# Lowering the Boom

# Lowering the Boom

## Critical Studies in Film Sound

Edited by

**JAY BECK AND TONY GRAJEDA**

UNIVERSITY OF ILLINOIS PRESS

Urbana and Chicago

© 2008 by the Board of Trustees
of the University of Illinois
All rights reserved
Manufactured in the United States of America
1 2 3 4 5 C P 5 4 3 2 1
♾ This book is printed on acid-free paper.

Library of Congress Cataloging-in-Publication Data
Lowering the boom : critical studies in film sound / edited by
Jay Beck and Tony Grajeda.
p.   cm.
Includes bibliographical references and index.
ISBN-13 978-0-252-03323-0 (cloth : alk. paper)
ISBN-10 0-252-03323-X (cloth : alk. paper)
ISBN-13 978-0-252-07532-2 (pbk. : alk. paper)
ISBN-10 0-252-07532-3 (pbk. : alk. paper)
1. Sound motion pictures. I. Beck, Jay II. Grajeda, Tony
PN1995.7.L69       2008
778.5'344—dc22       2007045199

# Contents

# Acknowledgments

As a truly collaborative project from its very inception, this volume has benefited from the tireless efforts and invaluable support of many friends, colleagues, and mentors. Our debt of gratitude is enormous, and we would like to single out a few people without whom this book would not have been possible.

The origins of this project date back to when we discovered our mutual interest in film sound at the 1999 Society for Cinema Studies conference in West Palm Beach, and those early discussions on coediting such a book evolved over a number of felicitous events. Perhaps the most significant was the "Walter Murch and the Art of Sound Design" conference, organized by Rick Altman and Leighton Pierce at the University of Iowa in March 2000. Not only did it provide a forum for exchanging ideas about film sound between academics and practitioners, but it was also an initial sounding board for several of the essays gathered here. In addition, the annual conferences of the Society for Cinema and Media Studies and the Modern Language Association, the Chicago Film Seminar, the University of Iowa's Sound Research Seminar, and the Sounding Out symposia all served as public venues for nascent versions of several of the essays in this collection.

We would like to thank all our contributors for their patience and commitment throughout the long process of working together across great distances to see this work to completion. Specifically, we would like to thank Debra White-Stanley for the title and Paul Théberge for suggesting the cover design. We also would like to thank Nataša Ďurovičová, Jason Middleton, Patrice Petro, and Rob Spadoni for their help, direct and indirect, in getting this collection to print. Special thanks to Shaun Frentner for indexing the collection and to the DePaul University Research Council whose Competitive Research Grant underwrote the process.

For their guidance and support over the years, Tony would like to thank Charlie Keil, Tony Conrad, Peter F. Murphy, Michael Walsh, Andrew Martin, Bernie Gendron, Paul Dickinson, Anahid Kassabian, Bob Miklitsch, Maureen Turim, David Shumway, and Michael Ryan, as well as his colleagues at the University of Central Florida.

For their collegiality and free exchange of ideas, Jay would like to thank Christopher Babey, Timothy Barnard, John Belton, Dana Benelli, Marisa Carroll, Cho Young-jung, Ofer Eliaz, Clark Farmer, Gregory Flaxman, Christian Keathley, Franck Le Gac, Allison McCracken, Gregg Millman, Angelo Restivo, Joe Wlodarz, Susan Yell, and Prakash Younger, as well as his colleagues at DePaul University. A portion of Jay's essay appeared in an earlier version as "'Rewriting the Audio-Visual Contract': *Silence of the Lambs* and Dolby Stereo," *Southern Review* 33.3 (2000): 273–91.

We both would like to express our gratitude to Joan Catapano for her early interest in and continued dedication to this project. We would like to extend our appreciation to the anonymous readers for University of Illinois Press, as well as the fabulous staff at the press, especially Angela Burton, Breanne Ertmer, John P. Hussey, and Marla Osterbur.

A very special thank-you is due to Rick Altman who has contributed immeasurably to both of our educations and to the field of film sound.

Finally, we wish to thank Victoria Grajeda, Cecilia Cornejo, and Gabriela Muñoz for their unwavering encouragement, inspiration, and love.

Lowering the Boom

# The Future of Film Sound Studies

## JAY BECK AND TONY GRAJEDA

In his introduction to a 1999 special issue of the film journal *iris* dedicated to "The State of Sound Studies," Rick Altman argues that "the growing field of Sound Studies" has finally matured to the point of exceeding the specificity of cinema: "Stepping out from under the protection—but also the anonymity—provided by cinema, work on sound has over the past few years benefited enormously from the growth of popular music studies and from the development of new technologies joining sound and image" (3). For Altman, the academic study of sound is, as the subtitle to his introduction asserts, "A Field Whose Time Has Come." Echoing Altman's sentiments from a different perspective, radio scholar Michele Hilmes opens her recent book review essay, "Is there a field called sound culture studies? And does it matter?" by answering her own speculative title: "I pose the two questions above in the face of mounting evidence that the study of sound, hailed as an 'emerging field' for the last hundred years, exhibits a strong tendency to remain that way, always emerging, never emerged" (249). Perhaps summoning Walter Ong's observation on the paradox of sound itself—that it "exists only when it is going out of existence" (32)—Hilmes bemoans the fact that work on sound as an object of study appears to remain forever in a state of becoming, never quite arriving.

Yet the apparent elusiveness of the study of sound may be more a symptom of the academic production and ordering of knowledges than some perceived ontology of the subject. Hilmes's essay itself goes on to demonstrate that a good deal of scholarly work on sound *has* emerged over the past few years. Indeed, as Hilmes, Altman, and other scholars of sound have noted, sound studies now traverses several academic fields: from film sound (Rick Altman, Michel Chion, Don Crafton, Amy Lawrence, Sarah Kozloff, Steve Wurtzler, Gianluca Sergi,

Charles O'Brien) to radio sound (Michele Hilmes, Susan Douglas, Allison Mc-Cracken, Allen S. Weiss), audio technology (Emily Thompson, Jonathan Sterne, David Morton), cultural analysis of sound (R. Murray Schafer, Friedrich Kittler, Douglas Kahn, Philip Auslander, John M. Picker, Michael Bull and Les Black), and music studies (Claudia Gorbman, Jeff Smith, Tim Anderson, Daniel Goldmark, Timothy D. Taylor, Mark Katz). This steady increase in scholarship from distinct disciplines currently coalescing around the question of sound implies that the subject has most certainly arrived, and this notable shift over the past decade places sound studies at the vanguard of academic discourse. The challenge, nonetheless, remains for these disparate fields pursuing the theory and history of sound not only to cohere around their object of study but also to articulate cross-disciplinary methodologies and analytical approaches.

Between Altman's anticipation of the promise signaled by the current state of sound studies and Hilmes's guarded reflections on its continued marginalization in the world of academic scholarship, we would note the rather paradoxical progress of the field to date, one in which, as Jonathan Sterne asserts, "a vast literature on the history and philosophy of sound" can be said to exist, "yet it remains conceptually fragmented" (4). While the somewhat inchoate project of sound studies inaugurates an emergent field still in formation, beset by fragmentation and awaiting theoretical cohesion, a more specific dilemma confronts the relatively stable field of film sound studies that has encountered some difficulty in establishing itself despite having so many vocal proponents, both academic and professional. In part this is due to the dominance of the image in cinema studies, preventing sound from being recognized as more than a simple "add-on" to the image. As Michel Chion has ably demonstrated in *Audio-Vision,* the challenge film sound studies faces is to develop a theory of sound and image relations during an era when image theory has become entrenched across multiple disciplines. The irony, of course, is that film has always been an audio-visual medium—a point made doubly clear through Richard Abel and Rick Altman's *Sound of Early Cinema* and, most recently, Altman's *Silent Film Sound*—yet the hegemonic relationship between image and sound in academe keeps sound studies in an asymptotic relationship to image studies. Altman, in particular, has done a great deal to foster the development of film sound studies, and his 2000 conference, "Walter Murch and the Art of Sound Design," was the genesis for several essays in this collection. As an intellectual endeavor, the conference provided a forum for academics and practitioners to exchange ideas and find a common ground for discussing film sound. This tradition has carried on in Philip Brophy's Cinesonics conference in Melbourne, Larry Sider and Diane Freeman's School of Sound conference in London, and the Sounding Out symposia around the UK; despite the renewed efforts of numerous theorists, critics, and practitioners, however, film sound still exists in the shadow of the image.

Laboring in this shadow is not a problematic exclusive to film sound studies. All studies of sound, as is the custom, typically decry the hegemony of vision and the privileging of sight historically endemic to Western culture. Yet the recognizable body of work on sound in film studies—dating back to the *Yale French Studies* special issue on sound in 1980—has laid the groundwork for an epistemology of sound-image relations that has the capacity for dialogic productiveness with the parallel, often overlapping project of sound studies. As such, the continuing expansion of film sound studies points to the theoretical and analytical significance of sound with regard to a range of crucial issues, including the apprehension of aesthetics and cultural forms, historicizing temporal and spatial phenomena, accounting for myriad creative and technical practices, positing a history of listening and subjectivity, and a host of other aspects of the audible reverberating across academic disciplines. Addressing many of these concerns within the first collection on cinema sound since Rick Altman's *Sound Theory/Sound Practice* in 1992, *Lowering the Boom: Critical Studies in Film Sound* contributes to recent developments in film sound studies by reconceptualizing film studies to place sound on equal footing with the image. To achieve this we draw from a wide variety of scholars who construct a necessary historical step in advancing sound studies. The essays gathered here not only amplify unheard and often ignored historical work but also represent new theoretical approaches and perspectives, while highlighting a number of films that are receiving for the first time scholarly attention in terms of their sound.

The work produced by the authors collected in *Lowering the Boom* would not have been possible without the significant growth of original scholarship in film sound studies over the past decade. A number of contributors to this volume, for example, have benefited from Claudia Gorbman's English translations of books by French filmmaker and sound theorist Michel Chion: in particular *Audio-Vision: Sound on Screen* (1990; trans. 1994) and *The Voice in Cinema* (1982; trans. 1999). One of the lines of inquiry developed by Chion that has had impact on *Lowering the Boom* can be found in *Audio-Vision,* where Chion takes issue with the stubbornly held assumption that our senses of hearing and sight are easily divisible, capable of being objectively isolated rather than thoroughly articulated. In order "to demonstrate the reality of audiovisual combination—that one perception influences the other and transforms it," Chion's work "formulates the audiovisual relationship as a contract—that is, as the opposite of a natural relationship arising from some sort of preexisting harmony among the perceptions" (xxvi). In other words, for Chion, sensory perception is socially constructed. His approach to film sound and issues of auditorship inform a number of essays in this collection, both with regards to theorizing sound as well as putting such ideas into practice through the analysis of particular films. Although his writing is often challenging—especially his use of neologisms and sudden jumps from one historical period or national context to the next—there is great lucidity as well. His larger project is to make his readers aware that film sound is more than simply a supplement to the image and that there is no inherently "natural" relationship between the two in cinema. Part of our work as sound scholars entails locating these seemingly "natural" relationships and exposing how they have been artificially constructed. Building on Altman's and Chion's work, and other recent developments in the increasingly coterminus and mutually inflected fields of film sound and sound studies, *Lowering the Boom* makes a distinct contribution to these fields in a number of ways. Corresponding to the organization of this volume into five sections, *Lowering the Boom* offers new theoretical and historical approaches, attends to a reconsideration of genre through sound, explores the relation between film sound and cultural studies, and, finally, provides pedagogical models of film sound through several case studies.

## New Approaches to Film Sound Studies

The first section in *Lowering the Boom* features four essays that address the expanding field of film sound theory and its significance in rethinking historical models of film analysis. Taking the perhaps most frequently read analysis of a film's sound as its object—David Bordwell and Kristin Thompson's work on *Un condamné à mort s'est échappé* (A Man Escaped) (France, 1956) in their canonical text *Film Art: An Introduction*—John Belton offers a new form of sound analysis that unhinges the terminology of sound theory from Bordwell and Thompson's definitions borrowed from industrial practices. Belton instead turns to Merleau-Ponty's philosophy of phenomenology to explore how this theoretical approach can open up what had previously appeared to be a closed text, thereby suggesting a model of analysis with repercussions far beyond the specificity of his particular reading. In his essay "The Phenomenology of Film Sound: Robert Bresson's *A Man Escaped*," Belton argues that Bresson's cinematic approach, especially his sound recording and mixing, seeks to put "essences back into existence." As such, Bresson's sound style tends to isolate certain sounds—heightening their discrete textures and qualities in a descriptive rather than interpretive mode—as a way of directing our attention toward objects, people, and places. By reducing sound to its most minimalist state, Bresson pares away the "noise" of traditional representational practices in the cinema, as Belton contends, in order to capture the unrepresentable essence of phenomenal reality.

Taking a different perspective on interpreting Michel Chion's "audiovisual contract," Arnt Maasø proposes an examination of sound space and the mediation of voices and sounds in film and television. Sound space is commonly discussed in film analysis and has also at times been among the most commonly debated topics in practical filmmaking. However, analytical terminology and methodology making meaning of spatial aspects in film dialogue, and mediated voices in general, are largely missing. Whereas scholars commonly supply transcriptions of significant visual spatial aspects—even when the analysis examines sound—none have shown similar methodological rigor and transparency when it comes to sound space. Through an analysis of examples from film and television, Maasø suggests that three interrelated levels of analysis are needed to study mediated voices: the intended vocal distance implied by the spoken voice itself, the microphone perspective, and the intended earshot of the idealized listener. Importantly, the interplay between these three levels unravels the spatial circumstances of "crooning," "whispered shouts," "pillow talk," and other mediated practices that cannot be sufficiently described by talking about "microphone perspective" or "tone of voice" alone.

Also supplying a multimedia theory of sound and image relation is Paul Théberge, who proposes a new model of analysis through the study of silence and its functions in film and television. In his essay, Théberge departs from the tendency of most academic work on sound to divide the soundtrack into more or less discrete analytic categories of dialogue, music, and sound effects, distinctions largely based on the technical construction of the soundtrack in film production and postproduction. Rather than reproducing such divisions, Théberge argues that a more integrated sense of sound design allows us to understand each of the sonic elements in their relation to one another as well as their cumulative contribution to the narration. To achieve this goal, Théberge posits an analytic approach to "silence" in order to simultaneously acknowledge the essential balance among all three conventional categories of the film soundtrack and to suggest that the *absence* of any one of these sonic elements may be just as significant as the audible presence of particular sounds. Though recognizing our cultural ambivalence to silence, the essay nonetheless considers relational, structural, stylistic, and generic silences through a number of examples drawn from classic cinema, the films of contemporary directors, as well as episodes of television drama. Théberge advises us to hear through such silences and encourages us to listen to the expressive possibilities of the absence of sound, those moments when silence creates meaning and speaks as loudly as sound.

Following on the trope of sound design, Jay Beck proposes a methodology whereby questions of hearing film sound are best theorized through how a film posits an idealized auditor, one who hears the sound of a film as an aggregate experience rather than as distinct elements of dialogue, music, and effects. By examining Jonathan Demme's *Silence of the Lambs* (USA, 1991), Beck demonstrates the ability of Dolby Stereo to render a sensorial accretion of acoustic details in the soundtrack mix. The polyphony of the film's mix operates on a series of acoustic "planes," thereby allowing the auditor to distinguish not just the individual sounds but also their location in the diegesis. In the film, FBI agent Clarice Starling traverses a previously unseen space where sound is used as a means of replacing the standard narrative convention of the establishing shot and a way to link the audience to her subjectivity. Restoring the spatial orientation lost in point-of-view cutting, sound functions as a new form of spectatorial anchor in the film by creating a stable acoustic space where the director can dispense with established patterns of shot/reverse shot. Beck notes that in doing so the film offers a new way of engaging the spectator that relies on the haptic ability of the surround-sound mix to activate audio-visual interactions not available in monophonic cinematic presentations.

Our second section historicizes cinema sound practices and their reception with three essays that fill in some of the major gaps in historical treatments of film sound. Contributors here address the uneasy transition to synchronized sound and the introduction of "the talkies" in Czech film culture, the mid-twentieth-century introduction of stereophonic sound for wide-screen cinema, and the practice of sound design as seen through the filter of cultural history. In the first essay, Petr Szczepanik explores an underexamined period in Czech national cinema. Szczepanik adopts the model of intermediality advanced by early cinema scholars André Gaudreault, Tom Gunning, and others in order to broaden our understanding of various factors contributing to the production of sound in early Czech film culture. Focusing in particular on the transitional period of the late 1920s and early 1930s, Szczepanik analyzes a number of arguments and tropes that structured film sound–related discourses traversing institutional, industrial, technological, and aesthetic concerns. Against a retrospective point of view that posits the conversion to sound as a short-term deviation from the dominant development of classical film, Szczepanik instead offers a prospective approach that takes the intermedial and cultural relations in account, thus providing a more thorough historical treatment of the transition during cinema's "uncertain" future.

In "Sounds of the City: Alfred Newman's 'Street Scene' and Urban Modernity," Matthew Malsky examines the function of the song "Street Scene" and traces its historical evolution from its eponymous origins in 1931 to *How to Marry a Millionaire* (Jean Negulesco, USA, 1953) where Twentieth Century-Fox used it to showcase the introduction of stereophonic sound for its CinemaScope wide-screen format. Malsky reads "Street Scene" not only as a sonic signifier of American modernity but moreover as an audible text that expresses and creates the subjectivity of female characters, precisely during a period marked by the postwar patriarchal crisis over issues of labor, leisure, and suburbanization. For Malsky, the use of stereophonic sound in *Millionaire* amplifies a dramatic shift in the representation of both the city and its female characters, portending and reflecting an anamorphic deformation of literal space into a fantasy space that is coded as crypto-suburban.

Rounding out the section is James Lastra's "Film and the Wagnerian Aspiration: Thoughts on Sound Design and the History of the Senses." Lastra's contribution branches out from the regular boundaries of film studies to examine the *longue durée* of sound design as a concept by starting with Walter Murch's inauguration of the term in the late 1970s and listening closely for echoes of its predecessors. Making the leap backward from Coppola's *Apocalypse Now* (USA, 1979) to the 1856 of Richard Wagner, Lastra examines the representational status of sound and the evolution of sound systems through modernity. Starting by interrogating the inclusion of sonic effects in the performance of music, Lastra follows the changes in representational strategies first brought on by the development of recording apparatuses and later refined by the perfection of synchronous, amplified cinema sound. He demonstrates how in the uneven history of representational technologies the Bazinian teleological argument of greater realism was regularly short-circuited. Instead of the technology capturing and replicating acoustic events without a trace of mediation, recording apparatuses were effectively constructing auditors. Skillfully weaving together an extended history of acoustic reproduction, Lastra's essay explores the deployment of image and sound to create prosthetic sensory experiences, ones that provide a perceived diegetic "realism" that is aesthetically preferable to actual experience.

The third section of the collection turns its attention to the question of sound and genre as our contributors examine both familiar and unfamiliar genres, from the relation between narration and image in documentaries and a rather unconventional film musical to the understudied and undertheorized areas of avant-garde film sound and animation. The section begins with Barry Mauer's examination of the history of Luis Buñuel's *Las Hurdes* (Land without Bread) (Spain, 1933), tracing its original release as a "silent" film screened in exhibition with live narration to an eventual print with sync-sound of music and voice-over narration. "Asynchronous Documentary: Buñuel's *Land without Bread*" focuses on the relationship between sound and image as a structural tension where Mauer notes how Buñuel retained the asynchronous quality that marked the use of live sound for his "finished" film. By pitting the tracks of image and narration against each other, as Mauer argues, *Land without Bread* not only reveals the manipulations involved in film construction but also trains spectators to adopt a subjective vision despite the repressive voice of narration.

Nancy Newman's "'We'll Make a Paderewski of You Yet!': Acoustic Reflections in *The 5,000 Fingers of Dr. T*" examines the 1953 film musical—based on the work of Theodor Geisel (Dr. Seuss)—as both a genre film and as an unconventional musical that anticipates the breakdown of the Hollywood studio system. Taking into account the contradictory reactions to Geisel's "vicious satire" of piano teachers, the essay considers how the film enacts a familiar tension in the Hollywood musical between the disciplined, exclusionary musical culture of Europe and the accessible, participatory activity that constitutes American popular song. To the extent that the film attempts to synthesize these two paths into a less threatening form of vernacular modernism that embraces both the specialization of classical training and the spontaneity of innate musicality, Newman suggests that *The 5,000 Fingers of Dr. T* effectively falls outside Rick Altman's generic definition of the Hollywood musical. Moreover, with its musical numbers forming a fantasy space in which a young boy searches for his musical identity—a search that is quite explicitly staged as an Oedipal drama—Newman argues that the psychoanalytic drive of the film helps explain the ambivalent reactions that its initial release provoked, one that narrates an aesthetic experience of both the pleasures and the terrors of the sonic realm.

A third close analysis of a director's approach to film sound is found in Melissa Ragona's essay, "Paul Sharits's Cinematics of Sound." According to Ragona, the project of thinking beyond representational conventions of the soundtrack as mere accompaniment to the image was at the center of Sharits's work during the period of structuralist filmmaking. Whereas his contemporaries Tony Conrad, Hollis Frampton, and Michael Snow also used sound to discover new translations of the cinematic frame, Sharits actually drafted a theory of sound and hearing in film he called *cinematics*. Informed by structural linguistics, the postserialist compositional work of Iannis Xenakis, and Wittgenstein's philosophy of mathematics, Sharits's work interrogated the epistemology of silence embraced by some avant-garde filmmakers, and sought to push sound past its purely rhythmic arrangements into the realm of nonrepresentational performativity. Ragona's essay examines several films by Sharits in order to grasp how his concern with the materiality and durational force of film's recording technology functioned as his primary compositional mode. Indeed, as Ragona contends, the work of Sharits reveals a history of avant-garde filmmaking inspired by experiments in music and sound, one that suggests the possibility for how sound could be an analog to vision.

In "'Every Beautiful Sound Also Creates an Equally Beautiful Picture': Color Music and Walt Disney's *Fantasia*," Clark Farmer argues that abstract animation represents an unusual case in cinema history because it explicitly takes an aural medium, namely, music, as the model for the visual. Farmer traces one aspect of that model—the long and fitful history of trying to understand color in relation to music—as manifest in what has generally come to be called "color music," across three distinct but overlapping moments: precinematic, cinematic, and postcinematic. The precinematic moment (dating back to the 1730s) covers attempts to build "color organs," instruments that would allow a user to "play" colors in conjunction with musical performances. The cinematic moment centers on abstract animators who emerged from modernist art movements and used film to add a temporal dimension to the graphic arts. Among these artists, the idea of a simple correspondence of sound and color was largely abandoned, though they were influenced by the resurgent interest in synesthesia, as well as attempts to create a rationalized system of color. The postcinematic moment in many ways represents a return to the precinematic moment as digital technologies are used to "perform" a regularized transformation of musical input into a visual output, which suggests that the abstract relation of color to music is an increasingly common aspect of audio-visual experience.

Our next section brings film sound into dialogue with cultural studies, and explores issues of gender, race, ethnicity, and sexuality as they are informed by and inflected through a film's soundscape. The authors here consider the various ways in which identity formation and subjectivity are constituted via sound, whether through musical expression, scoring, voice-over narration, ambient noise, or narratives that foreground audio technologies. In "'A Question of the Ear': Listening to *Touch of Evil*," Tony Grajeda considers the narrative setting of the border in Orson Welles's *Touch of Evil* (USA, 1958) as a critical metaphor, exploring the ways in which the film enacts the border-crossing movement of identity formation through a struggle over language, translation, and accented voices. In particular, Grajeda addresses how the film blurs the border between "American" and "Mexican" not only by drawing attention to the hybrid or "half-breed" subject but also by disguising and displacing the conventional treatment of the voice, in effect audibly "mixing" identity and thus unsettling the otherwise rigid category of ethnicity. While *Touch of Evil* provocatively engages the border-crossing trope of identity and difference, it also stages an encounter at the border between the field of vision and that of hearing, both diegetically and with regard to spectatorship. Through a close reading of Welles's disorienting use of sound in the film's famous surveillance sequence, Grajeda emphasizes the extent to which the film foregrounds the technological mediation of the "listening apparatus" and renders both visible and audible the production and reception of sound.

Extending the theoretical set of concerns arching across this section but shifting the discussion to more current work, Debra White-Stanley examines how postmodern representations of World War II recycle the aural techniques used in World War II films, only to effectively contain and manipulate the female voice. Recent war films, especially *Pearl Harbor* (Michael Bay, USA, 2001), *Saving Private Ryan* (Steven Spielberg, USA, 1998), and *The Thin Red Line* (Terrence Malick, USA, 1998), use aural devices such as voice-over narration and asynchronous representation of the image of women to signify gender through memory and absence—a sound container emptied of the female voice and filled with the voices of male soldiers and exploding weaponry. Against this gendered containment of the voice, White-Stanley searches for female agency in the rubble of the war film soundtrack, sifting through ambient noise, extradiegetic music, and voice-over narration. As she argues, the war-film soundtrack registers the cost of war and the persistence of militarism through the association between female characters and nationalist musical motifs, the voice-over narration of the soldier, the use of ambient noise to signify the power of modern warfare, and the portrayal of the engulfment of the female voice by the sounds of war.

Robert Miklitsch, in "Real Fantasies: Connie Stevens, *Silencio,* and other Sonic Phenomena in *Mulholland Drive*," recognizes David Lynch's 2001 film as a complicated audiovisual text, one marked by not only a hybrid character of realist counterpoint to fantasmatic surrealism but also a self-reflexivity riven with questions of gender and desire. Offering a close reading of its musical set pieces within a complex system of sound aesthetics, Miklitsch proposes that *Mulholland Drive*'s hermeneutic structure of dream-reality can be heard as an audible expression of Lynch's vexed relation to classical Hollywood cinema, and asks whether the film stages a postmodern pastiche or points to an inside critique of the Hollywood dream factory. By examining how Lynch utilizes sound strategies to function as the linking device between the disparate narrative elements in *Mulholland Drive,* thereby foregrounding the centrality of sound itself to his filmmaking, Miklitsch emphasizes both the degree to which *Mulholland Drive* exemplifies Lynch's continuing project of extending surrealism into mainstream narrative cinema (and television) as well as the significance of the director's work in understanding the contemporary uses of sound in film.

In our final section, we offer four case studies that provide close analyses of sound use in marginalized and noncanonical texts, featuring such disparate examples as industrial promotional shorts, punk rock films of the 1980s, Abbas Kiarostami's *Close-Up,* and contemporary films influenced by the narrative logic of video games. First, Paul Grainge presents an illuminating study of the packaging and marketing of sound technology for exhibitors and audiences through the corporate branding of Dolby Laboratories. After five years of modest success with their theatrical sound systems, in 1982 Dolby Laboratories produced their first promotional film for public exhibition. Evocatively titled *listen . . . ,* the film was intended for screening as a short prior to theatrical features. The short was produced with two promotional functions in mind: to embed and naturalize expectations regarding sound quality and to enable theater owners to market their investments in exhibition technology through the Dolby trademark. By examining *listen . . .* and other promotional shorts, Grainge discovers how

Dolby Laboratories constructed expectations about the cinematic uses of their surround sound technology in the rhetoric of their own marketing. For Dolby, the process of "selling sound" was as important as the development of the sound systems themselves, and the salient quality of acoustic "spectacle" was marketed as a form of sensory promise delivered by the sound systems. Through this study Grainge deftly constructs the link between Dolby sound systems as representational technologies, the industrial economies of contemporary Hollywood, and the relationship between sound and spectators.

Proposing an antinostalgic relationship between music and images, David Laderman listens to how the narrative punk rock films from the late 1970s and early 1980s signal a shifting cinematic sensibility through their representation and integration of musical performances. These films redefine rock music performance as a dystopic instance of postmodern spectacle, where a "playful anxiety" emerges when the performing body encounters technologies of reproduction. Constructing his own route of analysis through "(S)lip-Sync: Punk Rock Narrative Film and Postmodern Musical Performance," Laderman positions (s)lip-sync as an effect of disjuncture on both a narrative and a stylistic level between performer and performance. Many punk films seem to conventionalize (s)lip-sync where the performer stops lip-syncing and the spectacle of performance is laid bare. These punk (s)lip-sync moments challenge the classic representational coherence of the performing body—and more general traditional narrative film and music relationships—to self-consciously expose the apparatus of the film medium itself. Therefore, by reading moments of (s)lip-sync in these punk films as symptomatic of the broader cultural and political tensions of the postpunk, pre-MTV early 1980s, Laderman demonstrates how they perform the struggle of rock music's modernist rebellion against hypercommodity culture.

David T. Johnson's essay explores the interaction between sound and image strategies in Abbas Kiarostami's *Nama-ye Nazdik* (Close-Up) (Iran, 1990). The narrative hinges on the trial of Hossein Sabzian, a man accused of impersonating Iranian director Mohsen Makhmalbaf, and the film moves back and forth between reenactments of the actual events—played by Sabzian and other participants in the case—and direct presentations of the trial itself. Through this dialectic Kiarostami forces the audience to do more than just weigh one version of truth against the other; instead the director invites the viewer to probe the very nature of truth. Johnson points out that much of the film's self-conscious cinéma vérité style, designed to purport to observational objectivity, is in fact faked on the level of the soundtrack. Like many viewers, Johnson finds this form of cinematic guile distressing and investigates the ethical implications of Kiarostami's stylistic choice. Within his analysis of the film Johnson discovers that in revealing the artifice behind acoustic manipulation, Kiarostami signals how most audiences hear sounds referentially in relation to visualized sources. By deconstructing that relationship, the director is demonstrating how the regular practice of favoring image over sound has made totally observational cinema impossible. But as a direct response to this undercutting of acoustic objectivity, Kiarostami opens up the audience to a phenomenological experience of reality. Ironically, the result is that only by learning to hear the cinema more critically can we learn to experience and interpret the very film that seeks to give us this lesson.

Anahid Kassabian closes out the collection with her essay "Rethinking Point of Audition in *The Cell*," which examines sound strategies in Tarsem Singh's film to show how they challenge our conventional methods for analyzing film sound. As Kassabian demonstrates, the film undermines widely held assumptions guiding the role of sound to support traditional forms of narrativity (such as character development and forging identification mechanisms) not only by refusing to subordinate sound to image but also by blurring the presumably distinct sound registers that typically segregate dialogue, music, and effects. More specifically, the film puts into question some of the suppositions assigned to point-of-audition sound, the auditory means by which a film engenders spatial and subjective perspectives through certain characteristics of sound and is often used to stabilize and center a spectator aurally. By disrupting point-of-audition cues that otherwise serve to underwrite the fantasy of a unified, discrete subject, the film issues a different logic altogether. As Kassabian argues, *The Cell* (USA, 2000) audibly conditions a new form of subjectivity as it signals a shift in classical narrative cinema, one based not on the novelistic form of narrativity but rather on the principle of iteration, the aural logics of dance music, and the nonlinear narrative forms of video games.

## Alternative Paths

In organizing this collection according to familiar discursive constructs in film studies we do not want to leave readers with the impression that the essays are governed exclusively by a single analytical approach. Indeed, much like the condition or "nature" of sound itself, which often has the capacity to permeate boundaries and elude stable descriptive containers, the essays in *Lowering the Boom* are characteristically multidimensional in scope. For example, theoretical work on film sound has over the years typically separated around models based on either semiotic and psychoanalytic approaches or materialist and technological-industrial ones. Contributors to this volume, however, have selected previously ignored or largely underutilized models—not only phenomenology but also anthropology, cognitive psychology, mass communications, and social theories of modernity—along the way calling upon a diverse range of philosophers and critical theorists. Additionally, a number of essays take up theoretically oriented discussions on sound and subjectivity, auditorship, and ways of listening that are conversant with feminist theory, critical race theory, and cultural studies.

Similarly, whereas our history section aims to fill key gaps in the existing historical accounts of film sound, other historically based essays examine such unfamiliar areas as "color music" with regard to early cinematic development, avant-garde sound theory from the 1960s, punk films in the 1970s, sound technologies of the 1980s, and recent World War II films. Beyond our specific section on sound and genre, other essays in the collection consider such genres as suspense, film noir, war movies, and melodrama, without directly making questions of genre central to their analytical focus. Given the theoretical and analytical contingencies of much of the work gathered in Lowering the Boom, one hears a number of conceptual discussions resounding across the existing categories of the collection. In what follows we offer a few "alternative paths" through this volume, focusing on ideas such as the "acoustic auteur," technological change, "audio sensationalism," and film music as sound in order to suggest additional ways of listening to contemporary work on film sound.

One key alternative path through Lowering the Boom concerns the question of authorship as it cuts across several of the essays. Since the heyday of structuralism in the 1970s and subsequent developments in film theory, the "cult of the auteur" has generally fallen out of favor in contemporary film studies. Yet our collection demonstrates how sound theory reactivates the practice of auteurist models of analysis and how more careful attention to the use of sound in cinema once again opens up the debate surrounding authorship. However, this move is not merely a reinscription of the auteur theory, since sound work often problematizes the traditional notion of auteurism, given the degree to which the "author" of a film's sound cannot necessarily be assigned (or reduced) exclusively to just the film's director, or its composer, or any single member of its sound team. Although there are several American directors whose work is closely associated with the idea of creative film sound—most notably Robert Altman and Francis Ford Coppola—it is generally in conjunction with very prominent "acoustic collaborators" in the form of sound recordists, supervising sound editors, rerecording mixers, or sound designers—here, respectively, Jim Webb and Richard Portman or Nat Boxer and Walter Murch. Due to the hierarchical structure of entrenched labor organizations in the film industry, this tradition has become almost reified in American cinema where several directors have become associated closely with an acoustic collaborator: Steven Spielberg and Gary Rydstrom, Robert Zemeckis and Randy Thom, Jonathan Demme and Skip Lievsay, Steven Soderberg and Larry Blake. However, there are rare examples of genuine "acoustic auteurs" who utilize film sound in new and innovative ways to advance the art of motion picture storytelling. A few mavericks still exist in American cinema—most notably, David Lynch and Terrence Malick—yet the bulk of cinema sound work is predominantly the domain of a large team of sound practitioners working with little contact with the production staff or, often, the director. The result is that film sound is still a vital site for investigating the issue of authorship in cinema, especially in light of the difficulty for a single figure to emerge as an acoustic auteur within current industry practices.

Within *Lowering the Boom* several contributors engage the issue of acoustic authorship, and debates about auteurism can be read across many of the essays. Working backward, David Johnson's essay on Abbas Kiarostami's *Close-Up* addresses questions of the director's intentions regarding the film's sound by embracing the presumed "inadequacies" of the recording equipment. The resultant effect is a canny use of sound that simultaneously creates a documentary aesthetic while challenging the audience to consider how the soundtrack is actually carefully manipulated to create the desired meanings. Even though Johnson examines only one of Kiarostami's films, similar techniques can be heard across his body of work, thereby positing the films' acoustic techniques as the result of a single sound auteur and directly referencing the conditions of filmmaking in Iran.

The argument about an acoustic auteur is easier to defend in the hermetic avant-garde films of Paul Sharits, and Melissa Ragona's essay surveys sound practices across the breadth of the filmmaker's works. By theorizing his approach of cinematics, Sharits laid out an audio-visual strategy for making each of his films a self-referential "structural-information system" that ties together words, images, and sounds ("Words" 33). Other such examples of sound authorship are read in relation to Orson Welles (Grajeda), David Lynch (Miklitsch), and Robert Bresson (Belton) to further advance the question of how the auteur may be reconsidered through film sound.

Another way of listening to the various voices in *Lowering the Boom* is through the parallel paths of technological change in film sound, and what James Lastra calls "audio sensationalism." The former historicizes the production of sound through the development of audio technologies, whereas the latter offers a particular history of reception with regard to a listening subject. Technological change and its importance to our work on sound link a number of essays across sections; for example, Petr Szczepanik, Jay Beck, and Paul Grainge all suggest that technological change throws sound practices into relief and thus merits our careful listening. Despite the many historical accounts that emphasize the impact of sound technologies—especially in relation to the transition to synchronized sound in the late 1920s—few actually consider the aesthetic effects of these technologies. Rather than simply acceding to technological determinism—an approach that dominates many general histories of film sound—each author deploys a different methodology to study the impact of technology on cinema sound. Szczepanik postulates that the emergence of film sound in Czech cinema was a highly discursive process informed by not only cinematic technology itself but also such disparate cultural influences as the 1920s Czech avant-garde, linguistics, phonography, and radio. In his examination of Dolby Stereo in relation to sound design, Beck argues that the coincident appearance of both the new sound format and the new labor practice hides the fact that the former reified the traditional divisions of sound labor at the same time that the latter was attempting to dissolve those barriers. And Grainge's essay on Dolby Laboratories' promotional films continues the argument by demonstrating how the corporation designed its marketing strategy around a brand identity that foregrounds the spectacular nature of sound.

In counterpoint, Lastra's notion of "audio sensationalism"—in which a technological spectacle of sound forms a crucial part of a history of sensory experience and subjective engagement—invites connections with essays in this collection that also emphasize cinematic moments when sounds insist on being heard differently. Matthew Malsky's account of the stereophonic use of Alfred Newman's "Street Scene" in *How to Marry a Millionaire* and Paul Grainge's essay both provide concrete examples of how sound has been used as a form of acoustic "attraction." In addition, moments of audio sensationalism can be considered as historically significant to the development of film sound practices and to the advancement of academic discourses. Relevant here are David Johnson's primary-scene example from *Close-Up*, Tony Grajeda's analysis of the surveillance sequence at the close of *Touch of Evil*, and each of the instances of slip-sync described by David Laderman.

One last example of an alternative route through *Lowering the Boom* is marked by a discussion of music. Scholarly work on film sound over the years has often been dominated by attention given to music, from the production and function of the film score to the place and purpose of diegetic pop songs. Apart from the well-established domain of musicology, such work has tended to fall within either studies of content analysis or treatments of the ways in which film music performs emotional labor for an often unsuspecting spectator. In *Lowering the Boom*, however, it is not our aim to reverse the conventional ratio of privileging music over sound—a ratio sustained by past and current scholarship; rather, we have included work that conceptualizes film music not as music, per se, but as a form of film sound. For example, Robert Miklitsch's examination of the musical set pieces in *Mulholland Drive* treats the film's use of music in an atypical fashion, with music performing within an ensemble of sound effects. For Miklitsch, the conventional distinction between music and effects, in which the former assumes narrative priority over the latter, becomes blurred in the complex arrangement of Lynch's sound design. Similarly, Anahid Kassabian's essay regards musical sounds in *The Cell* less as compositional figures than as aural layers folded within an unusually dense model of sound design. She explores how the film dissolves boundaries among music, speech, and sound, leveling the standard hierarchy of tracks to create an affective cinematic world, one in which the accustomed means for subject positioning through sound has been inhibited or disrupted altogether.

Although a few of our essays treat film music in somewhat familiar terms (Malsky, Newman, White-Stanley), still others challenge more traditional ways of thinking about music in cinema, especially in regards to the relation between sound and image. Clark Farmer's essay, for instance, offers a historical sketch of "color music" and cinematic efforts to create "visual music" in animation, highlighting those synesthetic moments when a correspondence between image and sound is taken to be not merely causal or illustrative. David Laderman's essay, meanwhile, scrutinizes the portrayal of disembodied lip-syncing within "live" performance sequences in punk films from the early 1980s, films that for Laderman stage a postmodern form of performative excess and spectacle that rupture the classical convention of synchronized sound-image relations. Several of the essays in *Lowering the Boom* thus contribute to the extant body of work on music in film sound studies, yet such contributions aim to position film music within the larger mix of aural elements constituting a film's entire soundscape.

## "Dark Corners" Revisited

Many of the essays in this collection start to sweep out what Rick Altman calls "sound's dark corners": the understudied and undertheorized areas of film sound history. In his 1992 collection, *Sound Theory/Sound Practice*, Altman describes "sound's dark corners" as previously neglected areas of sound studies "that call most urgently for investigation" (172). Some of these dark corners include Third World cinema, documentary, music in and on film, animation, short forms, and idiosyncratic auteurs. As previously noted, *Lowering the Boom* addresses a number of these underexamined areas by including work on documentary, animation, and short forms, along with several essays on forms of music in and on film.

With regard to idiosyncratic auteurs, our contributors focus on the work of Lars von Trier, Peter Weir, Jonathan Demme, Luis Buñuel, Paul Sharits, Orson Welles, and David Lynch. Though volumes still need to be written on the international and global perspectives on film sound, this collection makes a contribution by offering essays on early Czech film culture, Iranian cinema, and Norwegian television, as well as work that deals with issues of language, culture, and translation in Buñuel's *Land without Bread* and Welles's *Touch of Evil*. *Lowering the Boom* joins with prior works in expanding international perspectives on film sound; nevertheless, a great deal of work remains to be done. In particular, the creative sound work of contemporary global directors such as Aleksandr Sokurov (*Russian Ark* [Russia/Germany, 2002], *Solntse* [The Sun] [Russia/France/Italy/Switzerland, 2005]), Carlos Reygadas (*Japón* [Mexico/Germany/Netherlands/Spain, 2002], *Batalla en el cielo* [Battle in Heaven] [Mexico/Belgium/France/Germany, 2005]), Takeshi Kitano (*Sonatine* [Japan, 1993], *Hana-Bi* [Japan, 1997]), Tsai Ming-Liang (*Dong* [The Hole] [Taiwan/France, 1998], *Tian bian yi duo yun* [The Wayward Cloud] [Taiwan/France, 2005]), and Julío Bressane (*Dias de Nietzche em Turim* [Brazil, 2001], *Filme de Amor* [Brazil, 2003]) all point toward an efflorescence of auteurism in relation to acoustic expressions of cultural identities that still needs investigation.

The historical accounts of the technologies of sound film in this collection, along with Lastra's discussion of "audio sensationalism," also resonate with related concerns in the wider currents of sound studies. With the modernization of sound since the mid-nineteenth century, a particular nexus of production (mechanical and electrical audio technologies), phenomena (a spectacle of sound), and reception (the conditions for a listening subject) has marked our historical understanding of sound from the emergence of the phonograph (Gelatt, Sterne), to radio broadcasting (Douglas, Hilmes), "hi-fi" culture (Taylor, Grajeda), and the recent digital "revolution" (Théberge, Katz). Although *Lowering the Boom* contributes primarily to the specific field of film sound studies, many of the essays here offer points of contact with the cultural study of sound across multiple disciplines.

The potential overlap among these fields is likely only to increase due to the recent acceleration of new media technologies, ongoing developments in the political economy of media convergence, and what Arjun Appadurai has termed the "mediascapes" of global cultural flows. Thus, the future work of both film sound studies and sound culture studies is bound to explore the production of altered soundscapes generated by new technologies of creativity, transmission, storage, and reception, all of which can be heard across a vast landscape marked by the mobility and deterritorialization of diverse populations and the concomitant sounds of globalization. Such developments demand further studies not only in cross-cultural communication (particularly the translation of languages and their role in film and other media) but also in the cultural specificity of audiences, texts, exhibition spaces, and the variegated contexts for reception. Finally, new (and not-so-new) "mediascapes" call for increased attention to the very conditions for a listening subject, whether differentiated by place, race, class, gender, age, or certain psychic and affective relations.

Another underexamined area concerns new work on theories of film sound. The near total dominance of the visual in classical film theory has inadvertently foreclosed salient pathways for examining cinema as an audio-visual form. Though several nascent theories have begun to emerge, film sound theory has historically lagged behind visually based film theory. A case in point can be found in *Post-Theory: Reconstructing Film Studies,* where editors David Bordwell and Noël Carroll take issue with what they posit as two large-scale trends of thought in cinema studies since the 1970s: "subject-position theory" and "culturalism" (Bordwell, "Contemporary Film Studies and the Vicissitudes of Grand Theory" 3). As forms of what Bordwell and Carroll call "grand theory," film studies informed by psychoanalysis, apparatus theory, ideological analysis, and a number of other theoretical discourses has, in Carroll's words, "run out of gas" ("Prospects for Film Theory: A Personal Assessment" 38). Proclaiming that the "doctrine-driven" approaches of subject-position theory and culturalism are exhausted if not actually dead, Bordwell and Carroll ignore the fact that in its overt analysis of the image, framing, and editing, grand theory itself had largely over-"looked" a major aspect of the cinema—namely, film sound.

The study of film sound theory, historically marginalized and thus underdeveloped in cinema studies, has only recently started to evolve, and it offers numerous possibilities for advancing, revisiting, and revising current feminist, Marxist, psychoanalytic, queer, and apparatus theories. Yet even these inaugural moves to establish sound theory as a discipline run the risk of being abandoned in the wake of "post-theory." For example, if, as Bordwell and Carroll claim, subject-position theory is exhausted, should we assume that the nascent theory of "auditorship"—in which a spectator-auditor is sutured by way of sound—will also prove to be a dead end? Therefore, we recognize work on sound as a clarion call for a *return to theory,* one that allows for a number of innovative and original approaches to theoretical perspectives that have otherwise been regarded as defunct.

If film sound studies can open up new avenues in the study of film theory, so too can it function as a powerful means for reconceptualizing formal analysis. As Rick Altman demonstrates in his essay "The Material Heterogeneity of Recorded Sound," even the basic terminology used to describe film sound places it in a subordinate position to the image. Though several authors in this collection challenge this analytical inequity through close readings of and attentive listening to films, it is still necessary to educate a new generation of film scholars (and to remind current scholars) that sound makes an equal—and perhaps occasionally larger—contribution to a film's meaning as does the visual. Once film sound has attained equal standing with the image in film analysis, it becomes possible to revisit established cinematic texts to reevaluate how they generate meaning and to rethink the current canon of cinema history. This equal integration of sound and image in film analysis holds great promise for understanding the formal operations of cinema, as well as connecting these concerns to cultural issues of sound and difference to address race, gender, and sexuality through film sound. Moreover, there are great strides still to be made in understanding sound's rendering of the body in cinema through both acoustics and the diegetic construction of space.

Equally pressing is the larger position of film sound in film history. Certainly, specific aspects of sound history have been examined in detail—such as early film sound practices, technologies, and the transition to sound—yet there are numerous overlooked or lost histories of film sound. For example, even though it is one of the most common techniques used in filmmaking, there has never been even a brief study of Foley, the act of performing sound effects during postproduction to match the action of the picture. The practice owes its nominal title to Universal sound technician Jack Foley who, despite never receiving sound credit for any of the films he worked on, was responsible for developing an entire approach for adding sounds to motion pictures. Due to his cinematic anonymity, there is very little documentation about his life and his work in Hollywood, and most of what has been written is culled from long-circulating industrial apocrypha and oral histories with no verifiable sources. Yet in direct opposition to its hazy origins, the practice of Foley has become a cornerstone of modern film sound and deserves a complete history.

In addition, other than extant trade manuals, a few essays in a Museum of Modern Art (MOMA) "Dawn of Sound" catalog, and a brief mention in Don Crafton's excellent history of American cinema's transition to sound, there are no critical histories of the early sound film shorts and their impact on the development of film sound practices. In the 1989 MOMA catalog celebrating the restoration of dozens of early sound features and shorts by the Library of Congress, both Charles Wolfe and Richard Koszarski hint at the work that has yet to be done in establishing the sound shorts as integral to the emergence of film sound and aesthetics. In a parallel history, aside from Michele Hilmes's examination of the industrial interplay between film and radio in *Hollywood and Broadcasting: From Radio to Cable* and Rick Altman's essay on "deep focus sound," there has been little that examines radio's influence on sound aesthetics. These historical "silences" are but two of a host of potential areas for historical research, including the history of room tone or "MOS," the development of dubbing practices, the interrelationship between phonography and cinema, differing national cinema audio practices, multilingual versions, and industrial and economic approaches to sound history, including studies of sound unions, new roles in film sound postproduction, multichannel theater sound, multiple sound formats, digital editing, and other emergent practices.

Finally, perhaps one of the richest areas for research resides in the study of film sound and genre. Nearly all of the work that has pursued the study of sound and genre has been conducted exclusively in the domain of the musical, and, resultantly, almost all of the work on sound in musicals has emphasized film music. This, of course, creates a paradigm where other structural functions of sound are often ignored in order to concentrate on the relationship between music and the genre. Therefore, not only does there need to be more work done on sound *as* sound in musicals, but there is also room for examinations of how film sound functions in other genres. For example, despite the plethora of books that acknowledge the importance of the voice-over in film noir, there have been few scholars who examine the structural function of the voice in the genre and, importantly, its link to radio drama. Likewise, there is a dearth of work on the relationships between point-of-audition sound and horror films, sound effect construction for science fiction films, and the use of stereophonics in rock-and-roll films. Each area provides a new method for studying the intersection of film sound and genre and for advancing the field of genre studies in productive new directions.

In the years since Weis and Belton's *Film Sound,* Chion's *Audio-Vision,* and Altman's *Sound Theory/Sound Practice* helped to define the field of film sound studies, we find today a field still very much in the making. Therefore, we believe that *Lowering the Boom* provides a much needed bridge between film sound studies and the future of sound studies by building on existing scholarship while staking out new territory in the theory and history of cinema sound. By embracing a number of innovative and original approaches to film sound that have otherwise been neglected, we aim to raise the stakes of film studies by demonstrating how nearly all films go "unheard" in academe, and we seek to redress that oversight by opening the ears and eyes of film scholars, practitioners, and students to film's true audio-visual nature.

**PART I**

# Theorizing Sound

# 1

# The Phenomenology of Film Sound

## Robert Bresson's A Man Escaped

### JOHN BELTON

Maurice Merleau-Ponty defines phenomenology as "the study of essences" (*Phenomenology of Perception* vii). It is a transcendental philosophy that
looks at phenomena in terms of universal essences, which are arrived at through
a process of reflective contemplation of that particular thing under scrutiny, but
a contemplation that is guided by an attempt to discover that which is invariable
about a particular phenomenon. If we are to grasp any phenomenon "wholly and
purely," we must necessarily lay hold of what is essential and unchanging about
it. These essences, of course, are abstractions. But these abstractions differ from
the kinds of abstractions generated by other philosophical systems or by science.
These systems seek to explain or analyze a phenomenon by placing it in a new
context, a movement that effectively transforms it. For example, science renders
all matter in terms of different elements in the periodic table, but that table does
not really account for the matter that it systematizes. A topographical map can
accurately describe a certain geographical area, but it neither resembles what it
describes nor conveys its essence. In both instances, the description renders the
phenomenon it seeks to describe via a model that replaces it. The abstraction
replaces the real.

In attempting to describe our experience of phenomena, phenomenology refuses to separate abstract essence from concrete existence, or, as Merleau-Ponty
says, it "puts essences back into existence" (vii). For phenomenology, the abstraction remains embedded in the phenomena; it is not lifted out and given new
form. More than anything else, phenomenology is characterized by a manner
or style of thinking. It is an approach to experience that seeks to describe rather
than to explain or analyze (viii). As Merleau-Ponty wrote, "To return to things
themselves is to return to that world which precedes knowledge" (ix). For this

reason, phenomenology rejects systematic analysis, scientific and analytical reasoning, psychoanalysis, and other forms of rationalization. It is an approach to phenomena that permits them to preserve their essential mystery. This is precisely the style of thinking found in the narration of the films of Robert Bresson.

Merleau-Ponty has described film as the phenomenological art form par excellence. It gives us objects and things as they are grasped in the consciousness of the narrator presenting events to us ("Film and the New Psychology" 58). In this sense, sounds and images are perceptions of things, marked as perceptions, but nonetheless perceptions that more or less directly posit the things they represent. Robert Bresson's approach to filmmaking—in particular to sound recording and mixing—is phenomenological. It attempts to get at the essences of people, places, and things. The following essay about *Un condamné à mort s'est échappé* (A Man Escaped) (France, 1956) attempts to describe the way Bresson's soundtrack "puts essences back into existence."

The words that we see at the very beginning of the film tell us that the film is based on a true story, explaining in the first-person voice of the narrator that "this story is true. I give it as it is, without embellishment." At once, realism is doubly evoked; not only is the story based on fact, but the telling of it is "without embellishment," that is, plain, direct, realistic. In effect, the film is introduced as a documentation of the central character's experiences and thoughts—a record of his perceptions. "Without embellishment" is also an apt description of Bresson's stylistics themselves, conveying a sense of the way he pares down events to the minimum amount of data necessary for their telling. One of Bresson's favorite maxims was, "One does not create by adding, but by taking away . . ." (48).

The film is based on an account by André Devigny of his escape from Fort Montluc in Lyons, which took place just a few hours before he was to be executed. The film is rooted in a certain authenticity. Bresson consulted with Devigny during the writing of the script, and Devigny was present during filming. Bresson insisted on filming in the exact locations where the events originally took place, at Fort Montluc in Lyons, and even used the actual ropes and hooks that Devigny had fashioned for his escape, which had been preserved in a nearby museum (Murray 68).

But the film is also highly abstract, an attribute that is generally opposed to realism. As Bresson admitted, "I wanted all the factual details to be exact, but at the same time I tried to get beyond basic realism" (quoted in Ayfre 8). Abstraction from the real exists on the level of both plot and style. The hero, Fontaine, has a universal quality; he represents a larger social type—an everyman-as-prisoner-of-war in occupied France. This universality is conveyed through the film's title: it refers to a *man* condemned to death (who has escaped). Fontaine's anonymity—his status as "man"—becomes part of his universality.

At the same time, the performance style of Bresson's actors is abstract as well. The performances are not fully fleshed out but somewhat minimalist, to say the least. François Leterrier's "embodiment" of Fontaine involves the repression of traditional modes of character expressivity. His body functions as a more or less blank surface through which we have access to his thoughts. In fact, Bresson calls his actors "models," a term that suggests their status as material objects that provide a ground or foundation on which the artist can construct a character. Models are nonactors. According to Bresson, actors should be replaced by models:

> No actors.
> (No directing of actors.)
> No parts.
> (No learning of parts.)
> No staging.
> But the use of working models taken from life.
> BEING (models) instead of SEEMING (actors). (1)

"BEING" is paradoxically the consequence of not-being—that is, of not being an actor. It is the product of a series of negations ("no actors . . . no directing . . . no parts," and so on). Bresson's "models" empty themselves out of the traits, details, nuances, and affects that fill traditional performances, opening up a space that Bresson can fill with the being of the character. In phenomenology, essence is often arrived at negatively—by identifying what something is not.

Fontaine is also something of an abstraction in a religious sense. Though the prison in which he is confined is clearly a prison, it resembles and functions somewhat like a monastery—and Fontaine, as inmate, takes on the role of monk. Its occupants have "withdrawn" from the world, renounced all material and corporeal desires, and devote themselves to spiritual pursuits, including meditation. One of Fontaine's fellow inmates is even a Protestant minister, who cites scripture and provides religious counsel to the hero. Fontaine's imprisonment by the German Occupation forces resembles that of Joan of Arc by the British (Bresson's *Procès de Jeanne d'Arc* [Trial of Joan of Arc] was made only five years later). His ordeal as a prisoner becomes a kind of passion. Though sentenced to death, he escapes. The central character's progress is toward a spiritual rebirth, a kind of Christ-like resurrection.

It is not entirely without significance that the only music used in the film is Mozart's Mass in C Minor. The Mass punctuates Fontaine's "passion." It shapes our reading of the film, transforming it from factual history to figurative allegory, from a literal account of a man who escapes a prisoner-of-war camp to a meditation on the paradoxes of Christian salvation. The term *transformation* requires some qualification. The Mass is literally nondiegetic; it has no source within the space of the narrative. Nondiegetic music is generally commentative in nature—it speaks from the space of the narration and conveys, from outside of the diegesis, an attitude toward what is seen. But in terms of its relationship to the film's images, Bresson's use of Mozart's Mass problematizes traditional distinctions between diegetic and nondiegetic. Here, those distinctions break down and become useless. The music reveals the essence of the images—the truth that lies beneath their surface. In this sense, it is diegetic. In other words, the music is not a *filter* through which we perceive the action but a core reality within the images themselves.

The film's secondary title is "Le vent souffle où il veut (The Spirit [or Wind] Breathes Where It Will), which is taken from the third chapter of St. John's Gospel. The film works out, in a secular fashion, a crucial religious paradox. The will of God works in mysterious ways; it saves some but not others. Orsini's attempt to escape fails, and he is shot. Fontaine's attempt succeeds. At the very moment that Orsini is shot (shots heard off-screen), Fontaine is reading aloud the words of Jesus to Nicodemus—"the spirit breathes where it will." Even Bresson acknowledges the religious dimension of the film, telling interviewers, "I would like to show this miracle: an invisible hand over the prison, directing what happens and causing such and such a thing to succeed for one and not for another. . . . The film is a mystery. . . . The Spirit breathes where it will" (quoted in Murray 68).

Everything that happens to Fontaine is, like Grace, both predetermined and the product of his own will. His escape is predetermined, a given that is stated as a fait accompli (in the past tense) in the film's own title. Yet it is also the end result of his own will, his determination to escape, that constitutes the bulk of the film's action—that is, the cutting through the oak door, the fashioning of ropes and hooks, and so forth. When Fontaine attempts to escape in the film's opening sequence, he fails. Why? Because he was not ready. He had not sufficiently prepared himself to receive God's grace. Like Orsini, he took matters into his own hands rather than waiting for the right moment. It takes the remainder of the film for Fontaine to prepare himself.

The film is similarly paradoxical; it is both abstract and realistic at the same time; its style might be described as "abstract realism." For example, concrete details—realistic elements—repeatedly reveal their abstract essence. We hear the footsteps of the prison guard in the corridor outside Fontaine's cell, or we hear his keys or nightstick hit against the staircase banister, but we don't necessarily see the guard. The guard becomes a universal essence; he becomes "Guard-ness" (with a capital G). Whenever Fontaine looks out his window, we hear the off-screen sounds of street noise and train whistles, signifying the world of freedom outside the prison. During the escape sequence, we repeatedly hear the off-screen sound of a train whistle, which is often followed by the rattle of the train on the tracks. Unseen, the train becomes an abstraction—"Train-ness." As noted above, the sound also represents the world outside the prison, the Idea of Escape. Like Fritz Lang, Bresson uses off-screen sound to create an abstraction, but, unlike Lang, the concrete nature of Bresson's sounds grounds them in reality. Lang's abstraction moves in the direction of giving reality to the unreal. Bresson's moves toward an intensification of the essential reality of things.

Placing the film on a spectrum of different realisms will help reveal its unique brand of abstract realism. The realism of Roberto Rossellini and Jean Renoir is that of everyday reality, full of an abundance of detail. Bresson is not like Emile Zola, who overwhelms us with detail to create the illusion of a full reality, with complex relationships between characters and things in the world around them. The strategy of Renoir and Zola is to embed their characters in as fully realized a universe as possible, to present their actions as bound up with it or even determined by it.

Bresson's film has a realistic appearance. It employs nonprofessional actors, realistic settings, naturalistic lighting, even a quasi-documentary photographic style. But rather than creating a proliferation of detail to suggest a certain material weight to the world seen on-screen, Bresson strips away detail, paring each shot down to a single detail or action. Rather than adopting the deep-focus staging of Renoir or Orson Welles, where several actions or events take place at the same time, Bresson focuses, often in close-up, on one event at a time; he concentrates on the fashioning of a chisel out of a spoon, the bending of a window frame into a hook, the braiding of cloth into ropes. This paring down, coupled with close-up cinematography, has the effect of intensifying what we do see, investing objects and events with heightened meaning.

What occurs on a visual level takes place as well on the soundtrack. Though Bresson uses a great many sound effects, they tend to come *one after another*, rather than on top of one another, as in traditional sound mixing. In other words, we get one sound at a time rather than a proliferation of different sounds, as we get in Renoir or Welles. (By the same token, there is no dialogue overlap, as there is in the sound mixes of Welles or Howard Hawks.) For example, when Fontaine is trying to bend the window frame into a hook, he accidentally drops it, and it makes a loud clattering sound. He immediately hides it under the mattress of his bed, and we hear the footsteps of a guard who comes down the hallway to investigate; the footsteps stop at Fontaine's door, and we can see the guard peering in through the peephole. Sound functions here to direct our attention toward objects and to create suspense, but it also reduces the entire scene to a drama played out between one sound and another, between metal on concrete and footsteps. Sounds become extensions of the characters—of Fontaine and the guard. This is a kind of intensification of the status of these sounds; the sounds are there not merely to suggest time or place or to create atmosphere but to bear the weight of the drama.

One way of describing the aesthetics of the soundtrack is to characterize it as elliptical. It relies on a disciplined practice of omission—on one sound followed by a gap (the three dots of the ellipsis). As André Bazin has pointed out, the ellipsis functions as a central stylistic device in the narratives of neorealist films ("An Aesthetic of Reality" 35). Since it is literally impossible for the filmmaker to show us everything, he or she selects some things and omits all the rest, acknowledging the omission by underscoring the fragmentary nature of what has been selected. The most obvious employment of sound ellipsis in *A Man Escaped* is Bresson's use of the Mozart Mass, which is presented in brief fragments—each fragment standing in for a larger whole. But Bresson's succession of solitary sound effects functions similarly: each sound signifies both itself and all the other sounds that it was impossible to include. This reading of Bresson's sound style takes its cue from his overall narration, which is grounded in elliptical construction. Each scene in the film functions as a fragment of a larger whole—a fragment that is separated from every other fragment by elliptical dissolves or fades. Together, they constitute a larger, unrepresentable whole—the whole of Fontaine's experiences in prison.

There are a few exceptions to this elliptical (one-after-the-other) sound style. But they are exceptions that tend only to confirm that style. During the escape, as Fontaine and Jost walk on the roof, the gravel crunches under their feet, and the noise threatens to betray their presence to the guards below. They stop and wait for the sound of the train to rumble by, using it to cover their own sound. Later, Fontaine uses the noise of the train to conceal his descent from the roof and to cover the noise of his murder of the guard. In all these instances, the sound effects multiply the intensity of the event. Again, it is a drama played out between two sounds that have extraordinary status for us. The soundtrack remains minimalist; though two sounds are greater than one, they are still pared down from the full range of possible sounds.

But what is most striking is the way the characters use the noise of the train to cover their own. It is as if they intuitively understand the aural nature of their universe, of Bresson's sound style. Given the axiom of one sound at a time, they can only efface their own sounds, which might betray them, by concealing them beneath other sounds, which become, for the guards, the only sounds. In other words, Fontaine seems to know the rules that govern sound in his world, and this knowledge enables him to bend those rules to his own advantage.

But to return to the question of realism, Bresson does not give us much information about his characters; they are not realized through detail, individual traits, mannerisms, or specific information about their personal history. Or rather, their personal stories are kept minimal. Blanchet, arrested for helping out a Jew, has given into despair; Terry has a daughter; Orsini has been betrayed by his wife. But we never learn why Fontaine is in prison, what he did for a living before or during the war, or what his role might be in the Resistance. Nor do we even know where he is going after he escapes; he merely walks off into the night, into an undefined, unseen space shrouded in fog and smoke. His motives for joining the Resistance or even for his struggle to escape are never explored; they are just givens. Compare his opacity with the transparency of Rossellini's Resistance workers in *Roma, città aperta* [Rome, Open City] (Italy, 1945); the priest Don Pietro and the communist Manfredi, as well as the democrat Francesco, openly discuss their beliefs, even debating over their merits with one another at times. Whereas Rossellini's characters are defined socially, politically, economically (via class), and historically, Fontaine remains without family, social class, or political position; it is not even clear whether he is Catholic or Protestant, though I suppose most of us assume that he is Catholic. Like everything else in the film, he is defined elliptically, through the spaces left blank in the information that the film gives us about him. Even though this is a true story dealing with a precise period of history (note the title "1943"), the hero is not really embedded in history. History may have been a point of departure for Bresson—after all, the hero is in jail as a result of the war and the Occupation—but the action of the film, imprisonment and struggle to escape, is presented in such a way that it transcends history; it becomes a kind of universal struggle.

Bresson's characters are not defined by traditional means; they do not possess the basic elements of motivation seen in conventional films—greed, lust, jealousy, or guilt—elements that are generally considered to supply psychological depth or roundness to characters. We understand Bresson's characters in terms of their will, energy, obstinacy—traits that apply more to spirit than to anything else. As Amédée Ayfre says, "There is always something fundamental and mysterious in them which escapes us"; they refuse to provide us with the sort of information that can enable us to sympathize or identify with them. "This is why, even in their most extreme confidences, they never fundamentally reveal anything but their mystery—like God himself" (12–14).

Bresson's actors do not interpret their characters for us, as traditional actors do. In fact, he rejects "acting," which for him is, in essence, false. Acting is pretending; it can only be perceived as such on the screen. His actors do not try to make their characters accessible to the audience. They function as bodies against which their words are read. Their gestures and facial expressions are minimal and nonrevealing; more often than not, gestures are purely mechanical—walking in single file down the prison corridors, stooping to empty their slop buckets, washing their hands in the washroom; that is, as with Fontaine in the chisel-making, door-cutting, rope-braiding sequences, they are engaged in depersonalizing activity, in an action that takes on the nature of a ritual or routine. Their individual will is subsumed within a more universalized, ritualistic will.

The film's musical underscoring illustrates the incorporation of the individual into the collective through quasi-ritualistic activity. As David Bordwell and Kristin Thompson have noted, the Mozart Mass appears eight times in the film (381). It plays once at the beginning under the titles and once again at the very end, as Fontaine and Jost walk off into the night. This is hardly unusual; it structures the film, marking its beginning and signaling its end. But it also appears three times as the prisoners descend the stairs to the courtyard to empty their slop buckets. We also hear it when Orsini attempts to escape, when he is taken from his cell to be executed, and when Jost arrives and Fontaine debates as to whether he should kill him or take him along. Leo Murray has written that the music functions as a "sign of Providence . . . as well as of communion" (78). In other words, it marks moments of divine intervention, where the course of events is determined, and marks points of human contact, where one man communes with another.

As a formal device, the Mass structures the film, linking each event that it underscores to one another. It suggests that there is something essentially similar taking place in the scenes when the slop buckets are being emptied and in other scenes, such as when Orsini attempts to escape, when guards take him away to be shot, and when Jost arrives to complicate Fontaine's own escape plans. The latter three instances suggest a pattern of predetermination—the music's association with Orsini's escape attempt and execution suggests that when it accompanies Fontaine's encounter with Jost, Fontaine's efforts will similarly fail. But Fontaine succeeds where Orsini did not. The pattern that is being worked out is that of Providence smiling on Fontaine but not on Orsini.[1]

The association of the music with communion certainly applies to the emptying of the buckets, where there is a ritualized sharing of experience. But I think that what is as important is the power of the music to reveal, even in banal, mundane activities, a certain grace and elegance; it reveals for us the beauty of the ritual; it brings out its essence. A ritualized form itself, the Mass enables us to see the routine nature of their activities as a kind of transcendent religious ritual. Religious essence is put into mundane existence. At the same time, its recurrence suggests that the film itself is the working out of a larger pattern; various sections of the film emerge as different stages or sections of a Mass that is brought to completion with the escape of Fontaine and Jost, where the music occurs for the final time. Previously, the music occurred in brief, incomplete fragments; it is truncated. At the end, it continues long after the end title, playing over black leader for more than a minute until it concludes. The section of the Mass used is the "Kyrie eleison" (literally, "Lord, take pity / have mercy on us"). At the end, it functions as a prayer that is seen to have been answered.

The music imposes a pattern both on the scenes that it accompanies and on the film as a whole. As a work of artifice, the Mass becomes an ordering of experience that is more or less predictable and reflects divine order. Heard in conjunction with the ritualistic nature of the events seen on-screen, the music suggests that these events contain an underlying order and predictability. They are not random, but organized, part of a larger pattern that shapes all experience and phenomena. In other words, beneath the apparent insignificance of events there is a larger logic or consciousness. It suggests a relationship between fixity and apparent randomness.

*A Man Escaped* is a description of its central character's experiences. The description is marked as subjective by the presence of voice-over narration; it is Fontaine who tells us his story. Bordwell and Thompson identify the voice-over as "external displaced diegetic sound" (377–79). That is, it is displaced in that it takes place at a time later than that of the images, it is external in that the voice-over is not understood as Fontaine's inner thoughts at the time of the images, and it is diegetic because Fontaine is a character seen in the space of the story.

Like traditional voice-overs, it functions to provide the viewers with crucial narrative information, indicating the passage of time, describing features of the setting, or conveying the central character's thoughts.[2] But the voice-over is, in many ways, nontraditional. For example, it frequently describes what Fontaine does as he does it. In this way, the voice-over narration is itself phenomenological in its attitude to the world of the film. It prefers to describe rather than to interpret or analyze.

Near the beginning of the film, just before he meets Terry, Fontaine describes the contents of his cell for us as the camera pans around the cell, showing us what he describes. Looking around the room, he notes: "My cell was less than three meters by two. It was furnished sparsely—a wooden [bed] frame with a mattress, two blankets. In a recess by the door, a sanitary pail [slop bucket]. And set in the wall, a stone shelf. The shelf enabled me to reach the window." Later, he describes his actions as we see him cut up a blanket into strips, fold the strips over four times, twist the strips, and then wind wire around them. In each instance, the voice-over is redundant, duplicating the image. But the effect of the sound-image conjunction is not simple redundancy—the sort of parallelism that Siegfried Kracauer condemned in his discussion of sound editing (*Theory of Film* 117). As Bresson himself notes, audio-visual redundancy is to be avoided: "What is for the eye must not duplicate what is for the ear" (27). Rather, it is a form of asynchronism whereby the sound and image are, in Kracauer's words, "interrelated in a contrapuntal fashion" (115). Or, as Bresson puts it: "Image and sound must not support each other, but must work each in turn through *a sort of relay*" (28). Seeing the contents of the cell and hearing those contents described constitute a complex phenomenological experience of Fontaine's cell—a relay whereby each category of description transforms the other. The space is described in two different ways—in word and in image. One description overlays the other, generating a multifaceted portrait of the cell.

At the same time, the brick-by-brick assemblage of descriptive detail reveals a larger truth that lies beneath the surface of Fontaine's descriptive observation. Fontaine's seemingly random observation of this space moves linearly from the bed to the stone ledge, which he then uses to look out the window. There, he finds Terry, an agent who helps him on his path to freedom (Terry provides a pin that enables Fontaine to remove his handcuffs and smuggles Fontaine's letters outside the prison). That which is seemingly random has a logic and order to it. In retrospect, we realize that the objects in Fontaine's cell—the bed frame, the blankets, the window, and even the sanitary pail—become tools that he "repurposes" in his escape (wire from the bed frame and strips of cloth from the blanket are fashioned into ropes; the window is turned into grappling hooks; garbage generated by his use of these materials is concealed in his sanitary pail and dumped into a cistern). In other words, this apparently "blank" reading of the contents of Fontaine's cell contains within it the means of his salvation and liberation.

Even though Fontaine's voice-over ought to possess a certain knowledge about the events it relates, it occasionally pretends ignorance. When a package arrives, thus forestalling a search of his cell for a possible hidden pencil, Fontaine asks, "Did it save my life?" Yet in speaking the voice-over, he is clearly alive. *A Man Escaped* is not *Sunset Blvd.* (Billy Wilder, USA, 1950); we presume that whoever is speaking the voice-over is still alive. After learning of his death sentence, he wonders, "Was the guard taking me back?" and "Would I return to my old cell?" These questions undermine his own supposed omniscience about things of which he should be aware. Though he presumably speaks from a position of knowledge, he seems ignorant of that which he should know. The nature of the voice-over here rearticulates the larger paradox addressed by the film as a whole; events, like divine Grace, are both preordained and unknown. Fontaine knows yet does not know.

At the same time, the voice-over seems to be associated with interiority. It is an introspective device that Bresson directly relates to Fontaine's imprisonment. For example, it is absent from the first scene of the film, when he tries to escape from the car that is transporting him to the prison, and it is presumably no longer necessary after the escape. The last words of the film are not voice-over but simple diegetic sound. And Fontaine simply utters Jost's name, and Jost expresses the somewhat banal sentiment, "I only wish my mother could see me."

In fact, Fontaine's imprisonment might be said to drive his voice inside. The regimen of the prison prohibits conversation among prisoners, who are repeatedly cautioned by guards to stop talking in the washroom. The dialogue, therefore, bears the stamp of this repression; it is generally hushed or whispered. And it is limited to bare essentials—a few words here and there. The verbal style, in other words, is exactly like the visual style and the way sound effects and music are used—sparingly. Even though the Mozart Mass appears eight times, only small bits of it are used. This is far from the wall-to-wall scoring heard in a traditional Hollywood studio film. Often, the communication between prisoners is so minimal that it is nonverbal; Fontaine communicates with neighboring cells by tapping in code on the walls. The prisoner across the way watches for the guards as Fontaine scrapes away at his door, clearing his throat to give the sign that all is clear.

Even when the prisoners seem free to talk, as when Fontaine and Blanchet talk looking out the window, their conversations remain short. Nonetheless, dialogue, even though repressed, has special status as a form of exchange between men. It is a means of establishing community. The prisoners, with the exception of Jost, do not talk with their German guards. When Fontaine hears his sentence, he says nothing; he merely lowers his eyes. (Eye contact functions similarly: the prisoners exchange looks with one another, but lower their glances when submitting to the authorities.) After Jost tells Fontaine his story, he asks Fontaine for his. Still not certain whether he can trust Jost, Fontaine merely says they have talked enough for the night and remains silent. In effect, Fontaine refuses to trust Jost and thus refuses to engage in substantial dialogue with him.

The prisoner setting isolates the characters from one another in separate cells and restricts their contact to a few prescribed moments—the descent into the common spaces of the courtyard and the washroom, moments that are themselves carefully regulated. This isolation on the story level is echoed on the stylistic level. Close-ups isolate characters and objects in space, detaching them from one another. The sound effects have a similarly detached quality. The film was apparently shot without sound, which was added later. Though these sparse sounds play a crucial role in establishing the surrounding spaces, they do not so much realize space as a continuum as suggest its discreteness. The separation of one sound from another contributes to this, as do the different textures that each sound effect has. For example, as he works on his hooks, Fontaine unscrews squeaky screws, breaks glass by stepping on it through a blanket, then dumps the broken bits of glass into his slop bucket. Each sound has a different quality; each material has its own distinct texture. The next day, Fontaine dumps his slop bucket out. The sound of water and broken glass as he empties out the contents of his pail is both composite and discrete. The sound combines the different sound textures of glass and water (successfully fooling the guards) yet also preserves their discreteness, creating an array of distinct aural objects for us (see Metz). That is, they remain distinct and isolatable.

The thrust of the film is toward contact, toward a defeat of the barriers that separate the characters. The film works against isolation. Moments of contact become "miraculous." Fontaine breaks out of his cell, crosses the corridor to the cell of a comrade, and speaks to him. It is a moment in which he has defeated the barriers that the institution has created. Like his escape with Jost, it is an act of transcendence over the concrete space in which he is entrapped.

The characters struggle against their enclosure, their isolation; that is why the two-shot of Fontaine and Jost walking off together at the end is so significant. Not only are they together, but they have entered a new space that enables them to move freely within it. As they walk away into the night, the camera follows them, framing them tightly from behind. Then, as they cross a bridge, the camera briefly tracks from right to left to follow them, then stops, and their figures recede into the background. The space appears to open up, facilitating their movement into it. Steam (from the engine of an off-screen train?) obscures them. The screen goes black. They have ascended, as it were, into a different kind of space. The constricting spaces of the prison—its cells, corridors, washroom, and courtyard—no longer exist for them. The real has given way to its essence, the essence of confinement is the cell, and the essence of freedom and escape is the spatially amorphous night.

The final images and sounds of the film function as a resolution of sorts. The film's images and sounds can be understood as both realistic and abstract. Within each image and within each sound, resemblance and essence coexist as layers of our experience of that image or sound or both. When the image fades to black, resemblance cedes to essence—the image is stripped of its superficial status as bearer of appearances and forced to reveal the mystery of its phenomenological essence. The black screen is the culmination of a process within the image that structures the very nature of the trajectory of the film's narration. The sound— the "Kyrie eleison" of Mozart's Mass—seemingly does the opposite. If blackness empties out the image, then the "Kyrie" fills the soundtrack. But the "Kyrie" at the end resolves the film's soundtrack, giving completion to the incomplete fragments of the music heard earlier. And the Mass, like the blackness, functions as the end point of a larger process that informs the progress of the soundtrack—a movement toward the revelation of the divine order that Bresson believes structures our existence.

The work of the film is a form of self-effacement whereby images and sounds— as representations—give way to the unrepresentable essence of phenomenal reality. Bresson reduces image and sound to a minimum, paring away the "noise" that accompanies traditional representational practices in the cinema. As a result, his films give us access to a level of existence of characters, events, places, and experiences that eludes most other filmmakers. He lays bare the soul that lies beneath the visible surface of phenomenal reality. As Bresson explained, "I was hoping to make a film about objects which would, at the same time, have a soul.

That is to say, to reach the latter (the soul) through the former (objects)" (quoted in Ayfre 8). It is these objects, presented in a certain order and in a certain style, that reveal the soul—the inner being of people, places, and things. Bresson captures the essences of people, places, and things and puts those essences back into existence. He is *the* phenomenological filmmaker par excellence.

## Notes

1. Bordwell and Thompson read the music in terms of narrative development, tracking "Fontaine's developing trust in the other men on whom his endeavor depends" (381).

2. For examples of these uses, see Bordwell and Thompson 378.

# 2

# The Proxemics of the Mediated Voice

## ARNT MAASØ

His voice was as intimate
as the rustle of sheets.
—Dorothy Parker (1893–1967)

In everyday communication a voice is not only a bearer of words and semantic meaning but also a wonderfully complex instrument to express a range of communicative features. A speaker, for instance, more or less consciously uses his or her voice to signify a communicative "closeness" to or "distance" from a listener—something the listener interprets without much effort or even awareness. When a voice is technologically mediated, the way a voice *sounds* is affected, and hence also the way a listener may interpret such aspects as the communicative distance between speaker and listener (both between characters inside a diegetic space and between a mediated speaker and the listening audience).

In the analytical work on mediated voices to date, we generally lack terminology to help us distinguish between different spatial aspects of the voice, in order to, for instance, discuss the above-mentioned expressive and communicative features. This article is thus an attempt to provide some basic tools in this regard. In order to do so, it must first address a few basic questions: How may one in a fruitful manner analyze the way a voice signals a communicative "closeness" or "distance" in relation to a listener? How does the act of mediation influence the sound of a voice and the perception of communicative distance? What are the differences between interpersonal (nonmediated) and mediated voices?

## Body and Voice as Signifiers of a Communicative Relationship

Of all sounds, the voice is the most important in human communication. Research on perception and cognitive processing indicates that recognition and processing of faces and voices appear more specialized than processing of other visual objects and sounds (see Spelke and Courtelyou, Eysenck and Keane, and also Maasø for an overview of research). As mentioned, the voice clearly communicates a wealth of information other than mere semantic and referential meaning, such as attitudes, emotions, and closeness and distance in relation to other subjects in communication. Such paralinguistic features even seem to develop earlier and perhaps are more fundamental than other vocal aspects. Collier, for instance, states that "the ability to judge emotions through vocal features develops earlier than the ability to judge emotions through facial expressions and body movements and may even be innate" (141).

The ability of speech to signify a communicative relationship between a speaker and listener rests on several conditions, including the spatial relationship between speakers, the limited *reach* of sound in space, and the ability of the human voice to vary the address—from an intimate whisper to a distant call. The relationships between bodies and spatial vocal expression are thus of central importance in communication. Social-semioticians Hodge and Kress even claim that, "Of all the dimensions of the semiotic situation, the most fundamental is the physical relationships of the [bodies of] participants in space" (52).

## Spatial Listening

Vision is commonly held to be a more precise judge of spatial characteristics in our environment than that of audition (see Julesz and Hirsh; and Maasø 15–38 for an overview of research). The faculty of listening is most precise with regards to horizontal direction but less accurate when estimating distance. Though listening may not accurately distinguish the difference between a person speaking at six and at seven feet away—which would be a simple task for vision—we can, of course, easily hear if a voice is spoken seven inches or seven feet away, and whether speech is intended for *one* close by listener or a large group.

When evaluating *spatial aspects* of the voice, three acoustic factors seem of special importance: volume or sound level of the sound, frequency characteristic, and the relationship of direct to reflected sound. The most important factor for judging the distance of a voice in everyday life is the sound intensity of the voice. Even though our everyday experiences do not tell us that there is a precise inverse relationship between sound intensity and distance (known as the inverse square law in physics), we nevertheless know that sound intensity drops with the increasing of distance. In everyday life we are therefore used to adjusting the sound level of our voices according to whom we address, speaking more softly when addressing a nearby listener than a more distanced audience. Likewise, we are all effortlessly taking account of sound intensity when interpreting whether a voice is directed toward us or not. Meeting people talking loudly on a cell phone may thus sometimes lead to a moment of confusion, before we are able to see that they are talking not to us but to an absent listener. A second important cue for judging the distance of a voice is the frequency characteristics. When a voice is close we hear more of the lower frequencies, because of the so-called equal loudness contours in psychoacoustics (see Handel), whereas a voice becomes "thinner" and more high-pitched when heard from a distance. A third important factor in everyday perception of voices is the sound of the room in which the voice is spoken. When we are close to a voice, the direct sound will dominate, while reflections from the walls, ceiling, and floor will dominate with increased distance. The amount of reflection will of course vary with the environment, from extreme amounts of reflection in a large marble-surfaced hall to almost none in a snow-covered soundscape.

## Earlier Analyses of Mediated Voices

The concept of sound space is commonly discussed in film sound analysis, and has also at times been among the most commonly debated topics in practical filmmaking (see Altman, "Sound Space"). However, the analytical terminology and methodology for making meaning of spatial aspects in film dialogue, and mediated voices in general, are largely missing, and hence also specific analyses of aesthetic practices concerning the spatial aspects of mediated voices. For instance, when Altman, Chion, Lastra, Klimek, and Truppin use the terms *close-up* and *medium close-up* sound, none of these authors indicate to readers how one might distinguish between the two terms (let alone other perspectives), or what role technology plays in relation to the spoken voice *itself* as a sign of a spatial relationship between a speaker and a listener.[1]

The lack of analytical categories in analyses of sound in film and television is especially striking when considering the wealth of analytical tools, concepts, and theories for dealing with visual space and the mise-en-scène of the human body in film and television. Outside of film and media studies, analyses of vocal expression certainly exist, such as studies of prosody and proxemics. Yet these describe only nonmediated voices and provide little help for understanding the difference between interpersonal and mediated voices. As a main goal in this article is to better understand the acoustical characteristics of mediated voices and the act of mediation, I will explore the role of mediation in the following, before proceeding to a presentation of the analytical tools suggested for examining spatial relations in mediated voices.

## Mediation and Proxemics

In face-to-face communication speaker and listener are by definition tied together in time and space. The earshot and sound of a voice in space are hence tied to the intensity of a voice, the distance between listener and speaker, the room's acoustical features, and the laws of physics regarding the diffusion of the voice in space. Thus, a softly whispered voice has a limited earshot and will rarely have a high ratio of reflected to direct sound. When sound is mediated, the laws of physics and everyday experience are largely suspended. With electroacoustic sound, the voice may rather be characterized by what R. Murray Schafer has called "schizophonia": a split (Greek *skhizein*) in time or space or both between the production of sound and listening (90). The microphone and loudspeaker can thus raise a softly spoken sound to a sound level far outreaching what is possible within nonmediated talking distance. Through the act of mediation sound can thus be broadcast over vast distances or be recorded for playback in a different time or space. The original loudness of a voice, the sound of the voice, and the spatial signature from the room of the speaker thus become individual parameters in mediation; parameters that are all important in evaluating distance and the relationship between body and voice in everyday life are therefore open to technological manipulation and aesthetic choice in mediation.

When anthropologist Edward T. Hall coined the term *proxemics* in the late 1960s, his neologism sought to emphasize the importance of spatial proximity and physical relationships between subjects in interaction and communication. Hall showed how distance and social behavior were interrelated and varied according to whom one communicates with (for example,, a stranger, acquaintance, friend, family, lover) and in what social context (at work, at home, in bed, and so on). Based on extensive studies of social interaction, Hall distinguished between four main proxemic zones that seemed of special significance when interpreting different kinds of behavior: an intimate zone (where subjects were within eighteen inches of each other), a personal zone (from around eighteen inches to four feet), a social zone (from four to twelve feet), and a public zone (more than twelve feet). Within these four main zones, Hall also discussed a "near" and "far" region of each zone. Since my main concern here is not to discuss proxemics as a research area in interpersonal communication but rather to use it as an inspiration for developing useful categories in analysis of mediated voices, I will not go further into Hall's work here (see Hall "Proxemics"; *The Hidden Dimension*; and also Baldassare for a further account of research on proxemics to follow Hall). For the purpose of this article, I trust that the concept of proxemic zones, boundaries for intimacy, and so forth makes sense on an intuitive level, and may bring readers to reflect on their own experiences with spatial relations in social situations. I believe most of us, for instance, have experienced some awkward situation when someone has trespassed our borders of intimacy. Such shared experiences made it possible for the creators of *Seinfeld* to give this phenomenon a new name—"close talkers"—with an immediate and broad comic appeal (see *Seinfeld,* episode 82, "The Raincoats," first broadcast April 28, 1994).

Compared to interpretations of the proxemics of visual representations of bodies in space, interpretation of vocal proxemics to a much larger extent depends on the voice *as spoken*. Thus, to analyze the proxemics of the voice as a question of microphone placement, and registration of direct and reflected sound, would greatly miss one important goal of this study. It would be comparable to discussing visual framing of bodies filmed in the dark: visual framing is of little interest for the study of proxemic relationships if one sees only darkness. Hence, if the goal is to understand the proxemic and communicative use of the mediated voice, one needs to pay attention to both microphone perspective—the perceived reach or earshot—*and* the use of the voice itself as signifiers of proxemic relationships.

The voice is an incredibly flexible instrument, capable of varying the sound level from a barely audible whisper around 25–30 dB sound pressure level (SPL) at one meter to a scream up to around 120 dB SPL at the same distance (see Handel, Holman). This vast dynamic span is more than most acoustic musical instruments can achieve. The voice can therefore quickly vary the intended earshot from an intimate whisper to a public call, signaling whether an utterance is intended for a listener either within proximity or at some distance. Whereas the physical body in interpersonal communication is tied to relatively slow movements in and out of proxemic zones, the voice may therefore cross several proxemic zones within a split second.

The above-mentioned differences between visual and aural space call for caution concerning the scope and precision of proxemics as a tool in the analysis of mediated voices. It is thus important to bear in mind that the boundaries and categories must be regarded as more flexible when analyzing mediated sound than images. Hall also acknowledged the difficulty in application of aural proxemics (in face-to-face communication), compared to touch and vision (*The Hidden Dimension* 119). Introducing Hall's analytical apparatus to mediated voices does not make the categories less fuzzy. Thus, the analytical tools presented here should be considered as but one of several approaches to the study of the mediated voice. Yet the lack of other tools and methods nevertheless seems to make the search for some common concepts of spatial relations a necessary step in the study of mediated voices.

## Three Levels of Analysis

Since mediation introduces a schizophonic split between different audible parameters important for evaluating distance in a proxemic sense, this needs to be reflected in categories and analytical levels applied to the study of the mediated voice. Whereas in interpersonal communication one could make do with just the signs and zones of proxemic relations suggested by the voice itself (as Hall does), mediation makes at least two more levels important: that of loudness and what loosely might be considered the "sound" contributed by the recording technology. I thus suggest using three levels of analysis, which I will be calling vocal distance, intended earshot, and microphone perspective.

*Vocal distance* is the term chosen to describe the way the voice in itself signifies a proxemic relationship, as it would be in interpersonal communication. For instance, a whisper is intended for a listener within touching distance.

In mediated talk, the volume fader achieves such an important function in influencing our perception of sound space that volume is recognized as a separate level of analysis called "intended earshot." The term *earshot,* however, does not mean the outer limits of an audible sound, as one might usually employ the term. The slightly awkward term *intended earshot* is instead coined to describe the primary earshot signaled by the volume of the voice. In other words, when a speaker addresses a large or distant audience, she will speak with a loud voice that will also be heard within distances closer to the speaker. The intended earshot in this case will nevertheless be regarded as directed at the more distant listener and thus understood as a public earshot. And even though a whispered voice may be (over)heard at some distance, it is nevertheless clear that it is meant for a nearby listener. Its low volume thus is a sign of an intended earshot. Because of the schizophonia introduced with mediation, loudness becomes a parameter possible to manipulate separate from the speaker's vocal distance, that is, the sound of the voice itself. The speaker may even intend the voice for a different proxemic zone than what the production team achieves. For instance, changing a voice with a vocal distance intended for a close-by listener to an extended earshot is often used as a dramatic or comic device in film and television, such as in the scene in *M*A*S*H* (Robert Altman, USA, 1970) where Margaret Houlihan's intimate pillow talk with Frank Burns is broadcast over the public-address system for the whole camp to hear, thereby providing Major Houlihan with the nickname "Hotlips."

The last analytical level used to describe the proxemics of the voice is called "microphone perspective." This level encompasses the mise-en-scène of the voice by the technological apparatus and acoustical characteristics important in judging distance (such as the direct-to-reflected sound and timbre), with the noted exceptions of volume and intended earshot. Though the choice of the microphone itself—and the placement of the mike in relation to the speaker—is arguably the most crucial part of the technological factors in mediation of the voice, a range of other technologies may also be important here, such as compressors, equalizers, added reverberation, and more. In order to not make the analytical levels unnecessarily complicated, however, these factors are all included under the term *microphone perspective,* with the possible danger of thus overstating the role of the microphone itself.

Whether a voice is recorded in a closet or a cathedral will affect the sound of a voice in important ways, influencing in particular our perception of distance.[2] To simplify the discussion and tie microphone perspective to a reference that is meaningful for most of us, the descriptions in the second column of figure 2.1 are to be regarded as an ideal type, based on the kind of reflections, reverberation, and absorption conditions typical for a living room. Likewise, though microphones may vary in spatial representations of a voice, the description is an attempt to indicate the distance of a nondirectional mike even though the particular voice in question may be recorded by a shotgun mike at two meters' distance or a radio mike below the chin of a performer. The important point is that increasing the microphone distance will achieve a somewhat thinner sound, whereas the amount of reflected sound to direct sound will increase.

During my work developing and adjusting the analytical tools presented here, I have simplified Edward T. Hall's categories slightly. As mentioned, Hall used four zones divided into "near" and "far" regions. For the two closest regions (the intimate and personal), such a level of detail seems unnecessarily inexpedient when dealing with sound, as our ears are poorer judges of spatial details in distance than vision and (especially) touch, where smaller movements make a bigger difference within the most intimate zones. However, I found the distinction of "far" and "near" to make sense for both the social and public zones, with the added benefit that the zones would then correspond more closely with terminology known within visual film analysis. I thus ended with a blend of Hall's zones adapted into the language of image framing: extreme close-up (ECU) for the intimate zone, close-up (CU) for the personal zone, medium close-up (MCU) and medium shot (MS) for the social zone, and long shot (LS) and extreme long shot (ELS) for the public zone, as described in figure 2.1.[3]

Figure 2.1 brings short descriptions of each of the three analytical levels proposed. Although each description attempts to explain how the vocal characteristics sound, describing the aural qualities of sound in such a way is a nearly impossible task. The figure is hence also available as a Web site bringing concrete examples from film and television selected to show concretely how I have judged these clips.[4]

## Snapshots from a Study

The following will bring small samples from a larger study where the analytical terminology presented here was put to the test.[5] I will mention some general findings related to two of the eight genres analyzed, and go further into an analysis of a short promo from Norwegian TV-2 toward the end of the chapter in order to highlight a few findings with more general theoretical interest.

Figure 2.1. Proxemic zones and analytical levels

| PROXEMIC ZONES | Vocal Distance SOUND / FUNCTION | Microphone Perspective[a] SOUND / FUNCTION | Intended Earshot[b] SOUND / FUNCTION |
|---|---|---|---|
| INTIMATE (ECU) ≈ 5–45 cm ≈ 2"–18" | Moaning; breath; whispering between lovers; intimate confession; confidentiality; exclusion of other listeners; "back region" behavior | Direct sound only; nonaudible reflections; proximity effect and "bassy" voices; "dry" sound; clearly noticeable mouth sounds (breath, click of tongue, etc.) | Normal earshot for whispering; loud speech is intrusive; screams can hurt |
| PERSONAL (CU) ≈ 45–120 cm ≈ 18"–4' | Soft indoor voice; personal conversation between 2–3 friends; shyness; closeness; non-exclusive | Barely audible reflections; a cut; sense of room tone; not particularly "dry" reflection or "bassy" speech; audible mouth sounds and breath, but not "up front" | Normal earshot for soft speech; whispering is clearly intelligible; loud speech is intrusive; screams can hurt |
| SOCIAL (near) (MCU) ≈ 1.2–2.2 m ≈ 4'–7' | Soft to regular conversation between 5–6 people; closeness; personal and social mode of address; community; nonintimate | Obvious mix of direct and reflected sound, though direct sound dominates; rich midtone sound; some audible mouth sounds and breath | Normal earshot for soft to regular conversation; whispering is heard, but is not necessarily intelligible; loud screams do not hurt, but are intrusive |
| SOCIAL (far) (MS) ≈ 2.2–3.7 m ≈ 7'–12' | From regular to partly raised conversation between several persons, or to a small group; community; extrovert; social; partly public | Marked increase in reflected sound, clear distance, and perspective; normal mid-tone sound; barely audible mouth sounds | Normal earshot in regular to raised conversation; whispers may be heard, but may be difficult to understand; loud screams may be intrusive |
| PUBLIC (near) (LS) ≈ 3.7–7.6 m ≈ 12'–25' | Voice raised to many listeners; lecture; public; community; self-confidence; authority; attention | Reflected sound dominates, but not at the expense of intelligibility; slightly "thin" sound; nonaudible mouth sounds and breath | Normal earshot for raised conversation and public speech; soft speech sounds weak; intimate conversation hardly intelligible, but may be heard |
| PUBLIC (far) (ELS) ≈ 7.6 m < ≈ 26' < | Shout across the street; speech to large audience; scream for help; public; authority; attention; power | Exaggerated reflections; noticeable "slap" echo may be heard above 12–13 m; thin sound with clearly reduced bass; reduced intelligibility | Normal earshot in very loud speech to the farthest reach of screams; soft speech difficult to understand; whispering barely audible |

Subtext: The descriptions of the different zones and levels in figure 2.1 are best understood when compared to examples supplied at the following Web site: http://www.media.uio.no/personer/arntm/zones/.

a. The relationship between the sound of the voice and the sound space may vary within the same zone. For instance, a voice with clear "mouth sounds" (breath and so on) would qualify for an ECU mike perspective even though the frequency spectrum is not very rich in bass. Similarly, a voice with clear "proximity effect," such as in many trailers and commercials, will also be regarded as recorded with a ECU mike perspective, although this may not contain any mouth sounds or other signs of an intimate mike perspective, since such signs of "imperfection" are very often edited out of commercials and trailers (Maasø 115, 164).

*Proximity effect* is a technical term describing an increase in lower frequencies when using a directional mike at short distances. Although this has often been regarded as a technical flaw, others have used it as an aesthetic effect to achieve intimacy in radio, popular music, and advertising, ever since the directional microphone was introduced in the mid-1930s. Frith argues convincingly that "crooning" as a popular music style is intimately connected with this microphone technique, achieving a type of communicative intimacy unheard of previously. (See also Holman 70 and Woram 88–90 for technical and acoustical accounts of the proximity effect.)

b. Note that *any* sound within the zones of intended earshot will be possible to hear at closer zones, and will be audible at least two zones outside (with the obvious exception of the far public [ELS] zone). In other words, the *boundaries* between zones are especially fuzzy within the category of intended earshot. Note also that the examples used in the description of intended earshot point to interpersonal communication—approximately twenty-five feet—in interpersonal communication, although this is quite possible with electroacoustic sound, as the sound clips accompanying the figure clearly exemplify. Since listening levels will influence the experienced earshot when watching television, it is important to listen at a consistent level when performing analysis. I have chosen to listen at a level where regular news speech is roughly at the same level as that of a conversation within a social proxemic zone in interpersonal communication.

*Figure 2.2:* Typical proxemic spans found in film dialogue and ads, trailers, and promos on television

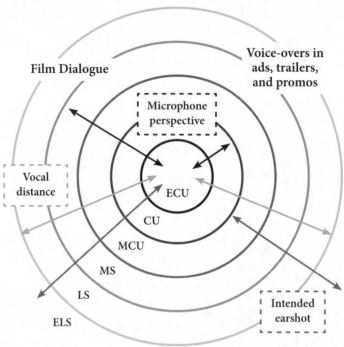

Subtext: Figure 2.2 gives a "bird's-eye view" of the proxemic range of *microphone perspective, vocal distance,* and *intended earshot* in two of the genres analyzed in Maasø (see 149–91, 269–307). The left-hand side of the figure shows distances found in film dialogue, whereas the right-hand side shows the vocal proxemics in voice-overs in ads, trailers, and promos. Abbreviations show the six proxemic regions, from ECU (extreme close-up) to ELS (extreme long shot).

There is quite a generic variation in the proxemics of the mediated voice as experienced in mainstream film and television. In general, the amount of least variation is found in the analytic level of microphone perspective. Here most of the television genres studied show an even narrower range of mike perspective than in film, with the most common perspective (in the limited material in question here) within the "personal" proxemic zone (CU). Some genres, such as promos, ads, and trailers, as shown, are very tightly miked, with little if any sign of the room surrounding the voice.[6] The study thus seems to confirm Altman's general claims for a uniform close-up and medium close-up perspective in film (at least according to the way the categories are defined here). This should come as no surprise since sound recording has become a highly standardized craft (as many have pointed to before), where the perhaps main goal during recording is to come as close as possible with the mike and get a "clean" recording.[7] Making the voice seem more distant is easily achieved in postproduction or live mixing by simply adding reverb or lowering the volume.[8] Removing reverb, increasing the gain (without also raising the general noise level), and making the voice sound closer in postproduction or live mixing, however, are much more difficult tasks.

The range of intended earshot varies more in both the genres showed in figure 2.2 as well as others analyzed. Yet the most common earshot across genres in the material analyzed is medium shot, that is, in the far social zone. Very rarely is the intended earshot intimate or personal. This would also be strange in media concerned with communication. A voice in interpersonal communication is often spoken softly precisely to prevent eavesdropping by anyone other than the intended addressee. Mimicking this is a fairly common narrative device in fiction film and TV drama (in order to conceal narrative information from characters within the diegesis and from the audience). This is nevertheless rare in film and virtually unheard of in the majority of broadcast genres.[9] When Altman uses the term *for-me-ness* to describe sound perspective in cinema, one might thus side with Sarah Kozloff and ask, "But who—or what—else would the sound be designed for?" (Kozloff 121).

Vocal distance shows the largest span of proxemic zones of the analytical levels in question. Though it is most common to keep speech within the social zone (MCU and MS), the voice regularly spans over all regions—from intimate whispers to loud shouts. Vocal distance also is the most divisive factor between the genres analyzed. I believe an important interpretative key here is in determining who is being addressed by the speaker. When one divides addressees into three groups consisting of viewers, studio audience or other conversation partners present (whether they are found within a fictional diegesis or mediated real-life conversation), and both viewers and the studio audience at the same time, the span in vocal distance varies relatively systematically in relation to these three main categories.

For instance, a voice addressing only the viewers at home (such as program announcers, voice-overs in ads, or certain news announcers) or one emanating from the diegetic space of the recording (such as in interviews, televised debates, or fiction film) usually displays a much narrower span in vocal distance within relative short segments than a talk show host addressing at once a studio audience, viewers at home, and guests. Such hosts often display a very dynamic vocal register, where one speaks within a social distance most of the time, but can both lower the voice to pull a guest into a more intimate mode or raise the voice to play to the studio audience, before returning to a general social zone again.

Fiction films arguably show the widest variety of vocal modes within a single text, from intimate pillow talk to loud screams of horror, depending on the subject matter and narrative. Nevertheless, many movies, especially mainstream Hollywood movies, display a much more personal vocal register than in most other genres. The vocal distance becomes much more personal than what would be possible in similar surroundings in (nonmediated) interpersonal communication. Much of this is presumably due to great control over the recording situation, and what I also suspect is an underlying ideology (or tradition) of vocal intimacy. Actor Carrie Fisher has perhaps touched on this in her humorous description of social relations in Hollywood: "You can't find any true closeness in Hollywood, because everybody does the fake closeness so well."

## The Schizophonic Shout

Nowhere is an intimate address more evident than in advertising and promos on TV. Curiously, this is also the genre with the most public earshot (see fig. 2.1). As more fully discussed in Maasø (105–48), advertising in Norwegian television (introduced in 1988) immediately led to an "arms race" for attention by advertisers. Yet the loud call for attention—itself an ancient rhetorical trope *(attentum parare)*—may today be combined with an intimate voice (concerning both vocal distance and microphone perspective), thanks to the schizophonia of mediation. This unprecedented combination of seemingly contradictory proxemic relations is worth a closer look, as I believe it displays some of the unique features of mediation.

Let me use a short promo for Norwegian TV-2 as an example here. The promo features a woman whispering the words "Se hva som skjer!" (Look what's happening!) while the same words are featured as a colorful graphical representation on-screen for the 1.2 seconds it lasts. When the promo was introduced in 1997, it served as a bumper between regular programs and advertisements, while at the same time branding the channel TV-2. Two aspects are of particular interest in this short promo. First of all, it displayed the largest gap between vocal distance and intended earshot *ever* in Norwegian television, as the whispered voice was also the loudest sound heard on TV until then (see an analysis of decibel levels in Maasø). Actually, the sound was so loud that within a few weeks of furious complaints to the television station, a new version appeared that was 9 dB softer than the original (in other words, the first version was nearly twice as loud as the second, as experienced by listeners). This brings us to the second interesting aspect of this promo: in the two different versions available, both the vocal distance and the microphone perspective are identical, whereas the intended earshot (volume) differs a good deal. This therefore makes an interesting case to study the interaction among the three analytical levels.

Both of these promos (available at http://www.media.uio.no/personer/arntm/tv2/) are simultaneously very *intimate* and very *public* at the same time. The promos thus invite us into an intimate communicative relation, and at the same time shout out this fact, thus making it possible to hear all the way from the kitchen or bathroom. This intimate shout is possible only through mediation, presenting a form of communication literally unheard of before the advent of mediation. When listening to the two versions mentioned, it becomes clear that this schizophonic split influences the way we perceive the individual aural parameters as well as the overall blend of them. Interestingly, the second (softer) version clearly sounds much *more intimate* than the first. As only the *intended earshot* varies in the two versions, this makes for an interesting hypothesis concerning the interaction of the analytical levels in mediated voices.

## The Schizophonic Average

Although a listener is able to simultaneously hear both the intimate and the distant aspects of the two promos, there is nevertheless also a *general* appearance in the two—a sort of average or sum of all proxemic levels interacting. Since these promos obviously vary in general appearance, this has led me to propose a hypothesis about proxemic relations that I call the "schizophonic average": the experience of the vocal proxemics of mediated voices is a result of what may be called a schizophonic average of all three levels of vocal distance, microphone perspective, and intended earshot. In the media these three levels will both interact in providing a *blend* of the three and, at the same time, allow each level to play an *independent* role. Hence, when there is a gap between the proxemic of the three levels, one can hear both the close and the distant aspects simultaneously, whereas the mix of the levels also provides a proxemic average of the segment or utterance in question. It is because of this average that the second version of the mentioned promo sounds more intimate, since the span between the proxemic regions among the three levels is less than in the first version of the promo.

Although the comparison of these two promos provides a particularly striking case (given the extreme gap between the three analytical levels), I believe the schizophonic average is a general phenomenon working across genres in film and television. This will, of course, have to be put to the test on a wider range of material than what I have analyzed elsewhere.

## Lowering the Boom?

The "schizophonic average" may also shed light on proxemic relations in a wider sense, outside of what is available for analysis. More specifically, it can help us understand historical changes beyond the reach of concrete analysis. For instance, although only a very limited amount of older television broadcasts are archived and available for analysis (at least in Norway), much is known about earlier re-cording practices. Interviews and document analyses show that there has been a profound change in microphone practices since the early 1980s. Before this time different kinds of microphones on booms were the dominant type of recording strategy in a variety of genres. During the 1980s, however, radio mikes became popular (partly due to increased camera mobility, including the use of cranes, and increased difficulties in avoiding boom shadows in lighting) and today dominate in most genres, except in single-camera productions in drama and film. With the diffusion of radio mikes follows a narrower dynamic range than with mikes mounted on booms, and a closer microphone perspective with less reflected sound, since the mikes are very close to the spoken source. Both the use of radio mikes and the practice of voice compression also bring out more of the intimate "mouth sounds" (breath and so on) that elsewhere I call the "back-region quali-ties" in the voice (see Maasø).

All in all, this has clearly led to an increased intimacy in microphone per-spective over the past couple of decades in television. Following the logic of the schizophonic average, this would then imply that the proxemics of the voice has become more intimate during the same period.

Over this same time frame, commercial television has been introduced in the Nordic region, as in many other European countries. On the commercial chan-nels, ads and promos make up roughly 20 percent of the total programming hours (compared to more than one-third on daytime U.S. television). Since ads and promos are aurally dominated by voice-overs (see Hirdman; and Amundsen et al.) and many of them display a very intimately spoken voice, even the vocal distance has become more intimate in television overall. One might thus say that television in some ways has raised the boom for good while achieving a more intimate voice than most booms ever did.

Raising one's voice in conversation could mean several things: that a speaker is uncomfortable with the situation and wants to signal further "distance" to the addressee, that she is excited or angry, or simply that the noise level behind the conversation is loud. Describing spatial relations and the proxemics of the voice can thus never take the place of further qualitative analysis. I nevertheless believe it is necessary to develop some terms and categories that reduce the enormous complexity and nuances *in* the mediated voice in order to facilitate communica-tion *about* the voice. I hope the vocabulary and categories presented here thus open up further discussion about the role of the voice in communication.

# Notes

1. The focus here will be on the sound of the voice. Although the interaction of vision and audition is of great importance when interpreting proxemic relationships in interpersonal as well as mediated communication (such as discussed by Chion), I will not deal directly with this here. The attempt of providing analytical tools to deal with the audible aspects of the mediated voices will thus need to be tested in audiovisual analyses of film and television.

2. Manipulating the level of reflected to direct sound, by adding reverb, is another fairly common dramatic device in film and television, for instance in *The Third Man* (Carol Reed, UK, 1949). In the sequence shortly before Harry Lime (Orson Welles) is shot trying to escape the police in the Vienna sewers, we hear a cacophony of distant voices, all with an extreme amount of reverb, thus also becoming an effective sign of the inner confusion and turmoil that the character is experiencing. Compared to real life in similar settings, however, the visual distant voices would not have the close-up vocal perspective such as many of these voices have, and would also sound "thinner" than the voices we hear.

3. After developing the categories for my research in Maasø (2002), I became aware of Theo van Leeuwen's then newly released (and excellent) book *Speech, Music, Sound*. Here van Leeuwen interestingly introduces similar categories for what I have called "vocal distance" (24–27), and also with Edward T. Hall's work as inspiration, although Van Leeuwen oversimplifies Hall's eight regions by presenting five (intimate, personal, informal, formal, and public). The "informal" zone in van Leeuwen's terminology, however, entails both what I call "medium close-up" and "medium shot" (that is, the near and far regions of the social zone). My own view is that it is too reductive to collapse the near and far regions of the social zone into one, precisely because this region seems to divide the more informal announcers and more formal news anchor, which in my view is a significant shift. More important, though, the main contribution of my own analytical terminology lies not in the adjusting and simplification of Hall's original categories but in the introduction of the analytical levels of microphone perspective and intended earshot in order to deal with the challenge of mediated schizophonia. Though van Leeuwen acknowledges the importance of loudness and microphone perspective, he does not incorporate them into his proxemic categories.

4. Listening for the vocal distance may be a good place to start, as the sound varies a fair amount within this level. The two other analytical levels require more training in what Chion calls "reduced listening" *(Audio-Vision)* in order to discriminate between the different zones, especially concerning the category microphone perspective. This has, at least, been my own experience, as well as that of a few students testing out these analytical categories in their own analyses.

5. The analysis sampled sound bites of fifteen to sixty seconds in length from different parts of programs in eight genres (news; sports; promo, ads, and trailers; fiction film; TV drama; talk shows; variety shows; and portrait interviews) broadcast on Norwegian TV-2 and NRK-1 during one week in March 1997. Slightly more than two hundred sound bites were analyzed (see Maasø 149–91, 269–307). The study was not designed to provide

findings in a strict statistical sense (which would have demanded a much larger and more representative selection); nevertheless, the material is sufficiently large and varied to give a sense of the typical *span* of proxemics one finds within these genres.

6. It is worth noting that voice-overs in ads, promos, and trailers are all recorded in acoustically "dead" rooms. More important, as the microphone is by definition not visible in voice-overs, microphones can be placed as close as needed, which leads to an ECU or CU microphone perspective in most of these recordings.

7. This is also one of the most consistent answers in my own interviews with twenty-five technicians and producers in television, also confirmed by observation of several TV productions (see Maasø).

8. Adding reverberation is also often used as a conventional sign for increased subjectivity or "inner thought" in many voice-overs in film and television. This also highlights the importance of interpreting vocal sound and distance in relation to the narrative and communicative context.

9. Notable exceptions are documentaries and other fact-based programming recording interpersonal communication, such as live broadcasts of political hearings and court cases where a witness may lower his or her voice, turn away from the mike, and address a lawyer. In many reality series contestants also try concealing conversations from the TV audience or other contestants present. However, this often fails, as radio mikes, booms, and concealed mikes tend to cover them very well (see Maasø [in progress] *Sonic Surveillance in Reality TV*).

# 3

## Almost Silent

### The Interplay of Sound and Silence in Contemporary Cinema and Television

**PAUL THÉBERGE**

Silence, absolute silence, is a rarity in contemporary cinema and television, so much so that it is difficult to even talk about silence in a meaningful way, let alone arrive at anything resembling a comprehensive theory of silence in cinema. Often, however, soundtracks can become *almost* silent, that is, sounds are reduced to simple room ambience or to a single tiny sound—the sound of an insect perhaps—such that, as Walter Murch states, "it feels silent, but it isn't" ("Touch of Silence" 96). In this sense, silences occur in film more often than one might expect, drawing our attention to particular sonic details or lending a special dramatic impact to the scenes in which they take place. There is also a kind of silence that is produced when, for example, music is allowed to dominate the soundtrack while dialogue and sound effects—the primary sonic modes of the diegetic world—are muted. This latter type of "silence" can function in a variety of ways in cinema, but it is peculiar, first, because the soundtrack is far from silent (in fact, an entire orchestra might be playing at full volume), but also because it suggests that audiences have become accustomed to hearing the soundtrack in a structural kind of way, as a set of relationships between the conventional divisions of the soundtrack into dialogue, sound effects, and music.

In this essay, I will explore some of the ways in which silence functions within the narrative structures of contemporary cinema and television. I will not attempt to be comprehensive in this regard: there are, I would argue, many ways in which silence is used and made meaningful within cinematic texts, certainly far more than can be surveyed in a single brief essay. What is more important for my purposes here is to develop an understanding of the ways in which silence works in relation to sound in film, the ways in which patterns of sound and silence emerge and contribute to the overall structure of the narrative. To this end, I will use the idea of silence in both a literal and a figurative manner, and as an element of narrative structure, in an attempt to analyze the dynamics at play in the construction of contemporary soundtracks. Ultimately, not only is the entire soundtrack the unit of analysis here—including dialogue, sound effects, music, *and* silence—but so is its relationship to narrative, to technology, to generic conventions, and to the culture at large.

## When Is a Soundtrack "Silent"?

Silence holds an ambivalent status in theories of film sound: on the one hand, it is conventionally held that absolute silence cannot be allowed to occur in film because it risks being interpreted, by audiences, as a technical breakdown, and, on the other, scholars have long argued that silence becomes possible, as a formal element within cinema, only as a result of the coming of synchronized sound. The fear of a perceived technical breakdown has its origins in the world of radio—a medium where sound is the only sensory modality and a silence of more than a few seconds might indeed confuse the audience—but has been repeated often enough, by filmmakers and theorists, that it has assumed the status of a maxim in cinema as well. However, though not denying the truth of this principle, Shoma Chatterji has suggested a more cultural foundation for this belief: "We are surrounded by all kinds of sound all the time. We're always surrounded by sound, as a result of which if there is silence we think something is wrong" ("Silence Juxtaposed" 105). Although Chatterji is speaking within the context of Indian culture, both in terms of everyday life in India and in the Indian experience of cinema, this almost phenomenological explanation would seem to be plausible for other cultures as well, but it should be noted that film audiences in India tend to be more vocal than their counterparts in the West, where long-standing traditions of concert hall and cinema decorum constrain audiences to remain largely silent.

In the West, our cultural understanding of silence is also marked by an ambivalence that in certain respects parallels that of film theory itself: on the one hand, silence is valued as a form of tranquillity, and, on the other, it is often the sign of abnormality, something to be feared. And film and television narratives have certainly contributed to our sense of both: consider, for example, the contrast between the quiet tranquillity of rural life and the danger and cacophony of the city as exemplified in the early minutes of the film *Witness* (Peter Weir, USA, 1985), or the cliché of so many television hospital dramas where the switching off of the sound of the heart monitor is the final admission of failure and death. In other instances, noise—even the constant background noise of modern communications media—can be thought of as preferable to silence: in the final moments of *Asphalt Jungle* (John Huston, USA, 1950), the chief of police turns up the volume on several radio receivers, resulting in a cacophonous montage of police communications; then, in switching them off again, he invokes the fear of a city without police protection, a city where dangerous criminal elements are allowed to operate in silence.

On a different tack, filmmaker Mike Figgis has suggested that the prohibition against absolute silence in cinema production may also stem, in part, from the way in which complete silence can place the audience in a form of direct confrontation with both the film and other members of the theater audience. Figgis deliberately experimented with complete silence in a sequence in *Leaving Las Vegas* (Mike Figgis, USA, 1995) in which Ben (Nicolas Cage) almost has a heart attack while drinking too much alcohol, falls, and is knocked unconscious:

> What I discovered when I started to tour with *Leaving Las Vegas* in America was that this moment in the film—in a crowded cinema, with a good sound system— was extremely uncomfortable when he's so distressed. And suddenly, it's so quiet in the cinema that you can literally hear everything, and you don't have the protection of this sound blanket of mush, or just ambient noise, or whatever, which we come to expect of a soundtrack. And I loved it . . . but it was even much more powerful than I thought it would be. . . . It's like that moment when suddenly you're talking animatedly and then everybody stops talking and you realize your voice is a bit too loud. (2)

It could be argued that audiences are always more or less conscious of their experience of film and the presence of other audience members, and sound plays a role in this self-awareness: humor, by invoking laughter in the audience, and horror, by soliciting screams, are genres that contribute to self-awareness in this way. However, as suggested by Figgis, the level of audience discomfort generated through the use of silence is quite different, extremely powerful, and has been exploited by filmmakers only on occasion, usually in instances where intensely dramatic or violent acts are depicted. For example, Michael Ondaatje has described such an instance in *The English Patient* (Anthony Minghella, USA, 1996) where Caravaggio (Willem Dafoe), who is about to have his thumbs amputated, pleads with his torturers, "Don't cut me": "[Walter] Murch [film editor and sound mixer on the production] makes the response to the line a total and dangerous silence" (x–xi).

Among those who posit that silence is, in fact, a by-product of the introduction of synchronized sound in cinema, silence need not be total, nor is it "a neutral emptiness. It is the negative of sound we've heard beforehand or imagined; it is the product of a contrast" (Chion, *Audio-Vision* 57). As such, "silence" is always relative, and relational to sounds heard in the context of the film itself. Relative degrees of silence have been used for specific effects in cinema for many years: for example, the general level of environmental noises can be attenuated, allowing for a focus on a particular sound (such as the ticking of a clock) (57–58), or ambient sounds can be suppressed entirely, thus giving the impression of entering into the mind of a character (89). With regards to the latter, we can think of selected *parts* of the soundtrack that are silenced (producing, in this case, what might be referred to as a "diegetic silence"), whereas other parts, such as music, continue to sound or are given special prominence (I will return to this type of silence in greater detail below).[1]

The possibility of reducing the soundtrack to near silence has been greatly enhanced over the past several decades by developments in the technologies of sound production and reproduction. The diffusion of Dolby technology during the 1970s was certainly a turning point in this regard, allowing for an expansion in both the dynamic and the frequency range of film sound. For Michel Chion, this expansion created not only the conditions for a greater fullness and density of sound production in film but also "a new sound space to empty, a new silence. This new silence, which reigns around isolated words or sounds, imparts a particular new intensity to certain scenes" ("Silence of the Loudspeakers" 150–51). According to Chion, this new sound space was not entirely inaugurated by Dolby technology—it is apparent even in earlier magnetic soundtracks—but it is greatly enhanced by Dolby.

Sound designer Walter Murch is in agreement with Chion on his point that the sheer reproductive capacity of new technologies necessitates a movement toward silence, toward a kind of emptiness: "One of the dangers of film, exacerbated in this brave new world of digital visual effects and six-seven-eight-channel sound, is that the real subject matter of the film can be crushed by the film's ability to represent it. As a result, I'm always searching for ways to get out of that rut and choose sounds that are off-axis and, ultimately, to find places where I can get to silence. And I do it not just for artistic reasons, but also for the physical relief that silence gives" ("Touch of Silence" 100). Not surprisingly, in his work, Murch has often exploited the expressive potential of the quieter realm of sounds produced through Dolby and more recent digital technologies:

> The desert is a vast space. When you're there, the feeling it evokes is psychic as well as physical. The problem is that if you record the actual sound that goes with that space, it has nothing to do with the *emotion* of being there. In fact, it's a very empty, sterile sound. . . . [T]he trick in *The English Patient* was to evoke, with sound, a space that is silent. We did it by adding insectlike sounds that, realistically, would probably not be there. . . . Also tiny sounds—as tiny as we could get—of grains of sand rubbing against each other. . . . We took those tiny things and made a fabric out of them. (quoted in Ondaatje 118; emphasis in the original)

It is intriguing to compare the quality of emotion suggested by Murch's approach to sonically representing the desert in *The English Patient* and that evoked by the grand orchestral gestures employed, for example, by composer Maurice Jarre in his score to *Lawrence of Arabia* (David Lean, UK/USA, 1962). Certainly, such differences should be attributed not to different uses of technology or even differences in the modality of sound versus music but rather, to how sound techniques and music can be made to resonate with narrative themes and the underlying ideological stance taken toward the desert in each of the films in question.

Dolby technology along with the standardization of multichannel sound reproduction, THX-certified sound systems in theaters, digital recording techniques, and other innovations have certainly created a world of new creative possibilities for the production and reproduction of both sound and silence in cinema, but these technologies may also have effects that are not only sonic in nature. As Mary Ann Doane has argued, sound ties the on- and off-screen spaces of the film to the metaspace of the cinema, the space of the audience ("Voice in the Cinema" 39–40), and so, too, does silence. "Dolby cinema thus introduces a new expressive element: the silence of the loudspeakers, accompanied by its reflection, the attentive silence of the audience. Any silence makes us feel exposed, as if it were laying bare our own listening, but also as if we were in the presence of a giant ear, tuned to our own slightest noises. We are no longer merely listening to the film, we are as it were being listened to by it as well" (Chion, "Silence of the Loudspeakers" 151). In large part then, this overlapping of spaces and the new level of "attentive silence" that is a by-product of Dolby technology are responsible for the degree of audience discomfort, mentioned earlier, that Figgis was able to produce in *Leaving Las Vegas*.

## Relational Silences within the Soundtrack

The conventional division of the soundtrack into dialogue, sound effects, and music, and the hierarchy that it implies, was not established immediately in response to the invention of synchronous sound reproduction during the late 1920s. As Rick Altman has argued, the achievement of what he refers to as the "multiplane sound system" took place over a period of almost a decade as the result of a complete reconsideration of the role of sound in film during the early sound period. Furthermore, I agree with Altman's assertion that film sound cannot be fully understood without due recognition and analysis of the relationships that exist *between* the various elements of the soundtrack and not simply the relationship of sounds to images (Altman, Jones, and Tatroe 341).[2] In his historical analysis of the development of Hollywood's multiplane system, Altman argues that during the early period of the conversion to sound, some films adopted an all-or-nothing approach, inserting musical cues, for example, only in spaces where there was no dialogue in order to ensure intelligibility. Later strategies pursued a subtle balancing of sound elements through the manipulation of volume levels, thus allowing for a more fluid relationship between on- and off-screen sounds, while maintaining a general consistency in the overall sound level.

What is remarkable, but not noted by Altman, is that almost as soon as Hollywood sound engineers achieved this new system of balancing sounds, a different kind of all-or-nothing approach appears to have emerged, one that pitted the various sound components against each other, not for functional reasons but for specific dramatic purposes. Indeed, I would argue that whereas the balanced multiplane approach to soundtrack construction certainly became the dominant mode of Hollywood sound practice, filmmakers continued to exploit a somewhat crude juxtaposition of sound components as a kind of special effect. Through such juxtapositions of sound, various relationships of presence and absence are created—what I shall refer to here as "relational silences"—and the contrasts thus produced are a powerful tool for creating meaning within film narrative.

My point can be illustrated through a comparison of the early sound version of Robert Louis Stevenson's *Dr. Jekyll and Mr. Hyde* (Rouben Mamoulian, USA, 1931), starring Fredric March, with a second version appearing a decade later (Victor Fleming, USA, 1941), starring Spencer Tracy. Mamoulian's 1931 version belongs to the period of early sound films before the establishment of the multiplane system when, partly in response to the continuous use of music in the silent period and partly in response to its particular uses in the early musicals, many dramatic films eschewed the use of background music entirely in favor of source music—music justified within the diegetic world—such as the opening point-of-view sequence where the hands of Jekyll (March) are shown playing the organ. Faced with the problem of representing, without recourse to music, the central dramatic moment when Jekyll first transforms into Hyde, however, Mamoulian created his now famous montage of voices, heartbeats, manipulated bell sounds, and the like, an innovative experiment in *musique concrète* more than fifteen years *avant la lettre.*

An examination of Fleming's 1941 production reveals an entirely different approach to sound: belonging to the period of classical Hollywood sound cinema, the film makes extensive use of dialogue in conjunction with sound effects and a rich background score composed by Franz Waxman, and, in this regard, the film exemplifies the conventions of a fully matured multiplane sound system. In the first transformation sequence, however, Fleming creates a dreamlike, Freudian scenario in which Jekyll/Hyde is depicted whipping a pair of galloping horses that then mutate into the two central female characters in the film. In representing this movement from the real world to that of the dream, Fleming cuts the dialogue and sound-effect tracks out completely—in a sense, silencing the conventional diegetic world—and brings Waxman's dramatic score up to full volume. Fleming's approach—a gesture that has become, by now, a cliché in the depiction of dream sequences, drug-induced states, and other moments when the audience is allowed access to the mental life of a film character (sometimes referred to as "metadiegetic" moments in a film)—relies entirely on the audience's understanding of the conventional relationships that exist between elements of the multiplane sound system and the roles they play within Hollywood realism, even while it appears to momentarily break with those conventions.

Although this use of what I would call a "diegetic silence" has certainly become something of a cliché in Hollywood cinema, it is nevertheless remarkable how often it continues to be employed and the extent to which it would appear to retain some degree of narrative utility. For example, Quentin Tarantino creates a slow-motion dream world around a sequence in which Vincent (John Travolta) shoots up in *Pulp Fiction* (USA, 1994) by concentrating on extreme close-ups, dropping out all diegetic sounds, and allowing pop music to fill the soundtrack. Likewise, when Almasy (Ralph Fiennes) carries the dead body of his lover, Katherine (Kristin Scott Thomas), from a cave in the closing moments of *The English Patient,* the desert winds and his cries of anguish are completely absent, submerged within Gabriel Yared's orchestral score and the dying words of Katherine in voice-over.

It is important to note, however, that diegetic silences—silences that are then filled by music or other nondiegetic sounds—are used not only to represent the inner life of characters, their dreams, fantasies, or moments of mental anguish, but also occasionally, in a somewhat different fashion, to represent any moment in which reality exceeds our expectations, when the real becomes the surreal. For instance, in the film *Kandahar* (Mohsen Makhmalbaf, Iran/France, 2001), music dominates the soundtrack in a sequence where the victims of land mine explosions hobble, on crutches, out into the desert as prosthetic limbs, air-dropped by parachute, fall from the sky.

Because it implies a break from the world of everyday sounds, a diegetic silence can also be used to bridge the gap between temporal and spatial moments in film narrative. In many road movies, or films that contain a significant number of road sequences, music offers a sense of time passing, of distances traversed. In *Rain Man* (Barry Levinson, USA, 1988), for example, the sounds of the road are sometimes muted so that the rhythms of Hans Zimmer's score or those of a pop song can lend a certain feel to the passage of time, a feeling of *durée* (see Gorbman 38). In one sequence, the song "Beyond the Blue Horizon" is first heard (apparently emanating from a radio or some other diegetic source) outside a roadside Laundromat, the music then comes to full volume and frequency range as we follow the silent car carrying the two main characters, in long shot, across a large, open terrain, and the song fades as the car arrives at another roadside stop; by leaving diegetic time and entering into musical time, what might be an entire day's journey is compressed into a single chorus of a pop song.

A more general type of musical bridging has become increasingly prevalent in contemporary cinema, in part as a result of the influence of music video on the world of film. In this instance, a number of events of relatively little narrative significance are often condensed into a single sequence with musical accompaniment but no diegetic sound. One senses this type of influence on a sequence in the film *Traffic* (Steven Soderbergh, USA, 2000), for example, when a series of drug busts is conducted with only the sound of the background score on the track.

Diegetic silences tend to be relatively obvious because they break so completely with the dominant representational mode of Hollywood realism. "Musical silences," on the other hand, are often more subtle and may go largely unnoticed in some instances, in part because we are not accustomed to hearing music all of the time in most films; of course, the overall quantity of background music is not so much at issue, but rather how musical silences occur, what takes place during them, and what they mean within the structure of the narrative. Even within film theory, however, there has been a tendency to pay more attention to the presence of music than its absence. This is perfectly understandable, but in the present context it is important to consider the potential for music to exist in almost any part of a film. With this in mind, the absence of music at any given moment may be as significant as its presence.

In an account of the reediting of Orson Welles's *Touch of Evil* (USA, 1958; reedit/restoration 1998), Walter Murch has described the timing of a diegetic musical cue—music played in the film by a pianola (a player piano)—that was not what had been specified by Welles in a memo, originally written to Universal after the studio had edited the film without his permission. When corrected, the cue functioned in an entirely different fashion than it had in the original mix: the sound of the pianola first enters while Quinlan (Welles) is outside a brothel, in effect beckoning him to enter; once inside, the music pauses and then stops entirely during the course of his conversation with Tanya (Marlene Dietrich), each time marking a shift in the character of their exchange. Unlike background music, which might add to the emotional content of such a conversation through melodic or dynamic emphasis, the more neutral character of the pianola music makes its impact felt through its starts and stops, what Murch refers to as a form of "dramatic articulation" ("Touch of Silence" 91–92). Clearly, the pianola music functions in the scene like background music—its origin in the diegetic world is merely the alibi for its presence in the scene—but it makes its presence felt more through its end points, its silences, than through the actual sound of the music.

Large-scale musical silences are also possible within film, but, again, it is only within the context of the film itself, or the context of the genre conventions in which it operates, that one can determine whether such silences are significant. In the film *Cast Away* (USA, 2000), director Robert Zemeckis and sound designer Randy Thom created an island world consisting of primarily the sound of wind, ocean waves, and creaking palm trees; background music is nowhere present in the soundtrack during the period in which Chuck (Tom Hanks) is stranded there (more than an hour of the film's duration) and returns only at the moment of his escape. The absence of music—that most collective of art forms—intensifies the sense of isolation, the distance from civilization that Chuck must endure.

At the local level, certain types of scenes—scenes of violence, for example—are often given special attention by filmmakers. In *Traffic*, a car bomb, intended for a criminal informant, kills a police officer: leading up to the moment of the explosion, music gradually fades up, pandemonium erupts as the would-be assassin is himself killed, and music then takes over the soundtrack almost entirely, with the explosion greatly reduced in volume and dialogue completely silenced. Later, when the informant is eventually poisoned, the scene is shot with only diegetic sound—a "musical silence"—lending a kind of stark realism to the victim's convulsions and the consternation of his police guards. As these examples suggest, the representation of violence, in sound *and* silence, can take different forms even within the context of a single film, yielding variations in the apparent degree of stylization, mediation, and "realism." Quentin Tarantino, in films such as *Reservoir Dogs* (USA, 1992) and *Pulp Fiction*, has pursued a highly individual, stylized approach to violence, ranging from a kind of gritty realism—tense, drawn-out sequences with no music at all and even very little dialogue or sound effects—to an ironic distancing, produced by pairing violent events with popular songs as source music.

In one of the few detailed studies of an individual filmmaker's approach to sound and silence, Elizabeth Weis has offered an insightful analysis of precisely the themes of murder, terror, and violence in the films of Alfred Hitchcock. For most filmmakers, according to Weis, silence is really a lack of dialogue: it might be used to emphasize the impotence of a character (an inability to speak), a terror beyond words, a sense of control (the cool efficiency and calculation of the murderer, standing in, in the case of Hitchcock, for the control of the director), or vulnerability and victimization (the murder victim's stifled scream, the scream to be filled in by the audience) (*The Silent Scream* 148–67).

Like musical silences, dialogue silences tend to be disguised in many films, in part because they often appear as a function of the script or to have been produced by actors in the context of performing a particular role or scene; in this sense, they appear "theatrical" in character rather than strictly filmic.[3] This is true even when momentary silences in dialogue are created, enhanced, or extended in duration through editing (as is the case in the example of the torture scene in *The English Patient*, described above). In some instances, the absence of dialogue may be more obvious but nevertheless produced as a function of the dramatic or performance context: after discovering the mutilated body of a neighbor in *The Birds* (Alfred Hitchcock, USA, 1963), for example, Mitch's mother (played by Jessica Tandy) runs from the house, mouth agape but unable to speak; she remains speechless, unable to describe the horror she has witnessed, for several scenes that follow (155). Occasionally, dialogue silences can be quite extended in duration, as in the case of *Vertigo* (Alfred Hitchcock, USA, 1958) where Scottie (James Stewart) follows Madeleine around San Francisco. And to the extent that animals can be said to have a "voice," it is interesting to note that after numerous attacks in which the screech and cries of birds figure prominently, the final attack scene in *The Birds,* when Melanie (Tippi Hedren) has wandered off to the attic alone, takes place silently, save for the sound of flapping wings (142).

A number of the examples of relational silence noted above— whether they are examples of diegetic, musical, or dialogue silence—are incidental in character, that is, they are designed to enhance dramatic moments in a film, whereas others have greater significance for the overall narrative structure or stylistic approach of the films in question. It is to this latter type of silence, one that is most important to the art of the soundtrack as a whole, that I would now like to turn.

## Structural, Stylistic, and Generic Silences

In her book *Unheard Melodies,* Claudia Gorbman introduces the concept of a "structural silence" (19), a concept that has since remained relatively undeveloped in her own work and elsewhere. For Gorbman, a structural silence occurs when a sound (music, in the case of her study) that has previously accompanied a particular event or sequence in a film is absent when similar events or sequences recur; the absence of sound is thus noticed by the audience and usually signifies some kind of shift in the course of the narrative. Understood in this way, a structural silence is essentially little more than a play of expectation created within the context of a particular film text. For my purposes here, I would like to extend this concept further to include instances of the use of silence that have other structural, stylistic, or generic implications.

Certainly, instances of the type of structural silence described by Gorbman occur with some frequency and can play an important role in the overall sound design of a film. In the case of Fleming's *Dr. Jekyll and Mr. Hyde,* once we have become accustomed to witnessing Jekyll's transformations accompanied by music (and no diegetic sound), Fleming chooses to stage one transformation in

complete silence: the scene in question occurs when Jekyll/Hyde is forced to reveal the secret of his psychological experiment to his colleague, Dr. Lanyon (Ian
Hunter); by highlighting the scene in this way, Fleming marks the revelation as
a significant turning point in the story. Used as an analytic tool in this way, the
concept of a structural silence is useful in moving us away from limited notions
of "sound design," thought of simply as the crafting of individual sound effects or
the creation of elaborate sound tapestries for individual scenes, toward a larger
sense where sound design is understood as a process in which the expressive
possibilities of sound (and silence) are planned as part of the overall thematic or
narrative structure of the work.

Audience expectations are not only created within the context of individual
film texts; they are also an integral part of the way in which film genres function.
As Steve Neale has argued, genres consist of systems of expectations that are internalized by audiences and become the basis upon which they can identify an
individual film as belonging to a certain genre and, by extension, what is possible
and likely to happen within the film. Neale uses the term *generic verisimilitude* to
describe this play of expectations and probabilities (31–32). *Dancer in the Dark*
(Lars von Trier, Denmark/France/Sweden, 2000) uses music and sound to bridge
the gap between the utopian tendencies of the Hollywood musical and the inevitable suffering and victimization of melodrama. Throughout the early parts of the
film, the mode of the musical genre is dominant. Indeed, no matter how dismal
her lot in life and despite the prospect of going blind due to a genetic disorder,
Selma (Björk) is never more than a production number away from apparent
redemption (the subplot in which Selma pursues her desire to play Maria in an
amateur rendition of *The Sound of Music* confirms these utopian hopes). Even
during a grisly murder scene, Selma and her victim are able to break into song.
However, as the film moves closer to its inevitable conclusion, Selma finds it increasingly difficult to resort to music as a form of solace: she finds the silence of
her prison cell to be oppressive and barely manages a last song and dance number
as she stamps her feet on the way to the gallows.

It is in the final execution scene that von Trier and sound designer Per Streit
finally close the door on the utopian hopes of the musical: on the gallows, Selma
sings her "Next to Last Song" but does so without the accompaniment of a nondiegetic orchestra, as has been the case in all the previous production numbers,
including the final march to the execution chamber; instead, Selma's voice alone
echoes in the reverberant space of the execution room (all other voices and sounds
in the room are muted—a diegetic silence reigns except for the voice of Selma).
In "staging" the voice in this particular manner,[4] von Trier lends the final scene a
kind of stark yet highly stylized realism that would be otherwise impossible within
the conventional codes of the Hollywood musical while, at the same time, retain-

ing the expressive character of the song mode. Furthermore, the reverberation of the a cappella voice is reproduced in full stereo sound and thus envelops the audience, wrapping them within the same acoustic space as the victim, enhancing their identification with her fate: most of the film's diegetic sound is reproduced in mono; only the production numbers and the final scene are reproduced in stereo, and, in this way, the implication of the audience within the musical portions of the film does not occur as the result of Chion's "silence of the loudspeakers" so much as through the activation of the stereo array at selected moments. But perhaps most important for my argument here is that in silencing the orchestra, von Trier creates not only a structural silence within the film but what I would call a "generic silence" as well: the merging of diegetic and nondiegetic sound is a hallmark of the musical genre, and the absence of the orchestra creates an empty space both within the context of the film narrative and within the expectations of the audience as dictated by generic verisimilitude. The silence thus becomes doubly significant, marking her song as the final shift from musical heroine to melodramatic victim.

In a more general sense, there are uses of silence that are specific to particular film or television genres. For example, there is a type of relational silence that is typical of crime films and television police serials: the interrogation room, essentially an isolation chamber within the police station, is doubly silent in the sense that it is both acoustically isolated and visually undecorated (the larger diegetic world is almost entirely reduced to the claustrophobic space represented on the screen) and thus lends its own kind of tension to the psychological intimidation and violence often employed in extracting confessions from criminal suspects. In addition, there is usually a separate room (the room behind the one-way mirror) where others may gather to listen (but remain unheard by those in the interrogation room) and look (but remain unseen); symbolically, this is also the silent space inhabited by the audience. Typically, music is not present during the greater part of most interrogation scenes, often entering only in the final moments when a confession is extracted or an allegation proved, or as a transitional device when a suspect describes their past actions: it is as if this musical silence is required so that emotions—all the shouting, intimidation, bribes, and tearful denials—can be expressed in their rawest form. In this way, a particular architectural space and the acoustic and relational silences that are a feature of it, become one of the regimes of verisimilitude that constitute the police or crime genre.

In television drama, the aesthetics of audio production are not unlike those of film except that there is often a tendency toward a greater stylization: given the need to create a unique identity for each television series, producers use audio to achieve a recognizable "sound" for the series—a sound that will be more or less consistent from week to week. The most obvious way that this is accomplished is through the signature tune and through background scoring. Signature tunes in television function much like jingles in advertising; in melody and arrangement they are designed to be immediately recognizable. Similarly, background scoring for television programs, which often use elements of the signature tune or its arrangement, tend to be remarkably consistent in their use of orchestration and general musical style (regardless of the number of different composers writing for the series), thus lending a kind of overall "tonality" to the program.

In some instances, the balance of sonic elements, and, hence, the implicit and explicit uses of relational silence in television programs, also lends them a unique character; the use of different types of "silence" can thus become an element of program style. In *The West Wing* (Aaron Sorkin/NBC, USA, 1999–2006), for example, a premium is placed on fast-paced dialogue: as the camera follows the various characters around the White House offices, they never seem to stop talking, strategizing, looking for the right public relations spin to put on political issues. The dominance of dialogue, and, in particular, the need to follow the subtle shifts in spoken content, puts a great deal of pressure on sound engineers to maintain intelligibility and places a heavy burden on audiences in terms of maintaining a significant level of attention. As a result, sound effects and music are kept to a relative minimum: in comparing sequences in any given episode of *The West Wing* with those taking place in a busy police station in a crime series, such as *NYPD Blue* (Stephen Bochco/ABC, USA, 1993–2005), or in a hospital drama, such as *ER* (Michael Crichton/NBC, USA 1994–), one finds an overall greater level of relational silence in *The West Wing,* sustained for longer periods of time. Whereas *NYPD Blue* or *ER* may contain significant dialogue sequences, they usually also contain significant levels of background chatter (sometimes referred to by sound engineers as "walla")[5] and sound effects; furthermore, dialogue sequences are balanced by more action-oriented sequences where sound effects, transitional music, and the like play an equal role in defining the sonic world of the series. The relative lack of opportunity for action in *The West Wing* and an almost obsessive emphasis on strategizing and political intrigue demand a certain level of silence: the characters are constantly moving from busy corridors to quiet offices, closing the door, as if to say, "Let's go someplace quiet where we can talk" (or, rather, "where only the audience can hear us"). Perhaps not surprisingly, when the rare musical cue occurs, it is often in the absence of dialogue—indeed, the entire diegetic world is usually silenced: for example, in shots from outside the Oval Office window where we see dialogue continue but do not hear it, or in music video–style sequences where action continues but only to the rhythms of a pop song. In these moments of "silence," the audience is allowed to momentarily disengage from the intense, highly stylized diegetic world of the series.

A particularly salient example of many of the uses of silence outlined above—total, relational, generic, and structural—within a single work can be found in Peter Weir's coming-of-age film *Dead Poets Society* (USA, 1989). On the surface, there is nothing terribly unusual about this film: it is a well-crafted, if somewhat sentimental, take on the coming-of-age genre, save for the suicide that marks the turning point in the story. However, the particular approach to the soundtrack taken by Weir and sound designer Alan Splet—an approach that in many ways resembles the all-or-nothing strategy outlined earlier—works in such a way as to underline the most significant themes and underlying tensions within the narrative (and, to a certain degree, the genre as a whole).

As is typical of the genre, the repressive, authoritarian, disciplined world of the boarding school is set against an expressive, irrational fantasy world of poetry and nature inspired by a romantic, renegade teacher-mentor; the playing out of this fundamental opposition results in the central conflict between the conventional, middle-class expectations of parents, on the one hand, and youthful exuberance and the need to find oneself, on the other. These key oppositions, and the narrative tensions they produce, are expressed in the film in a variety of ways: indoor versus outdoor scenes; full light and motion versus filtered lighting and slow-motion photography; and harsh, rational uses of speech and injunctions to "be quiet" versus the expressive use of language in poetry and theater. Most significant for my purposes here, however, is the way in which the film's themes and oppositions are represented through a set of relational silences, a play of absences and presences between the different components of the soundtrack. For example, throughout the film there is virtually no background music associated with the boarding school. The first significant use of nondiegetic music occurs when the boys escape from the repressive confines of the school into the nighttime fantasy world of the Dead Poets Society; as this transitional music rises to full volume, the sounds of the diegetic world gradually recede into silence.

Once established, the relational silences become a structure within which the narrative unfolds: in a series of back-and-forth movements between the various types of silence, increasing tension is created as the two opposing worlds move in closer conflict with one another, eventually leading to the suicide scene. Picking up this alternating movement at the point in the film when Neil (Robert Sean Leonard) appears in an amateur production of "A Midsummer Night's Dream," one enters the play as one might enter the fantasy world of the Dead Poets Society with a diegetic silence, the characters moving silently, in slow motion to the sound of background music. This is followed by Neil's closing soliloquy as Puck in the play, an eloquent, poetic recitation that is performed without sound effects or music and directed, or so it would seem, solely to his father, who stands silently at the back of the hall. An explosion of diegetic sound following the play—applause, euphoric shouts, and cheers—is muted by the mere mention that the father has come to take Neil home and Neil is reduced to silence.

The following sequence takes place at Neil's home, when an oppressive silence is shattered by an emotional argument between father and son (again with minimal sound effects and no music), resulting in the complete silencing of Neil, whose impotence before the voice of patriarchal authority is represented by his inability to speak his own most fervent desires. A complete dialogue and diegetic silence ensues as Neil retreats one last time to the fantasy world and, with music at full volume, silently prepares for his suicide. At this moment, Weir is able to extend the tension of the sequence by depriving the audience of one critical bit of information: we neither see nor hear the actual suicide take place. Instead, this moment of absolute silence is filled when the music is violently cut off and Neil's father bolts out of bed, having heard something but doesn't quite know what: for both the father and the audience, there is still the possibility that everything that has just preceded this moment was just a dream. With the music track silent, an intensely tactile sequence follows in which every sound effect—the padding of slippers on floors, the flicking of a light switch—is heard with the increased nuance that is characteristic of Dolby Stereo sound reproduction; in addition, the space between the sounds assumes a new kind of depth and emptiness (we hear the voice of Neil's mother from afar as she searches for him), and this in turn elicits Chion's "attentive silence" on the part of the audience. Ironically, upon discovering the dead body of his son, this intensely charged overlapping of acoustic spaces collapses for a brief moment as music enters and Neil's father's voice is heard in slow motion: the fantasy, the son's dream world, has become the father's nightmare.

*Dead Poets Society* does not impress itself on one as a "great" sound film in the conventional sense of the term: there are no climactic battle scenes, no elaborately constructed tapestries of sound that one might sit up and take note of. The soundtrack is, nevertheless, remarkable in the ways in which it creates a structure around selective uses of silence and the consistency with which it does so in relation to the narrative themes present within the film. This approach to sound design comes to full fruition in the climactic moment of the film when the dramatic tension is extended by actually suppressing the sound that accompanies the death of the main protagonist: in this sense, silence proves to be more powerful than sound.

## Conclusion: Fade to Silence . . .

Sound designer Randy Thom has argued that the art of sound design requires that one consider sound not simply as something that is "added to" or "synchronized with" visual images: "Sound has to be *designed-into* a movie. That is, the movie has to be designed for sound, not the other way around" (9; emphasis in the original). As the soundtrack for *Dead Poets Society* and other films discussed in this chapter suggest, movies can also be designed for silence.

In the study of sound design in cinema, the conventional analytic framework divides the soundtrack into dialogue, sound effects, and music. Such a framework is based, in part, on a technical distinction evident in film production and post-production: each sound element is produced, edited, and processed separately and brought together only in the final mix, or "dubbing session." The academic study of the soundtrack has tended to reproduce the distinctions that result from the early stages of this technical process by dividing the soundtrack into more or less discreet categories of knowledge. For the most part, this strategy has been extremely productive: we now have a substantial body of literature devoted to the study of music in film and a relatively smaller, sporadic, but nevertheless critical, mass of scholarship addressing sound and dialogue as well. What is still lacking as a result of this overwrought set of distinctions, however, is an integrated approach to the study of the soundtrack that allows us to examine the various sound elements in their distribution and relation to one another and to narration. That is, we do not have a framework that allows us to hear the "multiplane system" as the sound designer conceives it, as the mix engineer realizes it, and as the audience hears it: as a delicately balanced, elaborately interwoven set of sonic elements, each of which contributes to the overall narrative but does so in a cumulative and integrated fashion.

In this chapter I have, in part, attempted to elaborate an analytic approach that uses the notion of "silence" as a means of arriving at a more integrated framework for the study of film sound. Ironically, the particular use of "silence" suggested here begins with an acknowledgment of the plenitude of the film soundtrack, that is, its basic operating assumption is that any given scene or sequence within the narrative can, at least potentially, have all three conventional elements of sound (dialogue, sound effects, and music) present at the same time and that the significance of sound design lies essentially in the *balance* among these various elements, and their relationship to narrative and generic contexts, as much as in their individual makeup. Second, it suggests that the *absence* of an individual sound, or category of sonic elements, may be as significant for our analysis of film as the actual presence of any particular sound. Indeed, it is in attempting to grasp this play of presence and absence that the analytic approach suggested here forces one, at least conceptually, to hear through the silences, to hold all elements of the soundtrack together at all times.

Ultimately, as James Lastra has argued, sounds are not reproduced in cinema so much as they are "represented": the process of sound recording prestructures sound objects and stages them as representations of sounds ("Reading, Writing, and Representing Sound" 69–75). So, too, with silence: there is no such thing as a true, or absolute, silence, only a system of relative, structured silences—silences that are made to have meaning within the relational and representational context of the soundtrack itself. Understood in this way, silence is perhaps not an empty space so much as a space that is part of a representational system that can be fully understood only if we can arrive at an integrated way of analyzing it. Only then will silence speak as loudly as sound in cinema studies.

## Notes

Parts of this chapter were originally given as a paper titled "Toward a More Integrated Approach to Soundtrack Study," at a conference organized by the Sound Studies Group of the University of Iowa: "Walter Murch and the Art of Sound Design," 30 March–1 April 2000.

1. Throughout this chapter, I use *diegetic silence* and related terms to indicate the part of the soundtrack that has been muted. This approach differs from that of Claudia Gorbman, where the same concept would be referred to as a "nondiegetic silence," indicating that nondiegetic music continues to sound (18). There is, in fact, no widely accepted terminology for the use of silence in film, and I put forward these terms in the hope that the reader will find them clear, consistent, and useful.

2. Altman makes this assertion, partly, in response to Chion's insistence that sound-image relations are paramount and, for this reason, one cannot assume that the soundtrack exists as an entity unto itself.

3. I use the term *theatrical,* in this instance, in a limited sense: that is, to designate the type of pause often introduced by actors in response to a script in order to produce a sense of drama. However, there is a very different theory and use of silence in theater put forth in the work of Maeterlinck ("Silence") that concerns itself with the very inadequacy of language in the face of life's most profound experiences: according to Maeterlinck, our encounters with love, fear, and death can only be truly expressed through silence. In *Maurice Maeterlinck and the Making of Modern Theatre,* McGuinness discusses the impact of these ideas on later generations of playwrights; his discussion may have some pertinence in the world of cinema as well.

4. "Vocal staging" is a concept employed by Serge Lacasse ("Listen to My Voice") to describe the particular ways in which the voice can be represented, through panning, balancing, equalization, reverb, and other effects, in sound-recording practice.

5. For a discussion of the origin and uses of the term *walla,* see Juno Ellis, quoted in LoBrutto 215–16.

# 4

# The Sounds of "Silence"

*Dolby Stereo, Sound Design, and*
*The Silence of the Lambs*

**JAY BECK**

The name Dolby has long been synonymous with modern cinematic sound since the Dolby Stereo format was introduced in the mid-1970s; by the 1980s, it had become the standard for multichannel stereo releases around the world. But that standardization prompted a historical elision of multichannel sound before Dolby Stereo appeared and obscured decades of research and refinement in the art of utilizing cinematic acoustic space. Dolby Stereo ushered in a new regime of "listening" that reflected the hierarchical structure of the soundtrack's construction as well as the stylistic imperatives of narrative form. Because it was developed in piecemeal fashion from extant technologies, the resultant presentational effect commingled several different representational strategies. Though Dolby Stereo offered a new format and the possibility for widespread acceptance of stereophonic cinema, the technological restrictions limited its stylistic use and discouraged sound practitioners from exploring cinema sound's full potential.

This essay uncovers the competing representational modes present in Dolby Stereo, and posits the emergence of new modes in the hands of innovative filmmakers. The first section outlines the development of Dolby Stereo and its widespread acceptance as the standard film sound from the late 1970s through the 1990s. The second section explores the concurrent rise and fall of sound design, and how the possibility of a single guiding sound auteur was drowned out in the wake of Dolby Stereo. The last section examines how a few filmmakers resisted the prescriptive (and proscriptive) requirements of mixing for Dolby Stereo and resuscitated experimental modes of audio-visual representation. By means of example this essay concludes with a detailed analysis of sound use in Jonathan Demme's 1991 film *The Silence of the Lambs*.

## Dolby Stereo

In the late 1960s, the ability to rerecord and mix increasingly larger numbers of magnetic tracks became essential for postproduction work in music as well as on films, and Dolby noise-reduction techniques allowed more layers of material to be mixed without an increase in background tape noise. Acknowledging the need to provide high-quality sound from the low-cost monophonic optical sound track, in 1970 Dolby Laboratories experimented with the application of their A-type noise reduction to film in *Jane Eyre* (Delbert Mann, UK, 1970). The experiment demonstrated that films would benefit from the use of noise reduction, but Dolby Laboratories recognized that any attempt to improve cinema sound needed to involve the entire sound process from production through exhibition. In 1972, they introduced the Model 364 unit for decoding Dolby A-type monophonic optical soundtracks and the Dolby Model E2 Cinema Equalizer to take advantage of the increased dynamic and frequency ranges of A-type encoded soundtracks (*Chronology of Dolby Laboratories* 1–2). Although the few films released in the encoded optical format had sound quality that rivaled four-channel 35mm magnetic prints—a standard format in the 1950s but virtually obsolete by the 1970s—the monophonic Dolby track could not offer the same stereophonic sound available in the magnetic format (see Belton, "1950s Magnetic Sound"; and Beck, "A Quiet Revolution" chap. 2).

Stereophonic cinema sound never had been an interest of Dolby Laboratories until an outside researcher presented them with a fully functional stereophonic optical track. Ron Uhlig, an engineer at Eastman Kodak, was exploring the possibility of using Dolby noise reduction on split-channel 16mm optical tracks to compete with the burgeoning videotape industry. Working in collaboration with Uhlig, Dolby designed a 35mm stereo variable-area optical soundtrack known as Dolby Stereo, or Dolby SVA, that could provide high-fidelity, multichannel sound in the space of the standard monophonic optical track. At the heart of the system was a Sansui Matrix converter—a remnant of the quadraphonic music boom of the early 1970s—that utilized phase-change relationships to mix four channels of information into two Dolby noise-reduction encoded optical tracks ("Dolby Encoded"). This allowed for four full channels of sound to be decoded in playback, obviating the need for more expensive 35mm discrete-track magnetic prints. Though it proved a boon to distributors and exhibitors—Dolby Stereo prints were backward compatible and could be played on monophonic projectors, thereby making double inventories unnecessary—the technology did come with a number of presentational idiosyncrasies.

As expected, when presenting stereophonic music, the system functioned spectacularly. But dialogue and sound effects recorded in Dolby Stereo were not as satisfactory. Certain frequencies and moving sounds confused the Dolby Stereo matrix, which would erroneously send these sounds into the surround speakers. To combat such systemic artifacts, specially trained Dolby consultants were required to oversee the mixing of all films in Dolby Stereo. They required that the dialogue be mixed to the central channel to ensure comprehension and avoid phasing problems (Blake, "Mixing Dolby Stereo Film Sound" 3–4). Sound effects could be positioned anywhere in the left-to-right space of the screen, but moving effects were monitored carefully to ensure that their acoustic motion matched the on-screen motion of their source. And, as a by-product of mono-phonic downmixing, any information exclusive to the surround channels was lost upon monophonic playback; therefore, only "surround effects" were sent to the rear speakers so that no narrative information would be lost during downmixing (*Dolby Surround Mixing Manual* 5–1). Often this led to bifurcated soundtracks where music and certain sound effects (such as car drive-offs or plane flybys) burst free of the space of the screen while dialogue was locked behind the plane of the image.

Despite their own advances in 35mm optical sound, Dolby Laboratories rec-ognized that the 70mm six-track discrete magnetic format still provided the best sound quality available. In 1977, they improved the 70mm format with the addition of A-type noise reduction and rechanneled two of the six tracks to pro-vide room for "baby boom" low-frequency enhancement ("Bass Extension" 1). Simultaneously marketed as Dolby Stereo, alongside the 35mm SVA system, it was actually the 70mm system that was responsible for many of the earliest suc-cesses of Dolby Stereo. The Dolby Stereo systems were praised in part due to the popularity of *Star Wars* (George Lucas, USA, 1977) and *Close Encounters of the Third Kind* (Steven Spielberg, USA, 1977), and the "baby boom" aesthetic created a new audience base for Dolby Stereo.

Similar to the use of the 35mm Dolby Stereo surround channel purely for effects, the low-frequency enhancement channels in 70mm Dolby Stereo were often used to give the audience a visceral, Sensurround-like experience, rather than as an aid in creating a flat frequency range below 100 Hz. In a telling neologism, science fiction author and perspicacious film reviewer Ursula K. Le Guin christened the effect of Dolby's new low-frequency channels "decibellicocity." In her review of *Close Encounters of the Third Kind,* the author correctly notes that the soundtrack is often used to "whip up emotions, the same trick that's so easy to do with elec-tronically amplified instruments" (92). This "boom aesthetic" had far more to do with emulating the wideband loudness of home stereo systems and rock concerts than it did with realistic acoustic reproduction. Ultimately, the audience was left not with an impression of subtle sound editing and an increased dynamic range but an all-encompassing sensation primarily due to the extreme volume of the cinematic event. No reviews were written about the full-frequency reproduction of John Williams's orchestral score or the lack of distortion and ground noise; instead, the emphasis was on how Dolby Stereo, particularly 70mm Dolby Stereo, allowed filmmakers to "pack in" more sound.

A final change in the structure of Dolby Stereo occurred when sound editor and mixer Walter Murch and director Francis Ford Coppola decided that they would use the 70mm Dolby Stereo system as the primary release format for *Apocalypse Now* (USA, 1979). Coppola wanted a sound mix that emulated the quadraphonic musical recordings of the early 1970s, both in their multichannel location of the speakers as well as in their psychedelic sounds and style. According to Murch, Coppola "wanted the sound to fill the room, to seem to come from all sections of the room which had never been done before in a dramatic film" ("Designing Sounds for *Apocalypse Now*" 159). To achieve this, Murch needed a greater measure of control over the soundscape, as well as the ability to position sounds anywhere within a 360° area, and he reworked the 70mm channel configuration to place all of the low-frequency enhancement on one track, thereby freeing up another for use as a second surround channel. In doing so, Murch created five discrete channels for left, center, right, surround left, and surround right, with low-frequency enhancement—a template for contemporary digital 5.1 sound systems. This made it possible to position or move a sound anywhere in the 360° sound field, not just across the plane of the screen or between the screen and the back of the auditorium (Murch, telephone interview). Developing out of a principle of "sound montage," Murch expanded his role in the production and postproduction processes to that of conducting the film's sound from start to finish, a method that he referred to as "sound design" ("Sound Design" 246). In spite of his complex use of 70mm split surrounds to envelop the audience in the diegetic acoustic space of *Apocalypse Now*, it is commonly forgotten that Murch discarded the surround tracks entirely when preparing the Dolby Stereo 35mm mix because of the impossibility of re-creating his sound design and complicated effects in the matrixed format.

Hence, one of the biggest problems in studying Dolby Stereo technology is that one is actually examining a series of nested technologies and their subsequent aesthetics, a problem aggravated with each new model introduced in the 1980s and '90s. French filmmaker and theorist Michel Chion, writing in *Cahiers du Cinéma* in the early 1980s, put forth several loose doctrines on the nature of Dolby Stereo and offered insights into its aesthetic nature. Unfortunately, Chion was writing from a position of historical blindness. Most of the films referenced were released in discrete-channel 70mm, rather than the matrixed Dolby SVA format used on the 35mm optical prints, and Chion failed to separate the different mixing practices involved. Nevertheless, he does provide us with a number of interesting assertions that can be useful in considering theories of stereophonic cinema in general and Dolby Stereo(s) in particular.

In his first essay on the subject, "A Dolby Stereo Aesthetic," Chion identifies three properties unique to the systems. The first is the ability of the soundtrack to produce a "*hyper-realism* of the sounds as more precise, sharper and richer" ("Une esthétique dolby stéréo" xii; my translations, emphasis in the original). This effect is most certainly related to the system's attendant noise reduction and its ability to render sounds with an increased exactness. A second function of Dolby Stereo is the production of sound said to be more "present" and utilized "progressively" (xii). Put simply, the soundtrack tends to be filled with ever changing sounds, rarely pausing or becoming static. It is here that Chion identifies a latent potential in the Dolby system, the ability to reconstruct sonic "environments" rather than just events. Early examples can be heard in the rustling prairies in *Days of Heaven* (Terrence Malick, USA, 1978) or the acoustic miasma of *The Elephant Man* (David Lynch, UK/USA, 1980).

However, the most powerful observation in Chion's article is his evaluation of the changed status of stereophonic cinema in the Dolby 35mm system. Where the discrete-channel separation of the earlier stereophonic systems allowed for sounds to inhabit the lateral speakers, making possible the use of active off-screen elements, Dolby Stereo's limited separation can provide only a "passive off-screen" presence. According to Chion, the sounds are "no longer 'elsewhere,' but 'nearby,'" hovering in the wings, threatening to reveal themselves. He further states that "the sound has done poorly in finding its symbolic anchorage, its structural function in the mise-en-scène," because the "hyper-realistic aesthetic" of the sound makes plain the flatness of the visual presentation (xii).

Although I agree with Chion, in part, about Dolby Stereo's ability to render sounds with exceptional detail, I tend to part ways with his analysis of the spatial aspects of the technology. He is correct in noting that the separation in the flanking speakers is significantly less than in discrete systems, making the sounds seem less distant, but he does not correlate the change in side-channel effect with an increase in surround-channel signal. As an electronic by-product of a sound being channeled to either the left or the right speaker, a small amount of the sound will also be heard in the surround channel. This allows the sound to move in a manner uncommon to earlier discrete systems; sounds can move from the side channels into the surrounds and vice versa, or, more evocatively, sounds can be made to inhabit the space of the auditorium in counterpoint to other sounds on the screen. Dolby Stereo has the unique ability to construct acoustic *layers* on a soundtrack, and to render the sonic and spatial details of each layer with extreme accuracy. Yet it was never utilized in this way because this new acoustic layering required that the axes of cinematic action be perpendicular to the screen, and generally incompatible with shot/reverse shot editing and the 180° rule. Therefore, it took several years before filmmakers began to experiment with this effect, in part due to the near total control over postproduction sound mixing by Dolby consultants and the restricted ability of sound practitioners to experiment with sound design.

## The Predominance of the Vertical; or, "There Is No Sound Designer"

The title of this section takes its cue from Michel Chion's assertion that in the process of experiencing a cinematic presentation, "there is no soundtrack" (*Audio-Vision* 39). Chion's point is that the sounds in a film are always narrativized in relation to the images and are therefore perceived as subordinate to the dominant image track:

> By stating that *there is no soundtrack* I mean first of all that the sounds of a film, taken separately from the image, do not form an internally coherent entity on equal footing with the image track. Second, I mean that each audio element enters into a *simultaneous vertical relationship* with narrative elements contained in the image (characters, actions) and visual elements of texture and setting. These relationships are much more direct and salient than any relations the audio elements could have with other sounds. (40; emphasis in the original)

However, there is a blind spot hidden by Chion's position: the fact that the film industry has created a biased approach to the production and postproduction processes that perpetuates the idea of the dominance of image over sound in film. With an emphasis on dialogue editing and mixing over all other sounds in the picture, nearly all Hollywood films place primary emphasis on narrative acoustic elements via dialogue. This overarching drive for narrative comprehension has created a tacit agreement between the audience and filmmakers that all narratively significant sound elements will be heard. This rule of sonic signification mapped itself onto the divisions of dialogue, effects, and music, elevating dialogue to the top of the hierarchy and shaping basic film sound–studies terminology.

One major fissure in the rigid hierarchy of film sound appeared with Murch's notion of "sound design." This designation came into vogue primarily in the post-Dolby era of sound recording and mixing to define the individual in the production team responsible for the overall sound of the film. Analogous to the cinematographer and production designer controlling the look of the film, the sound designer is currently posited as the individual who constructs the overarching sound strategies of the film. However, unlike Murch's conception of a sound designer as an individual who can control and shape the sound strategies in both production and postproduction, major resistance on the part of the sound union IATSE Local 695 prevented this idea from becoming a widespread reality. Instead, the term took on a very different meaning in relation to emergent sound practices in the early Dolby era.

In a 1983 article, Marc Mancini evaluates the emergence of the sound designer and offers a similar definition by labeling the sound designer as an "aural artist" (361). The article examines the sound designer's presence throughout cinema history, observing that the previous designation of "supervising sound editor" can be read retroactively as a sound designer. He compares LucasFilm's Ben Burtt, best known for his sound effects work on the *Star Wars* series, to Jimmy McDonald, the vocalizer of many sounds in early Disney films, and to electronic sound-effects creator Frank Serafine. In doing so, Mancini helped to perpetuate an ongoing semantic problem with the term *sound designer;* specifically, his focus on the creation of sound effects helped shape the perception of a sound designer as a glorified sound-effects artist. Though effects creation is an important element in the industrial history of 1970s and '80s cinema, especially in regard to the increased role of the sound editor, it hides a much more important aspect in the history of the sound designer.

Mancini's ahistorical application of a contemporary designation to a variety of historical subjects opened up the synchronic nature of film terminology for debate. Is the role of a modern sound designer the same as McDonald's job in the thirties? And why does Mancini believe it is important to recuperate retroactively these liminal figures, and what is there to be gained from this venture? Mancini's article valorizes the role of sound designer without any critical analysis of how that role actually functions. Specifically, he haphazardly reads the function of sound design across film history without regard for the industrial and technological changes that subtend that history. It is precisely these presumptions that support my assertion that contemporary cinematic vocabulary is deeply anchored in the ideological biases of the studio system. This retroactive application of film sound terminology to historic subjects covers over points of cinematic evolution and conflict; specifically, Mancini creates an artificial teleology between past practices and present definitions, and shuts down a study of the fruitful period of experimentation existing before Dolby Stereo reified sound-editing and mixing practices.

By applying a simple tautological definition of a sound designer as a person who "designs sounds," Mancini does a disservice to evaluating the changing roles of film sound practitioners. In particular, the shift from a rigid hierarchical division of labor in the late classical Hollywood period to the central organizational role assigned to the sound designer in the 1970s is obscured by Mancini's presumed teleology. The fetishistic vogue for the director as "auteur" was possible only at the cost of obscuring or ignoring the highly segmented and specialized production labor force within the studio system. It was only because the labor system was so highly compartmentalized and divided that a director could theoretically dominate the style of a film. Due to the unbridgeable barrier between the divisions of labor in soundtrack construction, it was impossible to develop a "sound auteur" during the heyday of the studio system. Moreover, Dolby Stereo, with its specific demands for the centrality of dialogue mixing, ensured that the recording apparatus encapsulated and perpetuated the classical Hollywood divisions of labor. This "vertical" hierarchy of labor was inscribed in the apparatus, thereby giving rise to continuity between the hierarchical division of labor in Hollywood and mixing practices in the era of Dolby Stereo. The resultant sound practices remained tied to the narratively determined constraints of character intelligibility at the cost of a progressive and experimental audio soundscape.

Thus, at the time Mancini was writing, the misappropriated appellation of "sound designer" was one that disguised the fact that the "quiet revolution" of Dolby sound was not a revolution at all. The acceptance of Dolby Stereo as an industry standard only concretized the segregated system of labor that many directors were resisting in the 1970s. In the vertical relationship of sounds and the hierarchical segregation of sound technicians, narrative cohesion served as the dominant structuring principle, and the possibility of a central sound "auteur," or sound designer, faded away. The current continued use of the phrase "sound designer" tends to cover over the lack of change between classical Hollywood sound use and sound practices in the Dolby Stereo era. Therefore, there really never has been a central figure constructing the sound in Hollywood, and to paraphrase Chion's evaluation of the ongoing model of classical Hollywood sound production, it can be claimed, "there is no sound designer."

Perhaps a different way of looking at the question of "sound design" is to recognize that the true designers of sound are either those who are able to dictate what technology will be used or those who adapt the existing technology to suit their particular aesthetic needs. The former set includes individuals such as Twentieth Century-Fox president Spyros Skouras, who decreed in 1953 that all CinemaScope prints would have four-channel stereo magnetic soundtracks, or Steven Spielberg, whose *Close Encounters of the Third Kind* was released in Dolby Stereo to compete directly with the success of *Star Wars*. In the case of such individuals, the choice of technology generally preceded any serious thought about its aesthetic application. Often, the resultant sound aesthetics were restricted by the demands of the technology, leaving little room for experimentation or growth. Hence, the literal design of the sound was entirely predetermined and prescripted in relation to the narrative. However, in the case of the latter set of sound practitioners, those who adapt existing sound technologies to their own uses, it is possible to see the potential for a conceptual and creative deployment of sound design to emerge. Through the experiments of nascent sound designers from the 1970s—including Walter Murch, Ben Burtt, and Alan Splet—it becomes possible to see how the loosening labor structure of the film industry and the decline of the studio system created a space for "sound authorship." These individuals were predecessors of a progressive sense of sound design because they sought to craft the sound of the film by tweaking the technology to serve their individual needs. In the case of the Dolby Stereo era, it took well over a decade before sound practitioners began to work around the technical and stylistic limitations of the technologies to construct their own sense of interactive sound design.

## The Silence of the Lambs

Few films managed to utilize 35mm Dolby Stereo's expanded spatial capabilities as well as *The Silence of the Lambs* (USA, 1991). Director Jonathan Demme, along with rerecording mixer Tom Fleischman, sound editor Ron Bochar, and sound designer Skip Lievsay, regularly deployed highly creative patterns of sound use in order to advance the narrative. The film's focus on a young FBI agent, Clarice Starling (Jodie Foster), and her search for a serial killer known only as "Buffalo Bill" (Ted Levine) presented Demme and his sound team with an opportunity to use Dolby Stereo technology to replicate the subjective state of the central character in her quest. The penultimate sequence in the film serves as an excellent example for illustrating the acoustic construction of space in the post-Dolby era of cinema sound. In this scene, Agent Starling attempts to arrest the serial killer in his labyrinthine basement, and is forced to navigate the previously unseen space by following the shouts of kidnapping victim Catherine Martin (Brooke Smith). In the process of traversing the space, sound is constantly used as a means of replacing the standard narrative convention of the establishing shot and as a way to link the audience to Clarice's subjectivity. The sound of the sequence is indicative of Michel Chion's notion of the Dolby sound space becoming an "acoustic aquarium" where sound "magnetizes" the fragmented diegetic space into a concrete whole (*Audio-Vision* 69–71).

Chion bases his concept on how the improved definition of the Dolby soundtrack allows for a more faithful "rendering" of the accretion of acoustic details in the mix. Accordingly, this polyphony constructs a series of acoustic "planes" that allow the auditor to clearly distinguish not only individual sounds but also their location in the diegesis ("Quiet Revolution" 70–71). Through Dolby Stereo, the soundtrack expands the two-and-a-half-dimensional space of the screen (left-right, up-down, but only depth "into" the screen) to a full three dimensions by extending the diegetic space into the space of the theater. Two crucial effects emerge from this unique use of the 35mm Dolby Stereo surround channel: first, the listener is surrounded by the sound field while several discrete layers of sound are heard simultaneously, and second, the individual sounds are rendered with a clarity that allows them to direct the listener's attention through their spatial specificity. Therefore, the Dolby Stereo soundtrack effectively guides the spectator in the same way as the establishing shot—delineating a stable space that is subsequently carved up through shot/reverse shot deconstructions—while creating a new regime of acoustic attraction that destabilizes the narrative centrality of the frame.

According to Chion, "instead of establishing space (sound now does this), the image selects viewpoints onto it" ("Quiet Revolution" 79). It is precisely Dolby Stereo's capability of rendering a three-dimensional diegesis that offered Demme and his team the ability to edit unstable images that break the 180° rule, jump from close-up to long shot, or rapidly cut between impossible perspectives onto a stable acoustic space. In the penultimate sequence from *The Silence of the Lambs,* the space of Buffalo Bill's basement is made tactile in the acoustic dimension through Dolby Stereo, while the visual space is limited to the restricted perspective of Clarice's point of view. The smooth transition of the acoustic space, where only one layer of sound is changing at any given moment, compensates for the "jumping" of the camera from Clarice's perspective to direct head-on shots—a common stylistic trope in Demme's films—as she moves throughout the house. This is extremely important since the "point of audition" (see Altman, "Sound Space" 58–64), initially aligned with Clarice, subsequently reverses itself to construct Bill's point of audition later in the sequence. The result is a desired dislocation from the previously stable acoustic environment and a disconcerting alignment between the audience and the acoustic perspective of the serial killer.

The resultant soundtrack, especially with the expanded frequency range of Dolby Stereo and its spatial dispersion of sounds, buzzes with an intensity of its own. Chion refers to these myriad acoustic events as a "micro-rendering[s] of the hum of the world" whereby the "minor denizens of sound . . . breaths, squeaks, clinks, hums" each compete for the attention of the auditor and function by carving out the acoustic space (*Audio-Vision* 70–72). Yet isolating these individual sound sources becomes difficult since they coalesce into an acoustic "lump" (Chion and Brewster's term) where it is often difficult to dissect elements from the soundscape and identify localizable sound sources. This multiplicity of sounds in the modern soundtrack begs for a new theoretical approach for categorizing sound as a transsensorial, polyspatial "event." And instead of describing the individual, hierarchically divided elements of soundtrack construction, it becomes beneficial to consider the function of the overall film sound design.

It immediately becomes apparent in the Dolby Stereo era that the original divisions of dialogue, music, and sound effects rapidly fall away as clear-cut categories.

The roles of the sound effects editor and Foley artist have extended far beyond the inclusion of room ambience and footsteps. Modern film soundtracks demand a highly interactive process where a film's sound strategies need to be thought out during the earliest stages of production (see Murch, "Sound Design"; Thom). The acoustic depth of the multilevel soundtrack blurs the boundaries among dialogue, sound effects, and music, making it extremely difficult to ascribe any one sound to a specific category. The recent reemergence of the role of the sound designer in the wake of widespread resistance during the 1980s also begs for an understanding of how film sound's status is being transformed (see Blake's rebuttal to Mancini, "A Sound Designer by Any Other Name"). No longer can sound simply be thought of as existing in one of three discrete areas; instead, sound designer Skip Lievsay transmutes these sounds into a fourth element—spatial ambience—that effectively traverses all of the classical divisions of sound mixing and dissolves the uniform segregations of contemporary soundtrack construction.

The three primary divisions of sound in the film are derived from their classically defined and discursively constructed industrial designations. First, dialogue and ADR (that is, automatic dialogue replacement, also known as "dubbing" or "looping") editors are responsible for providing clear, discernible dialogue, then Foley artists and sound effects editors add hard effects to completed scenes, and finally the composer creates music after the images and dialogue are edited. In *The Silence of the Lambs,* Howard Shore's score is present for more than an hour of screen time, and it underpins all of the highly dramatic moments in the film, often guiding the emotion of the spectator through its suspended minor chords and nervous-sounding lower-register tremolos. The scripted nature of the music follows late-romantic compositional styles that ascribe emotional qualities to the music through familiar tonalities, chord progressions, and tempos. Despite this prior use of the score in the film, Shore's music actually drops out at a crucial point when Clarice enters the basement and does not return until the end of the sequence. Instead of eliciting emotional responses from the audience through scripted score music, Demme prefers to rely on the diegetic sounds, each carefully positioned to coincide with Clarice's point of audition.

The initial sound that guides Clarice, and the audience, through the space of the basement is the diegetic music playing from an unseen, distant source. The song "Hip Priest" by the Mancunian band the Fall is heard resounding throughout the basement, and as Clarice moves through the space, the music increases in volume as its reverberant qualities decrease. When Clarice moves closer to the sewing room—the room that presumably contains the speakers—reverberation diminishes, eventually disappearing when she opens the door. The music then reverses this trend, lowering in volume and increasing in reverberant qualities, when Clarice enters the next concatenated space and the narratively significant sound cue of Catherine Martin's voice takes precedence.

Perhaps the most important function of the source music is that it provides a temporal thread that allows the audience to make sense of the rapid cutting and reversals between point of view and head-on shots of Clarice. By using these unique shooting and editing techniques, Demme offers the audience neither establishing shots of the space nor the spatial familiarity of the 180° rule. Instead, the connective tissue that holds the sequence together is in the linear temporal flow of "Hip Priest." Because the song does not stop or cut during the sequence,[1] the audience does not perceive any temporal gaps in Clarice's spatial leaps between the shots. Rather, a sense of continuous motion across the basement is created through the diegetic music.

Once the source of the music is discovered, Catherine Martin's voice takes over as the mechanism for charting the hidden space. The dialogue between the characters completes the narrative function of guiding Clarice through the space while simultaneously acoustically rendering the spatial qualities of the basement. First heard at a low volume with a high level of reverberation, Catherine's voice guides Clarice to the room where she is trapped in the well. This is the only space that the audience has seen in a previous sequence, and therefore, accordingly, there is only a brief shot of Catherine at the bottom of the well to confirm Clarice's discovery. In keeping with the strategy of letting the sound lead Clarice and the audience, once the discovery is made and Clarice moves to the next adjacent space, Catherine's voice is no longer used as an acoustic cue. At this point, the primacy of the voice is supplanted by increased attention to Clarice's breathing that will serve as a crucial pivot point once Bill reemerges.

It should be noted that throughout the first part of this penultimate sequence Clarice's breathing is heard increasing in volume as she becomes more anxious. This sound is anchored in the plane of the screen, and it moves spatially wherever Clarice goes in the frame when she is seen in the shot. If the shot depicts her point of view, the sound carries over from her spatial position in the previous shot and does not change in volume or reverberant qualities. This observation is important because Demme's regular oscillation between point of view and head-on shots breaks the visual continuity of the sequence, yet the soundtrack creates a spatial consistency and lends the locale a conceptual "wholeness" in counterpoint to the shifting visual perspectives. Despite the fact that Clarice's breathing occasionally is heard from a position away from the location of her body, the combined impact of the other sound sources (diegetic music, dialogue, and crucially ambience and sound effects) adds up to mark the point of audition as Clarice's.

At this point, both dialogue and music, which were easily identifiable in the first half of the sequence, become murky with the increased presence of two other registers of sound: ambience and sound effects. These two areas are difficult to identify primarily because of the intense imbrication of acoustic details and the subtlety of the sound mix. Throughout the film, Demme and his sound team regularly integrate narratively significant hard sound effects with background ambient sound. Generally, any sound that can be visually "tagged," that is, identified by a profilmic source, is categorized under the heading of sound effects. However, other noises that are neither identifiable by their sound characteristics nor locatable through their on-screen presence are categorized as ambience. Although this evaluative framework does not do justice to the complexity of the sequence, or reveal the process involved in the creation of the sound mix, it does emulate the interpretative process involved in sorting out and making sense of the Dolby Stereo soundtrack.

The identifiable sound effects (dog barks, footsteps, door crashes, moth flutters, refrigerator sounds, and so on) operate by adding materiality to the objects and depth to the diegesis. Michel Chion uses the terms *materializing sound indices* (MSIs) and *elements of auditory setting* (EASs) to describe two different ways that intermittent sounds serve to concretize a filmic space, seen or unseen, through the accumulation of auditory details. Materializing sound indices are the specific acoustic qualities of given sounds that firmly anchor them to their physical counterparts, the objects presumably producing the sounds (*Audio-Vision* 114–17). The resultant "realism" of any sequence is partially contained in the aggregate number of these MSIs that render the space as diegetically "real" and concatenated, rather than a construction on a soundstage or in an editing suite. Thus, the rustling of Clarice's clothes, the scuffling of her footsteps, the creaking of the door hinges, and other microsounds combine to create the perception of a tactile space. These sounds are often classified as ambience for the sake of ease, but their effect is much more central. Although the audience is generally not conscious of the individual sounds existing outside of a lumped acoustic experience, distributed throughout the stereo field they pull at the borders of the frame to reveal the presence of a larger threatening, unseen space.

Elements of auditory setting punctuate the acoustic space through their discontinuous presence (*Audio-Vision* 54–55). Perhaps the dominant example of this is the barking of Precious, Bill's poodle, heard throughout most of the sequence. Despite the fact that the sound of the poodle is no longer narratively necessary once Clarice has discovered it is with Catherine, the barking continues until the point when the electricity is shut off and the image goes black. Of course, this brings up several questions, such as why the dog continues to bark after Clarice has told Catherine to keep it quiet and why it stops only after the lights go out. Perhaps one reason is to heighten the audience's sense of anxiety since Clarice is decidedly not in control of the situation despite her proclamations to the contrary. But another reason is that the increasing crescendo of EASs, such as the poodle's barks, creates a cacophony that extinguishes itself at the same time as the lights. The resultant effect is a double absence: the loss of visual material and

3M SelfCheck™ System

Customer name: Audureau, Helene

Title: Cinema and landscape / Graeme Harper &
Jonathan Rayner (eds)
ID: 30114014765716
**Due: 27-06-19**

Title: Cinema at the periphery / edited by Dina
Iordanova, David Martin-Jones, and Beln Vidal.

ID: 30114014687209
**Due: 27-06-19**

Title: Lowering the boom : critical studies in film
sound / edited by Jay Beck and Tony Grajeda.

ID: 30114015872008
**Due: 11-04-19**

Total items: 3
04/04/2019 15:13
Overdue: 0

Thank you for using the
3M SelfCheck™ System.

the perceived "silence" of the soundtrack, both of which serve to engender fear in Clarice and the audience.

In this last part of this sequence, after a few seconds of a black screen, Bill reveals his presence to the audience when the image returns, tinted in green and silhouetted, representing him observing Clarice through a pair of night-vision goggles. It is at this point that the soundscape shifts radically to deconstruct the acoustic space of the basement in order to align the point of audition with Bill to match his controlling point of view. Inverting the previous acoustic regime, spatial cues such as MSIs and EASs become less important in lieu of a heightened awareness of the few remaining sounds, primarily Bill's and Clarice's breathing. Previously, Clarice's breathing was heard as part of the dialogue, and her breathing and spoken voice were mixed in the soundscape to match her presumed position in the diegesis. Yet in this last section a crucial change occurs when the point of audition reverses from Clarice's perspective to Bill's to convey his immanent threatening presence. Clarice, panting and groping blindly in the dark, is kept visually and acoustically centered in the frame. This is done by channeling all of her breathing sound to the center speaker without including any reverberant details in the left, right, or surround speakers. Furthermore, Bill's breathing is heard through the surround speakers only, clearly aligning the audience with his point of audition. The effect creates an acoustic objectification of Clarice that matches her visual objectification; here Bill is in control of both the visual and the "acoustic gaze." Thus, the conjunction between the spatial location of the sounds and the binocular matte on the image firmly places the audience in Bill's point of view and point of audition.

In this final section the scarcity of EASs adds to the tension, especially in contrast to the crescendo that precedes it. The few sounds that we do hear—the sizzle of Clarice's hand on the furnace or her tripping and falling—function less as indicators of Clarice's presence in the diegetic space and more as reminders of the hidden threat in the room. The ambient sound in this section can be described as a general sound of air movement that serves as an acoustic "bottom" for the other hard effects. Importantly, the use of this minimal ambience creates a psychoacoustic sensation of "silence" primarily because of the markedly diminished number of elements in the mix. Sound designer Skip Lievsay commented on this intimate effect, saying, "It's just a miraculous way of pulling the audience in. We were able to take all sorts of sounds like jungle sounds, animals and raindrops—manipulate them with equalization and reverb and layer the components" (quoted in LoBrutto 266). Although the scene could have been constructed with just the breathing and Foleyed effects, the highly evocative background of the ambience prevents the section from crossing over from realistic to oneiric and the balance hovers somewhere between the two. Visually, the film reverses the previous point-of-view pattern for more than a minute of screen time—cutting between Clarice stumbling about blindly and a dusky image of Bill in the night-vision goggles—before Howard Shore's score music returns. The music coincides with confirmation that Bill is in Clarice's direct presence when his hand extends into the frame. The ominous tone of the score, constructed from low-register horns and percussion instruments, emerges from the din of the ambient sound and effectively masks it as the music builds to a crescendo while the confrontation ensues.

Here the music provides the dominant role on the soundtrack until the point when Bill cocks the hammer on his gun and the acoustic regime is transformed once again. This vital sound provides the clue for Clarice to discern Bill's location in the space, and it is heard as a very loud reverberant sound that does not match any realistic sense of the acoustic event. It can be argued that this is the moment when Clarice becomes aware of Bill's location and where the audience is aligned once again with her acoustic perspective. The psychoacoustic rendering of the sound as a highly reverberant, high-volume event simulates Clarice's heightened sense of awareness. This exaggerated sound is followed by the only slow-motion shot in the film as she spins around and empties her gun. The six shots fired are seen as flash frames, and a nonsynchronous sound of breaking glass is matched with a brief shot of sunlight streaming in through a cracked window. At this point, the soundscape returns to its original construction, with Clarice's breathing heard in a localizable space through the left, center, and right speakers while the score music continues. A few random EASs are used to confirm that the sounds are now anchored to the space of the diegesis and no longer internalized as a specific character's point of audition. This is reinforced by the first two-shot where Clarice is seen in the background, reloading her gun, while Bill lies bleeding in the foreground.

In this sequence from *The Silence of the Lambs,* Jonathan Demme and his sound team demonstrate a new way of engaging the spectator that relies on the ability of Dolby Stereo to activate audio-visual interactions generally not available in monophonic cinematic presentations. Although the effect is quite striking when viewing and listening to the sequence in isolation, the fact that it is activated through the artful interweaving of sounds and images with the larger narrative demands of the film makes the sound strategies exponentially more effective. By using this example from *The Silence of the Lambs,* I wish to emphasize how the sound team harnessed Dolby Stereo as an expressive device for original and creative sound work. Unlike the early Dolby Stereo soundtracks that were tethered to monophonic codes of representation, *The Silence of the Lambs* restores the role of sound design to create and spatially deploy effects that both augment the narrative and heighten dramatic effect. Demme and his sound team blend the auditory and the visual in such a way as to engage the audience more directly than either dialogue or images alone. First, the film's sound design functions as a unifying element to cover over the spatiotemporal breaks of the editing and the oscillation between point-of-view shots and their reverse angles. Second, it renders a perceivable space through the use of MSIs and EASs. And third, it constructs points of audition that coincide with the perspective of key characters. Moreover, the patterns of sound use in the film also control spectatorial positioning by constructing a stable acoustic space where the director can forgo established patterns of shot/reverse shot in favor of point-of-view cutting. By mobilizing the sound in such a fashion, the film provides a model for sound design in Dolby Stereo films while sidestepping the systemic limitations that have prevented such an application in the past. The conventions of Dolby sound mixing had, in fact, deafened audiences to the variety of potential audio and visual interactions. The fixed aesthetics that emerged from Dolby Stereo's rapid acceptance in the 1970s precluded such experimentation and industrial reorganization, and it is only after more than a decade that filmmakers and sound practitioners have started to challenge these limitations and rethink film sound's potential.

## Notes

A portion of this essay appeared in an earlier version as "'Rewriting the Audio-Visual Contract': *Silence of the Lambs* and Dolby Stereo," in *Southern Review* 33.3 (2000): 273–91. Thank you to Rick Altman and Angelo Restivo for their editorial suggestions.

1. In truth, the song does skip at the moment when Clarice enters the workroom, and a two-second musical passage from the prior shot is heard again. It is difficult to determine whether this is a mistake on the part of the music editor or if the shots were timed in advance to the music and a decision was made to extend the shot.

# PART II

# Historicizing Sound

# 5

# Sonic Imagination

## or, Film Sound as a Discursive Construct in Czech Culture of the Transitional Period

**PETR SZCZEPANIK**

Much has been written about the discussions around film sound among film directors and well-known theorists. Innovative studies have also been written on the discourses and role of sound technicians in shaping the norms of film practice and representation (Altman, "The Technology of the Voice"; Lastra, *Sound Technology and the American Cinema*). But not much has been written about the broader field of discourses surrounding the advent of standardized synchronous film sound in the framework of contemporaneous media, culture, and society—discourses that are much more heterogeneous and less connected to the actual film practice and technology than those produced exclusively by film directors or theorists. Nonetheless, they have shaped the way in which meanings were ascribed to a new technological form, determining the position within the cultural landscape of the period sound cinema was supposed to occupy.

A discourse analysis of the texts that were defining the emerging sound film and projecting cinema's possible futures and identities can teach us a lot about the general evolution of media in history. But in order to reach such an understanding, it is necessary to broaden the usual definition of a medium. Any medium should be defined as a hybrid on both horizontal and vertical axes. Horizontal hybridity entails blurring boundaries and mutual redefinitions among different media and cultural formations. Vertical hybridity refers to a medium's identity as a mix of technological devices, practices, discourses, and institutions. This implies that a medium cannot be defined as a purely technological invention or as any kind of delimited material device but can only be defined as a multilevel construct, which also includes purely discursive "inventions"[1] and objects, assembled from values, expectations, projects, fictions, and dreams circulating within contemporary discussions about the nature of a new medium. I use "discourse" here as a certain mode of speaking or writing about film sound, a mode that may be reconstructed from a limited body of writings and other historical documents. It is important to stress that a discourse does not describe an already clearly delimited and defined object of its interest; rather, it acts to itself construct this object, together with certain institutions and social practices. Every discourse thus delimits, defines, classifies, and evaluates its object and creates its boundaries and relationships toward other objects, thus potentially constructing alternative identities and futures for that object.[2]

Conclusions drawn from this analysis also reveal recurring topoi, comparable structures of experience and reactions of contemporaries relating to other new media in the past,[3] namely, that they all share a certain kind of catastrophic-utopian rhetoric, hesitate in calling the new medium by a proper name, and, uncertain which term to choose, oscillate from one to another. Along with conducting an argument over media essentialism, the discourses tend further to redefine the nature of what it means to be human and its relation to death. In Czech culture we can find these similarities especially between the period of the coming of radio and that of the sound film, understood as a new medium.

The discourses surrounding the advent of film sound in Czech culture shared one typical feature: a catastrophically utopian imagination. The debates and visions of filmmakers, critics, writers, and theoreticians were usually composed of two parts. The first one evaluated contemporary film production and concluded that it was too naturalistic, mechanical, and theatrical. The second one, my primary focus here, was usually much more charged with imaginative powers, often constructing the image of the possible futures of sound cinema. These alternative images of cinema's future and of its media identity were multiple, heterogeneous, and rather different from the development of the classical silent narrative cinema up to that point.[4] On the level of critical reception, seen from the prospective[5] point of view of its contemporaries, cinema was often freed from the dominance of fictional narrative forms as well as from the classical cinematic *dispositif.* Cinema was then frequently understood as a hybrid form—mingling theater, radio, gramophone, telephone, and even television. It was also conceived as a medium for nonnarrative purposes: as a scientific tool for analyzing physiological and mimicking processes of live speech; as a device suitable for a special archive preserving "audiovisual autographs" of politicians, scientists, artists, and the like for future generations; as a new kind of communication technology resembling recent television networks or interactive media or as a new kind of "opto-phonetic" poetry based on synesthesia.

The grounds of such utopian imagination could be found in the specific nature of the Czech avant-garde film movement of the 1920s, which, though quite strong and influential, was purely theoretical, very visionary, literarily oriented, and concerned with the potentialities of cinema while fundamentally alienated from any actual film practice of the time. A related reason is that of the strong presence of writers connected with the avant-garde in the field of daily journalism. Another important framework for discussing film sound was the scholarly domain of linguistics, namely, the so-called Prague Linguistic Circle, whose members and sympathizers considered film sound as a new opportunity for the study of and influence on the sound culture of the spoken word as such. During the late 1920s and early 1930s a further framework was provided by the radio and recording industry, undergoing at that moment profound transformation, expansion, and diversification, parallel to the transformation of cinema, and creating a context for the social understanding of film sound. Radio reached a new level of self-consciousness when it started to seek its own aesthetic specificity in the new radiophonic genres, working with noise and speech in more imaginative ways. New gramophone genres emerged as well when the medium started to be used for nonentertainment purposes again (presidential addresses and gramophone archives of sonic autographs). And it was at the intersection of phonography, radio, and early film sound that new ideas of media archives and their significance emerged.

The profound transformations that cinema underwent during the period 1927–34 on all levels were defined not only by technological development but also by the discourses (comprising reviews, theories, advertisements, business decisions, patents, and so on), practices (different ways of creation and consumption), and institutions (established media, arts, sciences, legislation, and other social and economic structures) surrounding it and inserting it into the world of everyday life (Lastra, *Sound Technology and the American Cinema* 13). From a prospective position of analysis, cinema stood on the crossroads of alternative possible futures, which implied different modes of social use and meaning and which differed significantly from the institution of silent cinema. In what follows I propose a description of five groups of arguments and tropes structuring the film sound–related discourses present in the Czech culture of the late 1920s and early 1930s. This description is far from an exhaustive account of the coming of sound as a historical process; it is intended instead as a case study and an experiment with alternative perspectives.

## Cinema Is No Longer Cinema—It Is a New Medium

The initial theoretical speculations of reviewers, directors, writers, musicians, businessmen, or technicians had to cope with a key question: was sound cinema still cinema, and if so, what should it be called? Prominent theater and film critic Otto Rádl wrote in 1930:

We talk about "silent cinema" and "sound cinema"—in both cases about "cinema," with a difference only in adjective. It is a pity that a completely new term was not created for the new invention, similarly distant from the original one as the American "talkie" is different from "moving." Because the difference between these two is so fundamental, like between Shakespeare's *A Midsummer Night's Dream* and a pantomime about "duped Peter." . . . "[T]alkies" are really *not cinema* any more. They are of course *not theatre* either. . . . The sound cinema is a *new art form,* which hasn't had a parallel till now. (65–66; emphasis in the original)

In contradiction with the traditional perspectives of the historical reflections on sound cinema that considered it to be inferior to the silents, Rádl identifies Shakespeare with the new medium, even though he certainly is fully conscious of the fact that the emerging form could not equal the genius of the playwright. What mattered for him, though, was not the purity of artistic technique but the potential of the new medium.

In contemporary discourses a number of terms appeared for sound cinema, most of them functioning as equal alternatives, diverging only slightly in referring to different aspects of the medium. Some of these terms implied specific ideas as to the medium's future or as to its links with another media. The most frequently used terms were *tonofilm, tónfilm, fonofilm, soundfilm (zvukofilm), talking film (mluvící film), spoken film (mluvený film), aural film (sluchový film), radiofilm,* and finally also *sound film (zvukový film).* Apart from these, there also emerged some special terms introduced by writers inspired by the aesthetic assumptions of the French and Russian avant-garde: Czech avant-garde's premier theorist, Karel Teige, wrote of *opto-phonetics,* left-wing critic Lubomír Linhart of *opto-phonogenia* or *photo-phonogenia.*[6] From these terms and their uses we can deduce alternative notions of media identity: would sound cinema limit itself to supplementing the silents with sound effects and music and thus maintain its internationality? Would it take the form of the mechanical opera or operetta with a dominance of singing and music? Would it become an extension of the phonograph or the radio, or would it define itself as a unique medium or even as a new art of audiovisual counterpoints or synesthesia?

A second vast terminological field emerged within the domain of advertising, where a new vocabulary served as a quantitative substitute for established genre terminology as well as for aesthetic terminology. As film technician, historian, and critic Karel Smrž noted, "The film business finally acknowledged that it was a business accustomed to express all the necessary values in percentages. Movies aren't anymore the most absorbing, the most touching, the most perfect and the most delightful, but 100% sound, talking, singing and colour" (425–26). Diversity of the quantitative terminology refers at once both to a new variability of the film commodity and to a new framework of reception: film was no longer evaluated as a unique artwork or as an original story, but rather as a more or less technologically advanced or "upgraded" version of a given "software."[7]

On the level of discourse we only rarely encounter what David Bordwell has defined as a "functional equivalent" on the level of style (see "The Introduction of Sound"). Especially in the early reflections on sound film practice (in 1929–32), there is a strong awareness of a historical change and a break, with the accompanying idea that sound cinema is no longer cinema but rather a completely different art or medium. The goal of this new medium should be to find and better define its "sound essence," which is far from materialized in the contemporary practice. This essence was often thought to be more clearly present in the field of another media: not only traditional ones such as theater but also the new electrical media such as radio or even the (incipient) television (see, for example, Frejka, "O zvukovém filmu").

The argument "cinema is no longer cinema" is thus related to the crisis of media identity, and is most visibly expressed in an unsuccessful search for a new noun with which to designate sound film. Interestingly, a few years earlier there was a similar debate concerning the Czech term to refer to radio or broadcasting. The neologism *rozhlas* was introduced with great success in 1924 by a broadcast journalist, and remains widely in use today.[8] The basic ideology concealed in the argument that sound cinema is a new medium is thus media essentialism, expressed in the demand that cinema should look for a new kind of "audio *raison d'être*"; there are significant similarities here with current debates concerning the so-called new media in the 1980s and 1990s.

## Mechanism versus Organism

First among the key oppositions structuring the contemporary discourses on the coming of sound was that of a machine and a human being, of mechanical production and artistic creation. The machinist nature of sound cinema was commonly emphasized, in particular in discussions of multiple-language versions (MLVs), which were referred to as "cans of words," that is, as strictly serial products, in contrast to traditional concepts of artistic creation. On the one side stood terms allied with mechanicity, on the other side those of full human presence and of acts of individual creation and expression.

The dichotomy between the mechanical and the organic is the link that connects early sound cinema to the history of phonography but also to that of early cinema. James Lastra and Tom Gunning have described the humanoid forms given to phonographs in the nineteenth century to compensate for the uncanny effect of the separation of hearing from the other senses (see Lastra, *Sound Technology and the American Cinema* 16–60; and Gunning, "Doing for the Eye" 19–20).

These human- and animal-like puppets and automatons were supposed to simulate the lost integrity of a human subject by neutralizing the horrifying effect of the split between voice and body, or of a voice still alive after the body is already dead. Similar themes can be found in the analyses of discourses around the sound performances and lectures of the silent cinema period, where the pianist or lecturer was put into a prominently visible place beside the screen (instead of being hidden in the orchestra pit or behind a partition, or definitely replaced by a mechanical device) to stand for a substitute organic identity of a cinematic apparatus, which was otherwise understood as too machinelike to be part of the culture (Tsivian, *Early Cinema in Russia* 85–87).

After synchronous sound reproduction became dominant, cinema in the technological sense came closer to the ideal of embodied human perception. Paradoxically, however, on the cultural level the more complete simulation of human sensorium instead strengthened the awareness that "what sound cinema does is to transform organic links into links which are mechanical" (Honzl 39). The reason for this idea not only lies in the imperfections and disturbances of the contemporary systems of synchronization, along with certain regressive features of film style, but also arises, with regard to the cultural level, with the uncanny, ambivalent effect of talking shadows. Contemporary critics often reflected on the borderline position of reproduced film speech: "The photographed speech of talkies is as much the product of a machine as it is the human creation and expression. But *what is predominant?* It is an art based on using language and therefore as much a kind of expression through words as an intentionality of physical-chemical process itself" (Vančura 40).

On one side, critics wrote about cinema's regression back to the style of theater, but on the other, sound reproduction foregrounded the mechanical and reproducible vein of cinema more than ever before. As one prominent theater critic and dramatist put it: "Modern cinema doesn't want to be a substitute for spoken drama—no way! Instead it attempts to create a world in its own way. . . . In cinema there are no more premieres, since it is based on perfect, endlessly multiplied reproduction. . . . Cinema instead, is precision itself, the microscopic precision of vision and the microscopic precision of hearing" (Tille 73–74).

A concern with the uncanniness of an all-too-faithful audiovisual representation of a speaking body and with the repulsiveness of something as familiar as an open mouth of a singer was, for instance, formulated by film critic and radio author Franta Kocourek: "How many times and for how long have we had to look into the scary gullet, opened in front of us bordered by more or less shining teeth and allowing us to look deep inside the entrails of tenor's or baritone's heroic oral cavity!" (97). Jiří Frejka, the avant-garde theatre director, noticed too the ambivalent borderline between the spheres of the human and the machine when he wrote about the image of a man in sound cinema: "*Machine against man.* This is the cinema . . . or rather, I would say that it is a mechanical imitation of a man. . . . But the most refined and mature part of the society will never content itself with this industry of human voice and music. . . . We are still not fully aware of how disturbing this feeling of a mechanized man in the cinema is! It is a man, a robot, which is much too similar to man" ("Divadlo a film" 1–2; emphasis in the original). In another article he describes a feeling of something inhuman that accompanies the filmic reproduction of a speaking body, with the help of a puppet metaphor: "It is often said that sound cinema provides a surprisingly perfect illusion, as if the voice were coming directly from the image, but in fact you have constantly the feeling that the human voice . . . is somehow coming unstuck from the image, that it is incommensurable with the image and doesn't adhere to it. Sometimes this turns into grotesqueness, and one can't help but think of puppet shows in which humans speak for the artificial dummies" ("První zvukový film" 2).[9]

When we consider the reflections of contemporaries expressing as they do a culture's anxiety relating to a fragmented human subject, we can conclude that, to its critical audiences, early sound cinema seemed to be, on the one hand, a more accurate and truthful representation of the human being, but, on the other, it was also reminiscent of robots, automatons, nonhuman dummies, or ventriloquists. This dialectic needs to be read as an expression of the contemporary culture's ambiguous relation to the art of mechanical reproduction—a relation growing in ambiguity in direct proportion to the apparent perfection of the reproduction.

But the sense of sound cinema's robotic nature was related not just to the image of a speaking body but also to the new mode of film production itself, especially to the parallel production of multiple-language versions. As elsewhere in Europe, so too in Czechoslovakia this was the issue that stimulated the most intense resistance of the cultural elites attached to the ideology of the original and a unique artistic expression against the serialization and mechanization in the field of cultural production. Paramount's Joinville studio, where several Czech-language versions and many other multilinguals were produced on the basis of American patterns, became a symbol of this "decline," and was called by one of the Czech visitors, film journalist J. A. Urban, "a factory manufacturing words and sounds accompanied by images . . . , a warehouse, in which cans with words are stored and prepared for the distribution to all parts of the world" (206). Although the MLVs were intended to create the effect of "liveness" and full presence on the part of the various national audiences, in the imagination of a critical discourse they figured as the very image of destruction of preceding concepts of artistic creation. Critics felt that cinema was losing its uniqueness and the process of filmmaking was subject to the strict regiment of shifting crews on the same set—

with the director becoming an imitator, whether of his own or someone else's work—which changed studios into factories.

## The Search for a New Synthesis: Synesthesia and Opto-phonogeny

Although synesthesia, that is, the phenomenon of one sensory perception provoking an experience within another sensory realm, was already a relatively common topic in discussions around silent cinema (and other arts), it took a different shape with the coming of sound. In the framework of the critical discourses in Czechoslovakia it provided a key backdrop for speculations about the ideal relationship between image and sound.[10] The idea of a synesthetic (silent) cinema was based on the assumption that particular image compositions and rhythms could stimulate corresponding (virtual) aural perceptions (which is why theorists of synesthesia sometimes refused live musical accompaniment). But with synchronous sound this had to change, because what was now at stake was the relation of the two simultaneously present sensorial channels rather than a mere evocation of one by the other.

The general presumption implicitly grounding a synesthetic theory of sound cinema was that it could become a new basis for coordination, synthesis, and perhaps even translation among different arts and senses. For example, in his theory of "opto-phonogenia," the otherwise rather sober leftist critic Lubomír Linhart reworked the idea of an opto-phonetic counterpoint and went on to formulate the concept of a "synthetic line" connecting autonomous and separate images and sounds into a "new phono-visual form," composed of "sound-images or images-sounds instead of simple sum of images and descriptive sounds" (50–51). This synthetic line unites the two autonomous channels on a third level beyond the common reality, where "their essences meet, freed from all ancillary expressive elements" (55). What finally emerges on this level is the resulting "opto-phonetic emotion."

Another formulation of the principle of audiovisual correspondence was based on the concept of abstract movement. Influential critic Jan Kučera claimed a special kind of absolute movement (in counterdistinction to the restricted movement in painting or theater) connecting music and cinema and making the crossing of boundaries between senses and arts in sound cinema possible (242). Much like Linhart's texts, Kučera argues that interactions and the reciprocal opening of the senses take place on a higher level of feelings and ideas. But what needs to happen with the talkies first is that words—which unlike music are not based on pure movement—have to get rid of their petrified semantic sediments, and awaken in themselves a poetic power of sound rhythms and intermingle with noise. Under such conditions it would be possible for emotions and "poetic ideas" to travel across the full spectrum of sensorial perceptions, opening passages from one to another without uniting them into a homogeneous whole.

The most explicit synaesthetic theory of sound cinema was advanced by architect, acoustician, and painter Arne Hošek in whose view the laws of synesthesia were scientifically proven and universally valid. He defined the fundamental basis of all arts and senses as an "abstract," "absolute" composition, an internal organizational principle guided by subconscious intuition. This composition was seen as having its source in the "common emotional ground, which could be called 'sens-commun'" ("Komposice, umění, film a divadlo" 277).[11] Music is closest to this principle, and therefore it could function as an "internal compositional link" between particular arts. Whereas the other arts have moved away from this principle by their mimetic tendency, sound cinema offered a new promise for the rediscovering of "sens-commun" in that it "links movement with image and music, matches time with space" (277–78). Cinema thus replaces music in its function as a vehicle of communication between arts and senses (see Hošek, "Poznámky k estetice filmu"). A similar though more science fiction–like project of the synesthetic cinema was proposed by avant-garde theorist Karel Teige, who wrote about a quasi-cinematic apparatus that would immediately translate any visual patterns into sound compositions ("Film II: Optofonetika").

The recurrent motif of synesthesia should thus be understood as one of the utopian visions of cinema's future brought to life by the arrival of sound. It was grounded in the belief that a mechanical connection of sound and image would restore through technological means the primordial unity of multisensory experience, which could in turn provide a universal language for communication among senses, dissociated by modern media and arts.

## Sound Cinema as a New Communication and Recording Medium

Sound cinema emerged not as a finished and firmly defined medium with a clearly delimited identity, but rather as an ambiguous technology, a hybrid or intermedia form, an object of—in Rick Altman's terms—a "jurisdictional struggle." In other words, several media and formations struggled to define its identity according to their own criteria and to subordinate it to their own laws.[12] Although this is more clearly the case for the American cinema with its important period of the Vitaphone shorts than for Europe (at least on the level of industrial practices, technology, and film style),[13] a similar process took place in Europe and specifically in the Czechoslovak cinema on the level of critical discourses anticipating and surrounding the coming of sound.

The end of the 1920s and beginning of the 1930s was a period during which the entire culture of reproduced sound in Czechoslovakia went through a dynamic transformation and diversification; this was also the case for the development of popular-music forms. The recording industry expanded not only by the quantity of its production but also by the range of genres and applications such as advertisements; sports recordings; exotic, campfire, and street songs; as well as presidential addresses and even hypnosis records. Among them were also materials related directly to cinema: recordings of the best-known Czech film stars (singing Czech, American, or German hits); so-called melodramas; dramatic sound plays composed of dialogues, recitations, songs, and music;[14] and phono archives, founded by the Czech Academy of Arts and Sciences, which focused on recording national musical composers, political speeches, actors' monologues, and performances of ethnographic importance.

The phonograph arrived as a technological device for the first cultural model for the recording of human speech, preserving it for future generations and reproducing it as a process unfolding automatically in time. As is well known, an important function for the phonograph in the nineteenth century was to record "sound signatures" of famous personalities (see Kittler, *Gramophone, Film, Typewriter* 27). The other anticipated usage was that of phonetic analysis and refinement of the spoken word, as pictured in G. B. Shaw's *Pygmalion*. Both of these issues returned with the arrival of the talkies, when cinema was again briefly considered to be principally a recording medium, even better suited for the archiving of speech than the phonograph, given its capacity to also record gestures and facial expressions. At least in the Czech culture the coming of sound thus stimulated speculations about such issues and also advanced specific film archival projects. But there were two distinct kinds of film archives: one based on recording the "portraits" of real people, the other focused on preserving the heritage of the silent film art, which had suddenly become endangered since soon only second-rate cinemas would continue to exhibit silent movies even while the public was ready to forget them very quickly. The first instance of such an archive was partially realized by the Czech Academy of Arts and Sciences in cooperation with the A-B film production company in 1931–36, though the project never fully reached its original goals (see Míšková; and Zeman).

A strong stimulus for establishing a sound archive of musical and nonmusical recordings came also from the first radio station, Radiojournal, where it became more and more necessary to use high-quality gramophone recordings as a supplement to live broadcasts when more complex program schedules emerged, when it became necessary to preserve expensive musical performances, and when the sound effects used in radio plays became more sophisticated. A radio archive of records was established in 1929. In the beginning it focused on a broad range of

recording activities such as the recording of political speeches, actors' performances, or ethnographic projects of preserving folk songs and dialects coorga-
nized in Czechoslovakia by the Department of Phonetics of Sorbonne University,
the Department of Phonetics of Prague University, the Czech Academy of Arts
and Sciences, and Radiojournal, with the technological support of Pathé (see
Patzaková 292–95).

The expansion of sound culture was related less to the gramophone than to the
radio, which was at that time undergoing an intense quest for its own aesthetic and
communication specificity. Though quite distinct institutionally (much like the
BBC, the Czech radio station Radiojournal was partly state owned and mandated,
since its founding in 1923, to fulfill principally a public educational function),
the radio served as a backdrop for discussing issues related to film dialogues and
the use of sound effects. Moreover, the discourses surrounding the emergence of
the radio and its search for its own identity showed some common features with
discourses concerning the talkies: a catastrophic-utopian imagination (stressing
immediacy, instantaneity, omnipresence, the unstoppable nature of electromag-
netic waves, their infiltration into everyday life, and their potential for intense
social impact and more), uncertainties relating to the proper naming of the new
medium, and the frequent comparisons and identifications of radio with established media (press, telephone, gramophone, theater).[15] After 1926, in the wake
of the period when radio was considered an extension of either the press or the
stage theater, discussions about radio specificity addressed an explicitly aesthetic
question of new radio genres: radio play, radio reportage, sound montage, and
so forth. Their specificity was considered to be located in the dynamic interrelations among spoken word, music, and real noises; it was commonplace to refer
to cinematic montage as a source of inspiration for such sound compositions.

Linkages among cinema, sound media, and new means of industrial production and transport were widely acknowledged already during the silent period—
especially in the circles concerned, since the mid-1910s, with avant-garde poetry
and manifestos, where all reproduction technologies functioned as the agents
and symptoms of daily and artistic modernity. In the second half of the 1920s,
however, connections between sound and moving images began to yield specific
visions of cinema's alternative futures, approaching it as a medium of new regimes
of communication and spatiotemporal structures. For example, film and radio
writer Kimi Walló labeled both radio and cinema metaphorically as "dreams that
whisper from the distance." By these "dreams" he meant the complex sensorial
experiences mediated by loudspeakers, which could combine music, the spoken
word, nature and city noises and above all were able to immediately transport
us back and forth between distant spaces or times. Film and radio sound was
thus supposed to work as a new kind of energy: it was seen as the sum of coor-

dination between electricity, magnetism, and photochemistry, and as such it was able to, in effect, dynamically render the visual space of film by providing instant connections that would bridge gaps between separate and incommensurable time-spaces. "With the stroke of a gong or the hoot of a siren, the protagonist of a radio play could appear in America, in hell or on the seafloor" (Walló 79). The metaphor of a "distant whisper" is, of course, quite prescient with regard to the later manipulation of volume in proportion to the narrative significance of a particular sound, pointing as well to the corresponding construction of subject position both in cinema and in broadcasting.

The critic proposing the most elaborate ideas concerning sound cinema as a new communications medium was Karel Teige, who had been writing about cinema sound since the first half of the 1920s. In his essay "Film II: Optofonetika" (1925) he described contemporary inventions in the area of synchronization systems for sound and the moving image, and formulated a hypothesis that "filmed events will be broadcast into our homes and projected there, though they are just taking place on the opposite part of the globe" (11). Teige assumed that optical, acoustic, and "electro-kinetic" waves vary only in frequency and that it thus would be "possible to translate acoustic, luminous or electric energy into another"; from this assumption he proposed a new "opto-phonetic art" based on the principle of translating light into sound and vice versa. This vision was to be made possible by a special apparatus connecting the film projector with a telephonic inductor capable of automatically transforming the luminous energy generated in the projection of images of the outside world into harmonious sound compositions: "Visual patterns, projected onto the screen, will produce a particular sound. A square would probably cause a different tone than a triangle and a different one than a luminous sphere. . . . It will be possible to combine the chords of such luminous geometrical forms. . . . Perhaps the music of the future will not be written on a score paper: instead the filmic light compositions will be projected. Crystals and stars will talk to us: in what tone, what language?" ("Film II: Optofonetika" 11).

Teige's imaginary apparatus, which drew in part on older experiments in the field of chromatic keyboards and abstract films, is thus constructed as a fusion of cinema, telephone, and radio (or television), and at the same time as a materialization of the idea of a new audiovisual poetic communication of man with the world. In his later writings, related more specifically to actual film sound practice, Teige saw the future of sound cinema in nonfictional modes of artistic and documentary forms, which again strikingly resemble today's television transmissions: "Synchronized words and synchronized sounds will double the documentary value of film newsreels. Sound films of world events! Sound films of daily news! . . . Or sound films of folk festivals, folkloric dances; an important tool of scientific research" ("Divadlo a zvukový film" 39).

For Teige, the main artistic potential of film sound was, however, in its au-tonomization in the form of sound poetry that would combine artificially created sounds not imitating any outside reality. Anticipating today's digital effects, he thus proposed the artistic use of film sound not as reproduction and representa-tion but as sound writing or simulation. In his opinion, sound film should con-struct "sounds so far unheard in reality, sounds only graphically inscribed into sound track. Sound poetry—not His Master's Voice!" (41). Drawing on Teige's speculations, we can see how film sound functioned as an intense stimulus of avant-garde visions of cinema's future, a future clearly distinct from that of a classical narrative film that was soon to return as the dominant form. Still, this deviation from the dominant line of development was present not only in writ-ings of avant-garde theorists but also in the much more sober newspaper articles of various critics, scientists, and even technicians. On the level of discourse, cinema returned, so to speak, to its early stages, where many alternative futures were awaiting the new, still undefined medium, among which narrative fiction was not the most important.

## Noise versus Speech, the Noises of Speech, and the Spoken Word as Autonomous Action

On the level of aesthetics, the most specific link between the radio and cinema was that of a new conception of speech and the related issue of noise. To fully un-derstand the preoccupation of cinema-related discourses with speech, it is neces-sary to broaden our perspective and take into account not only radio but also the development of the entire sound culture of the period, especially the soundscape of the modern city. One of the key changes under way during the late 1920s and early 1930s was the new intensity and pervasiveness of modern noises, which then partly defined the experience of modernity as such. As recent research has shown, technological (nonorganic) noise was a major public issue especially in large cities (through the noise-abatement movement), culturally related further to the new artistic and scientific experiments in the field of music, spoken language, and sound technology (see Thompson, *The Soundscape of Modernity*). The social awareness of technically produced and reproduced noises reached its peak at the moment of cinema's irretrievable conversion to sound and with radio undergoing a self-reflexive process of searching for its own media specificity.

In the realm of Czech culture this process could also be seen in relation to poetry: poems by prominent authors S. K. Neumann, Josef Hora, or Vítězslav Nezval (from the mid-1910s to the mid-1930s) provide metaphorical descriptions of the aesthetic and social impact of the sound of human masses in the streets, of factories, locomotives, electric trams, lightbulbs, cars, typewriters, telegraphs, phonographs, telephones, radios, and jazz music, alongside the glorifications of cinema and American film stars. The same could be said of the peculiar avant-garde genre of film writing, the so-called film librettos, or unrealizable scenarios, which were so typical of the purely visionary, impractical relationship of the Czech avant-garde of the 1920s to cinema. The clearest example would nonetheless be the history of radio with its nonfictional genres of live sports broadcasts and industrial reportages grounded in attempts to represent the logic of industrial production through a live montage of real noises.

From 1926 to 1930 these two genres had a strong potential to shift the aesthetic character of the radio from the domain of musical and theatrical transmissions toward montage compositions of extraordinary noises. In reproducing noises of exercising bodies, applauding mass audiences, workplaces such as coal mines or power stations as well as public spaces of everyday life such as cafés, *varietés*, train stations, and of course streets, these genres achieved a considerable dramatic impact and effect of reality. These experiences with nonfictional genres were then also transposed into the realm of fiction radio plays so that the most widely discussed expressive means of the radio play became the so-called sound backdrop, or background noise, which does not "have to have only the form of stage music so as to illustrate a scene in the usual way. It can instead 'photograph' even the most realistic sound of reality: the grumbling of engines, the shouting of demonstrators and sportsmen, the vocal timbre of an important politician or professor—as though the new radio reporters would obtain, devise and compose it" (Frejka, "Rozhlasový žurnalismus" 6).

Alongside these discussions about noise as artistic material, discourses also emerged reflecting on noise as disturbance of radio transmission. Noises produced by radio transmitters and receivers, but also by their physical environment—for example, weather, nearby electrical devices, as well as reckless manipulation of the so-called feedback on the part of the listeners—brought along several diverging perspectives on noise. Dividing them into categories, we can speak of disturbances in the realm of the aesthetic (developing a new sensitivity and terminology for distinguishing the noises of different kinds of breakdowns), the ethical (struggle against the irresponsible manipulation of radio receivers), the legal (lawsuits against the users of electrical devices producing intolerable noises or disturbing their neighbor's radio reception by feedback), and the political (international divisions of frequencies). These multiple issues and conflicts defined the radio as a complex social and ecological[16] system interacting with spaces of everyday life as well as with city soundscapes.[17] As for cinema, the technological noise of a projector or reproduction was not a big issue (except in specialized technical writings), and neither was the noise of the audiences, who had already been silenced in the course of the preceding evolution of exhibition practices[18] (with the exception of short-lived nationalist demonstrations against German-speaking movies in Prague in 1930); in other words, in counterdistinction to radio, cinema of that period was not an "ecological" medium, since its reception lacked the interactive and ethical dimensions typical of the radio. What was instead intensely discussed and also closely observed by critics and rank-and-file viewers was the problem of insufficient synchronicity and incommensurability of film personalities and voices.[19] Their nonsynchronicity could thus be considered a specifically cinematic disturbance of communication between the viewer and the fictional world of the diegesis.

But the end of the 1920s was also the period when noise was foregrounded as a general social and cultural problem: as a noise of the new international popular culture, disturbing (thanks to the new technologies) the values of the traditional national culture, as a noise of the vulgarization of speech under the influence of changes in the social and national structure of the population in the wake of World War I and the foundation of an independent Czechoslovakia. The changing sound culture of the spoken language caused by the mixing of social classes, the expansion of media and popular culture, and the impact of intense war experiences was studied with a great deal of attention by linguist Miloš Weingart, a member of the well-known Prague Linguistic Circle (along with Jan Mukařovský and Roman Jakobson), who called this vulgarization "argotization." According to Weingart, sound cinema as well as radio function as mirrors and symptoms of such argotization (marked, for example, by mispronunciations and faulty intonation in real social life or deficient differentiation of language expressions among fictional characters from distinct social backgrounds in film and theater), but could at the same time be used as a tool for scientific analysis of speech behavior (as an extension of the phonograph, but additionally permitting the analysis of corporeal expressions and of the social context of speech acts), thus leading to refinement and cultivation of sound culture (see "Zvukový film a řeč" and "Zvuková kultura českého jazyka").

The intense preoccupation with the sound quality of speech—quite understandable with regard to the escalating nationalism in a small country like Czechoslovakia, surrounded and permeated from inside by the powerful foreign-language culture of the Germans—soon prevailed also in the realm of radio-related discussions. From 1929 on it was more and more common to shift attention away from real noises and toward the sonic wealth of speech. Nonetheless, the vaunted radiophonic speech was not defined by the laws of written language; rather, it was conceived as malleable expressive material, capable of incorporating and to some extent subjugating the domain of real noises. Through careful manipulation of rhythm, timbre, word order, voice modulations, and dynamics or significant pauses, the spoken word should serve "not only as a symbol of various things, but directly as a live (plastic) image and live (rhythmic and dynamic) music!" (Stoklas, "Úvaha o hře rozhlasové" 1).

Many film critics, too, searched for the essence of sound cinema as a new medium within the realm of film speech. Curiously, it was more often a rather radiophonic concept of autonomous speech as a flexible expressive material than a synchronic or counterpoint aesthetic of sound-image relationships. For example, O. Rádl defined the aesthetic identity of talkies—which ought to differentiate themselves above all from a theatrical use of language—by the associations of sounds themselves and by emancipating the semantic power of spoken words from the actor's body: "Tomorrow we will hear in a movie theatre the words alone. You will make sure that more horrible than the terrified face of a dying soldier would be one syllable sounding pianissimo in a trembling tone with frightful gradation. . . . You will find out, how much exalting energy there is in words, how much sadness sounds in two syllables of a single expression" (68).

Similar visions were also advanced by two authors close to the Prague Linguistic Circle, well-known writer and film director Vladislav Vančura and stage and film director Jindřich Honzl. Both asserted that film speech should not imitate stage dialogue; rather, it should undergo the creative process of purposeful deformations so as to intensify its expressive powers. According to Honzl, sound and speech should develop their sign properties, stand independent against the image, and break the organic relations with the speaking body. Vančura emphasized that film speech had to become autonomous dramatic action in itself, and indeed the main structuring principle of film form.

What we can conclude from all these arguments and issues as formulated and defined by discourses surrounding the coming of sound is that, in the realm of a broader cultural context, cinema underwent a process of profound transformation, which brought with it the need to define its identity anew, bringing it in many respects back to the earlier stages of media history as well as connecting it in some respects to the changes relating to the present expansion of electronic and digital media. Although from a retrospective point of view the conversion to sound could appear to have been a short-term deviation from the dominant development of classical film, from the "prospective" point of view—and taking intermedia and cultural relations into account—we see the full heterogeneity and uncertainty of this situation. Early cinema, early sound cinema, as well as today's cinema permeated by digital technologies are all moments of film history opening multiple alternative futures in the face of a new technology to the point of suddenly being forced to look for their media identity in relation to preestablished forms and media.[20]

## Notes

A special thank-you to Nataša Ďurovičová for her editorial assistance with this essay.

1. On the notion of discursive inventions as an object of study of media archaeology, see Huhtamo 302–3.

2. I wrote the present article on the basis of an analysis of a broad scope of Czech trade cinema and radio journals, newspapers and general cultural or artistic journals, as well as theoretical, technical, and scientific books and poems. I gave priority to longer texts with theoretical or generalizing intentions concentrating on film sound or reproduced sound as such. The final number of these preselected and closely read texts was approximately one hundred. To further elaborate on the proposed conclusions here, however, it would also be necessary to study other documents of different kinds (business, legal, and so on).

3. On the notion of topoi in relation to media archaeology, see Huhtamo.

4. On general questions of cinema's possible or alternative futures as a historiographical concept, see Elsaesser.

5. On the concept of "prospective look" in film historiography see Altman, "Introduction: Sound/History."

6. The rival terms in English and French were listed by Rick Altman: *talkie, speakie, dramaphone, cineoral, audifilm, pictovox, phonoplay, audien* ("De l'intermédialité au multimédia" 38).

7. For a more detailed description of quantitative terminology in Czech film advertising, see Klimes.

8. The term *rozhlas* corresponds semantically to the English *broadcast,* but it had not existed before as a noun, only as a commonly used verb, *rozhlásit* (to announce).

9. A metaphor similar to dummy was that of ventriloquist, used in the same context by Karel Teige (see "Divadlo a zvukový film" 38).

10. The general importance of synesthesia for the avant-garde movement in plastic arts of that time could be seen in the anthology about the work of kinetic artists Zdeněk and Jöna Pešánek (see *Světlo a výtvarné umění*). By the way, kineticism could also be seen as one of the ideas of synthesis and coordination between the arts, which were so active during the early sound cinema era.

11. The term *sens-commun* was used in the original.

12. For discussion of the issue of the intermedia phase of early sound film's development, see Altman, "De l'intermédialité au multimédia," and on the problem of jurisdictional struggle, see Altman, "What It Means to Write the History of Cinema."

13. Comparison of the intermediality of early sound cinema in the United States and in Europe, particularly in France, was proposed in Barnier.

14. A Czech historian and collector of phonograph and gramophone devices and recordings considers the melodrama to be an important predecessor of talkies (see Gössel 113).

15. For a typology of early writings about radio, see Čábelová 147–80.

16. On radio as an ecological medium, see Cubitt 104.

17. I analyze the issue of radio noise in a more detailed way on the basis of Czech radio

trade journal Radiojournal in my article "Speech and Noise as the Elements of Intermedia History of the Early Czech Sound Cinema."

18. On this development in the United States, see Altman, "Film Sound—All of It."

19. On a similar situation in Germany, see Müller 291–312.

20. In his theory of "crisis historiography," Rick Altman proposes basic analytic concepts as well as an orientation time line for the systematic study of such moments of cinema history (see "Cinema Sound at the Crossroads").

6

# Sounds of the City

## Alfred Newman's "Street Scene" and Urban Modernity

**MATTHEW MALSKY**

Dear Al,
Do nothing but continue to use
"Street Scene" wherever it fits.

—Darryl Zanuck

## "Street Scene" Recycled

It is November 1953, and you have settled into your seat at the grand Loew's State theater in Times Square, Manhattan. The film starts with the brass and percussion strains of the famous Twentieth Century-Fox fanfare. On-screen, velvet curtains part as the strings swell, revealing the wide-screen splendor of a full orchestra in formal attire. In the background behind the majestic stage, Doric columns punctuate a nearly cloudless blue sky. "The Twentieth Century-Fox Symphony Orchestra's 'Street Scene' composed and conducted by Alfred Newman" in CinemaScope is about to begin. You revel in the glorious, enveloping stereophony in which "the sound [comes] naturally and clearly from the sections of the orchestra where it was produced" (Crowther, "CinemaScope"). After a full six minutes and a rousing climax, Newman turns to the camera and bows, then immediately—finally—launches into the main titles for the evening's main event: the premiere of *How to Marry a Millionaire* (Jean Negulesco, USA, 1953).[1] Neither Newman nor his music returns again. How can you make sense of this unusual concert prelude?

This opening sequence exceeded contemporary musical practice for the start of a classical narrative film: to underscore the opening credits, introduce thematic materials that will be used later, signal the film's emotional mood, and then fade into "inaudibility." There were resonances here with musical performance on early sound film that imitated established and popular forms. Or perhaps it even evoked the live-performance codes of first-run silent movie exhibition—more keenly felt because this CinemaScope premiere was held in exactly such a movie palace.[2] Thus, this representation of a symphonic concert functioned to buffer audiences from (even while it introduced them to) the new and unfamiliar cinematic experience of CinemaScope by utilizing a recognizable sonic model. By presenting stereo sound in a "cultured" context, Twentieth Century-Fox encouraged audiences to accept stereophonic music as prestigious while also "relying upon codes established earlier to ensure its reception as verisimilitudinous" (Belton, *Widescreen* 207). However, this technological showpiece did more than simply reference ancient film sound history and the high-status sonic experience of the concert hall. The musical choice of "Street Scene" made this opening into a recapitulation of the entirety of the classical Hollywood sound era. Although the form (that is, the technological mediation of sound) was new, the content (the music) was ripe with associations, and hailed spectators both through specific historical allusion as well as through ideological address. To more fully understand how this concert prelude might have been experienced, we need to follow its historical traces.

The music was originally composed by Newman for the opening sequences of an early sound-era film: King Vidor's gritty realist drama *Street Scene* (USA, 1931). The success of Newman's music fueled its wider acceptance. It quickly entered the combo, concert, military, and dance band literatures. It was transformed into a song with lyrics, published in a variety of instrumental forms, performed live, broadcast on the radio, and recorded frequently from 1933 on. This was already a well-known popular "standard" before it reappeared in the film noir *I Wake Up Screaming* (Bruce Humberstone, USA, 1941), and then in no fewer than six additional Hollywood films through the 1940s and on television in the early 1950s: *Do You Love Me?* (Gregory Ratoff, USA, 1946), *The Dark Corner* (Henry Hathaway, USA, 1946), *Kiss of Death* (Henry Hathaway, USA, 1947), *Gentleman's Agreement* (Elia Kazan, USA, 1948), *Cry of the City* (Robert Siodmak, USA, 1948), *My Friend Irma* (George Marshall, USA, 1949), and when the latter film was made into a TV sit-com on CBS (Richard Whorf, USA, 1952).[3] Its appearance in *How to Marry a Millionaire* was its last but most grand, and it capped a twenty-two-year history. In its longevity and (relative) stability, "Street Scene" offers an unusual opportunity to consider three distinct moments that provide a remarkable soundtrack to the shifting ways in which modernity was heard and imagined: the early sound era, the height of classical Hollywood sound practices, and the "frozen revolution" of 1950s magnetic sound.

## "Street Scene" and the City: How Modernity Was Articulated and Mediated

How can we account for this curious recycling over such a long time and in such diverse genres? Partial answers lay in the efficiency of the studio system that reused existing musical properties to cut costs, and even in the personal preferences of an influential studio head.[4] However, a more convincing answer was evident in the visuals at the start of every picture. All used "Street Scene" to underscore the establishing shots that followed the main titles: a panoramic view of the Manhattan skyline. These were all urban pictures, and more specifically all featured New York City as the locus of the drama. Although this was not at all unusual for Hollywood movies—the "Empire City" was (and continues to be) America's most identifiable icon and Hollywood's most durable alter ego—this specific recurring combination of location and music pointed to the privileged position that the concept of "the city" held in these films, and in social theory more generally. It was the persistence of the representation of modernity through this semiotic equation—the visual depiction of the bustle and excitement of New York City with the sounds of a distinctly American form of vernacular musical modernism—that Twentieth Century-Fox drew upon in recycling this music.

Modernity as a concept encompasses quite a few overlapping domains; moral, political, cognitive, and socioeconomic definitions have dominated contemporary thought. In a fourth major category, what has been called neurological, modernity is conceived of as a barrage of stimuli. In the writings of Max Weber and Georg Simmel through Lewis Mumford and Richard Sennett, the ties between individual subjectivity and the culture of cities has been a formative theme, claiming that the primary characteristics of modern city life are social-psychological in nature.[5] This conception of modernity should be understood as a distinct register of subjective experience characterized by the physical and perceptual shocks of the modern urban environment. From before the start of the twentieth century, the saturation of advanced capitalism—what has been called the "second industrial revolution"—could be gauged in terms of the rapidity of industrialization, urbanization, and population growth; the proliferation of new technologies and transportations; and the explosion of a mass consumer culture. These symptoms of the city transformed the very texture of everyday experience, and modernity was the individual's subjective response.

Cinema—and especially sound cinema—is an exemplary case of this type of modernity, one subsequently developed by Siegfried Kracauer and Walter Benjamin.[6] Cinema is constituted, on the one hand, through its status as an industrial-technological commodity. As with trains, photography, electric lighting, telegraph, and telephone of the early 1930s, sound cinema was experienced by its early audiences as a quintessentially modern stimulus (Singer). At the same time, classical Hollywood cinema, through the impetus of its narrative and its formal techniques of representation, provided an aesthetic mode of experience that reflexively reproduced modernity. For the audiences of Street Scene, I Wake Up Screaming, and How to Marry a Millionaire, to watch a movie was to participate in modernity both viscerally and cognitively. Miriam Hansen asserts exactly this point. Distinct from "high" or "hegemonic modernism," this vernacular modernism of interwar American cinema was created because:

> Hollywood did not just circulate images and sounds; it produced and globalized a new sensorium; it constituted, or tried to constitute, new subjectivities and subjects. The mass appeal of these films resided as much in their ability to engage viewers at the narrative-cognitive level or in their providing models of identification for being modern as it did in the register of what Benjamin troped as the "optical unconscious." It was not just what these films showed, what they brought into optical consciousness, as it were, but the way they opened up hitherto unperceived modes of sensory perception and experience, their ability to suggest a different organization of the daily world. ("Mass Production of the Senses" 70–71)

Cinema represented a specifically modern type of public sphere. This new mass public functioned as a discursive forum qua modernity in which individual experiences and subjectivities could be expressed on-screen, and also recognized and reexperienced by the viewing spectators themselves. Each in its own way, these films articulated a vision of this modernity, as well as offering its audiences a unique means for experiencing it. Specifically, this type of modernity came to be associated with the music of "Street Scene." This music carried with it such distinctive metonymic intimations of experience and place that it became a signifier of modernity. This music operated as one formalized element within a chain, an arbitrary musical signifier that added to the portrayal of the city as modernity's public sphere, part of the code of the sensorium, which gave spatial and sonic coordinates to the symbolic network of intersubjective relations.

However, all these films also emphasize the construction of particular subjectivities. Although modern, urban living in general characterized these disparate films, they all placed a particular emphasis on the portrayal of women's options of living in the American city. At the center of each of these film narratives was an exploration—within mass culture—of the modern challenges that faced young, urban, white, middle- and lower-income women—always focusing on their fulfillment through romantic opportunities and marriage strategies. The gold digger, the working girl, and the adult daughter striking out on her own— the choices these women made defined the modernity of the city and epitomized their position in the public sphere. My claim is that the music "Street Scene" as used in these different contexts reveals a shifting subjectivity in the tumultuous social transitions from the end of World War I to the years following World War II. In examining the soundtracks of *Street Scene, I Wake Up Screaming*, and *How to Marry a Millionaire*, we will find marked changes in the relationship between modernity and the urban public life as reflected in the status of female characters in these films. "Street Scene" exemplified the interrelationships between the sound practices of Hollywood, metropolitanism as a key term of modernity, and companionate marriage. Put differently, the soundtracks of these films both articulated and mediated the changing relationship between female subjects and the larger discursive field of signifying structures.

## Street Scene and the Sounds of the City, 1931

After the *Street Scene* credits are shown over a cartoon drawing of a moving van in front of a tenement house, we begin, as all proper tours of Manhattan should, with a view of the skyline from afar, where the grandeur and majesty of the city are impressive and overwhelming. However, the camera quickly pans downtown. Moving over the tops of shorter and more sordid dwellings, zooming in as if through a microscope, the details of a teeming and sensuous street-level scene come into focus: kids cooling themselves in an open hydrant, a workman laboring to chip ice from a block, a horse swishing flies away with its tail, a sleeping man on a fire escape. Finally, we settle on a microcosm, "the exterior of a 'walk-up' apartment house, in a mean quarter of New York" in the midst of a summer heat wave, the principal setting of Elmer Rice's Pulitzer Prize–winning play (*Street Scene* 3).

From the film's opening moments, Vidor succinctly evokes the city as a physical setting characterized by its varied palpable associations: textures, spaces, temperatures, movements, sights, and smells. But most of all, King Vidor's *Street Scene* is an urban world permeated with sound. "The noises of the city rise, fall, intermingle: the distant roar of the 'L' trains, automobile sirens and whistles of boats on the river; the rattle of trucks and the indeterminate clanking of metals; fire-engines, ambulances, musical instruments, a radio, dogs barking and human voices calling, quarrelling and screaming with laughter" (5). Vidor brings more to this production than his strikingly mobile camera style and an affinity for representing a quotidian city life sympathetic with Rice's play. Both were already evident in his 1928 silent *The Crowd*.[7] He also brings, with his early use of sound technology, an unusually sure and developed sense of continuity and space. He already follows the dictates that Rick Altman has described as "Hollywood's standard representational system" ("Sound Space" 64). For example, taking his cue from the original Broadway production, Vidor blankets the soundtrack not only with the dialogue and diegetic sound effects requisite to the story but also with a constant layer of random street noise.[8] Throughout the movie we hear the sounds of the city—almost exactly as described in Rice's stage notes—even if we do not always see their source. Throughout the film, these sound effects contribute significantly to our sense of the sonic oppressiveness, intensity, and enormity of this metropolitan world, even though only a microcosm is seen.

However, rather than the "realistic" diegetic sounds that permeate the rest of the soundtrack, the opening sequence described above is underscored with music only—Newman's "Street Scene"—and the music is notable not only for its very presence but for its style as well. Hollywood musical practices were still maturing in 1931, both technologically and musically, and musical underscoring, beyond the ubiquitous main-title themes, was still the exception. It was not until at least the mid-1930s—initiated by movies such as *King Kong* (Merian C. Cooper and Ernest B. Schoedsack, USA, 1934) with Max Steiner's wall-to-wall underscoring—that Hollywood films were saturated with music in the established hegemonic underscoring style (that is, a "classical" music that adapted Wagnerian-leitmotif techniques to Hollywood's narrative demands).[9] Instead, Alfred Newman's music for the main titles and opening sequence, the beginning of the "second act," and the closing titles of *Street Scene* clearly has a foot in two camps: popular and classical. With its ordered, symphonic forces, on one hand, and its Jazz Age feel, on the other, the music of *Street Scene* participates unabashedly in Gershwin's characteristic concert rhetoric of the 1920s—jazzy syncopated figures and ornaments; moving accompanimental response lines in contrary motion with the melody; "stinger" chords repeated in motor patterns; bold skips in the melodic line; pentatonic collections juxtaposed against expressive blue notes; a fragmentary, nondevelopmental, "modular" form; and a striking orchestration.[10] This crossing of the "great divide," the combination of high and low forms, was not accidental or uninformed. Newman, who grew up in New York City as both a classically trained piano prodigy and a Broadway musical director (he directed several of the Gershwins' early Broadway productions), was witness to the birth of this "American modern music" that "officially" began with the Paul Whiteman Orchestra's performance of "Rhapsody in Blue" in 1924. This was the musical equivalent of vernacular modernism, the very sounds of Ann Douglas's "mongrel" Manhattan.[11]

Yet the connection of "Street Scene" to a film narrative that articulates urban "neurological" modernity is anything but ineffable. This music assumes specific meanings through two forms of connotation. First, these musical sounds are closely synchronized to stereotypical urban activities on the image track. This correlation of "Street Scene" with not only the New York City skyline but also its "hubbub" is repeated in virtually all of the other films that used this music over the next twenty-two years. Though not originating diegetically from the cityscape on-screen, this music hardly suggests the psychologized register of later classical Hollywood nondiegetic soundtracks. In fact, Newman's music was more often criticized as being too concrete, and derided as heavy-handed "Mickey-Mousing." This is clearly evident in the "Street Scene" music orchestrated to mimic flies buzzing around a milk horse, the brashness of boys dashing around in the street, or a man snoring. As Kathryn Kalinak defines it, "Mickey-Mousing is a structural device that authorizes the non-diegetic presence of music as an emanation of the narrative itself. Its perfect synchrony with narrative action masks its presence so that the music can create certain effects on a semi-conscious level without disrupting the narrative credibility on a conscious level" (126–27). The musical representation of this cityscape is structurally synonymous with a "realistic" representation of the city's sounds, noises that might otherwise be "actual" sound effects.

Second, in a number of "experimental" works of the late 1920s and early 1930s, music was treated as an integral aesthetic and formal feature of realist urban cinematic representation. This rendering of modern public life in the city through music can be seen in Carl Mayer's idea of the city symphony (Kracauer, *From Caligari to Hitler* 182–89). *Berlin, the Symphony of a Great City* (Walter Ruttmann, Germany, 1927), a lyrical documentary of modern life in Berlin, was conceived through music both aurally (with a commissioned symphonic soundtrack) and visually (as "optical music" or "a melody of pictures"). *The Man with a Movie Camera* (USSR, 1929) relied on Dziga Vertov's detailed musical plan to give aural specificity to the modern dimensions of Soviet city life. In the narrative of *Street Scene,* the play Elmer Rice considered his most innovative, there were similar resonances between musical structure and the formal impetus that allows the realistic presentation of everyday life. Rice patterned the unfolding of the plot on musical organization and orchestration (*Minority Report* 237). Invoking sociologist Ferdinand Tönnie's distinction between gemeinschaft (community) and gesellschaft (society), Henry Jenkins has described *Street Scene* as "a plot structure appropriate for a Gesellschaft culture," an episodic narrative in which social relations are characterized as multiple, fragmentary, and impermanent ("Tales of Manhattan" 200). Music was the perfect metaphor to organize these social bonds that epitomized an urban, neurologically modern "society." For example, in one revealing scene, all of the main characters good-naturedly argue their musical preferences. Sam, a young Jewish intellectual, loves the seriousness of Beethoven. Mrs. Maurrant, who throughout voices her desire for some small measure of pleasure and understanding, adores Mendelssohn's "Spring Song." The cheerful Greta prefers the "walzer von Johann Strauss." Mrs. Jones, the resident busybody, says, "Well, gimme a good jazz band, every time." Thus, the clash of disparate musical stimuli—all available in the city—is thematized and parallels the stories of these people themselves. Until the plot coheres at the end, this tale of the city flits among these neighbors gossiping on the stoop, a married woman secretly meeting with her lover, a couple awaiting the birth of their first child, a fatherless family about to be evicted, and the political rants of an older Jewish socialist. This argument over musical taste, like the plot constructed of seemingly random narrative elements, mirrors the lack of cogency of the multitude of competing, chaotic stimuli of the city environment.

These swirling subplots of *Street Scene* finally cohere around the character of Rose, and it is for her that the implications of the sociosymbolic reality of the city are ultimately most pertinent. Rose (Sylvia Sidney in her first starring role) is a young woman living with her working-class parents, entering the adult world of work and courtship. All of this is to be negotiated in the public bustle in front of her building. By the end, she is confronted with choices that will shape her life thereafter. Over the course of the film, she is courted by two different men, each representing a different path. One is an understanding, empathetic, and upwardly mobile soul mate. The other is an older married man with the connections to put his mistress on the stage. This choice juxtaposes companionate marriage (but with delayed gratification, the difficulties of intermarriage, and the challenges of class and social acceptance that come with mobility) against the immediacy of the city's excitement and glamour (but at the expense of Rose's moral center). Rather than see Rose make her choice between these two men, we in the audience are forced to come to terms with the radical contingency of this narration of city life when Rose's mother is caught in flagrante delicto by her father and murdered. Rose, in turn, assumes custodial responsibilities for her younger brother and moves out of the city entirely. She rejects both men. But in participating in courtship rituals in the city—that is, the symbolic order of exogamy—Rose is offered several avenues of entrance into the world of ambition, culture, wealth, and experience. In short, this is the urban life that Manhattan seemed to offer young women like her. This is the city musically represented in the opening montage.

## Interlude: "Street Scene" in the City, 1931–41

Hollywood's colonization of the sheet music industry, New York's Tin Pan Alley, was crucial to the vertical integration of sound production, because, as Sanjek notes, "within the space of a year, talking pictures had become the principal means for plugging popular music" (54). First out of the gate, Warner Brothers bought the original Tin Pan Alley music house, M. Witmark, in 1928, and quickly acquired another dozen music companies. Metro-Goldwyn-Mayer (*Street Scene* was produced independently by MGM's Samuel Goldwyn) following quickly, purchased a controlling interest in the "hustling" New York–based Robbins Music Corporation later the same year. When in 1929 MGM released its first movie musical, *The Broadway Melody* (Harry Beaumont, USA, 1929), the songs immediately went into the Robbins Music catalog, and Robbins hired uniformed pages to hawk sheet music and recordings of the songs in MGM theater lobbies. It was Jack Robbins who in 1926 had assisted the "king of jazz," Paul Whiteman, to publish many of the arrangements of the "futuristic" music he had commissioned for his concert jazz band and dance orchestra. Many had been performed at his "Experiments in Modern Music" concerts at New York's Aeolian Hall. These "symphonic syncopated" scores and choral arrangements, sold principally to schools and bands, were featured in the Robbins catalog as "prestige" items along with MGM's new popular songs and compilations of newly composed motion-picture music (with themes by Erno Rapee, Hugo Reisenfeld, Domenico Savino, and Nathaniel Shilkret) (103, 107). Together, business partners Robbins, Whiteman, and Abel Green, who was *Variety*'s music editor, developed a business plan for marketing their publications that depended on close association with New York's most prominent bandleaders. The bandleaders' "'plans in editing and suggestions,' together with regular broadcasts of a song from one to three times a week and appearances in person at night clubs, hotels, presentation movie houses, and vaudeville theaters were essential in making a hit" (102).

"Street Scene" was a case in point. It perfectly suited the catalog as both a "symphonic syncopation" and a film score. In 1933 Robbins Music Corporation published no fewer than seven different arrangements. Versions existed for solo piano; for piano (or guitar) with voice (lyrics by Harold Adamson); in three different four-hand piano arrangements: by songwriter Vernon Duke, by house arranger Savino, and by jazz arranger Jacques Fray; for concert band (Erik Leizden, arranger); as well as a full symphonic version with piano and tenor banjo.[12] Though it has been impossible to confirm that Robbins hit its mark for regular broadcasts and live performances, there are traces that suggest that "Street Scene" was much in the public ear. On February 15, 1933, Newman conducted the Los Angeles Philharmonic in a special concert, which included his own "Street Scene" (F. Steiner 157). Anecdotal evidence suggests that Leizden's arrangement was a regular part of the Edwin Franko Goldman concert band repertoire, and may have been performed as part of their Central Park concerts series in the 1930s and 1940s.[13] Into the 1950s it was performed by the National Guard Band (Elkus). The *New York Times* lists several radio broadcasts between 1936 and 1949. Newman himself recorded it in 1940 as part of a 78–rpm album titled *Sweetheart Music*. Through the 1940s and '50s, big band and dance orchestra recordings were made by Les Brown, Benny Carter, Ray Anthony, Harry James, Stan Kenton, George Greeley, and Lester Lanin. Presumably, they were playing it at their concerts too.[14] All this is by way of saying that when the principal characters in the 1941 Twentieth Century-Fox film noir *I Wake Up Screaming* walked into a swank New York City nightclub and the dance band in the background struck up the music of "Street Scene," many film spectators would have known this music and would have likely been able to hum along.

## I Wake Up Screaming in the City, 1941

After an eight-year hiatus from cinema, "Street Scene" returned to underscore a more fatalistic and menacing vision of the city's modernity.[15] *I Wake Up Scream-ing* made the most extensive use of "Street Scene" of all of the films in which it appeared. It was used not just in the main titles; it was transformed by Cecil Mockridge into a feature-length, (mostly) monothematic classical Hollywood nondiegetic film score. The film opens similarly to *Street Scene* with Newman's music underscoring a line-drawing representation of a big-city skyline. However, the first hint that the previous impartiality and neutrality of the city's character— its open range of possibilities—is to be contradicted comes immediately with the cliché-ridden film noir visual style of the images behind the credits. Even in cartoon form, the cityscape is dominated by the ominous interplay of lights and shadows with buildings presented at unusual and defamiliarizing skewed angles. The music of "Street Scene" has been updated only insofar as its orchestration and tempo reflect the contemporary modernity of swing dance bands. How-ever, if there is any doubt as to the changed situation, the character of the city is quickly confirmed in the first sequence: still underscored with fragments of "Street Scene," a young newsboy among the morning rush-hour bustle in front of a skyscraper holds out a banner headline for us to read: "Beautiful Model Found Murdered." If in the strong familial ties within the tenement house *Street Scene* still bear vestigial traces of a gemeinschaft culture of reciprocal and stable social relations, the stories of the city in *I Wake Up Screaming* are even more thoroughly appropriate to a gesellschaft. This is a world of transitory and superficial social exchanges between characters who are without domestic lives or histories. From the start of the picture, suspicion and fear make all social relations problematic. Virtually every male character is potentially the secret, violent, and misogynic sexual psychopath; every woman could be his next victim. Yet this seeming chaos ultimately fits together deterministically by the end when the mysteries are re-solved. In retrospect, these ambiguities are portrayed as intrinsic to the city itself, as are the city's perils—especially for women.[16] The uses of "Street Scene" in the soundtrack of the two expository flashbacks that follow this opening confirm this changed character of the city and its relation to the narrative plight of the female protagonists.

In the first flashback, sports promoter Frankie Christopher, under police interrogation, recounts his first meeting with the murder victim, Vicky Lynn. After finding the young woman slinging hash in a lunchroom on Eighth Avenue, Christopher bets that he can engineer Vicky Lynn's Pygmalion-style transformation into a Café Society celebrity. As she enters the nightclub for her debut, a roomful of couples dances to a thoroughly diegetic version of "Street Scene" played by the musicians on the bandstand. By this point in the film, this music has crossed several boundaries for audiences. First, from its position in the world of popular music, it has crossed back into music suitable for a classical Hollywood soundtrack, and "Street Scene" wholly and effectively fulfills its obligations to these psychologized narrative codes. Though explicable in terms of marketing strategy for music in a vertically integrated Hollywood, the reemergence of "Street Scene" on film also represents the contiguity of Hansen's cinematic sensorium with the city's modernity. It would have been equally comprehensible for an audience to experience "Street Scene" as a live radio broadcast from a Manhattan hot spot as it would have been to see the same imagined on-screen. Second, with this scene the music leaches from nondiegetic cinematic space into the world of the story. In doing so, this shift from the subjective, omniscient underscoring to the objective diegetic sound of the club, an impossible continuity of the audible, subtly implies a manipulation of a world in which "subjects are spoken." The seemingly random events of the city are, in fact, malevolently organized by some unseen agency whose presence can be heard and understood only by the audience, the epitome of the urban world of film noir. Though this loss of autonomy and self-determination in film noir was often connected specifically with the loss of masculine authority, this example indicates that "Street Scene" has come to represent an ideology of social interactions in the city more broadly—the ineffectuality of heteronormative relations in a gesellschaft culture in which all individuals are in transition and struggle to form social bonds in the absence of enduring communal structures.

In the second flashback, the structural position of female urban subjectivity is bifurcated in two sisters and instantiated by the brief juxtaposition of "Street Scene" with another musical theme. In a flashback told to the police by her sister, Jill, Vicky Lynn recounts her first successful society outing. The flashback begins with a pair of "stinger" chords from "Street Scene" that serve as a sound bridge. Instead of more "Street Scene," the start of the flashback is underscored with generic and saccharin music: a galumphing two-beat bass with a simple and innocent melody as Vicky Lynn gleefully describes her success at the club. At the moment in the conversation when Vicky Lynn emphatically declares that she will take her fate in her own hands and grab what, through Frankie Christopher, the city offers, "Street Scene" ominously returns to overtake the saccharin tune. In what should be a moment of triumph and self-determination, Vicky Lynn articulates what we in the audience already know to be an impossibility—since we know from the start of the film of her imminent death. When she says "I've finally come to my senses!" the soundtrack turns from banal to menacing with the introduction of "Street Scene" because we realize that her relationship with the symbolic order—as personified by the city itself—is always already fundamentally compromised.

Jill's experiences throughout the film present an alternative rite of passage of a young woman new to the city, and likely the only one that could succeed (that is, not lead to death). When first introduced, Jill is underscored with a distinctly naive rendition of "Somewhere over the Rainbow." This theme occasionally interrupts the steady association of "Street Scene" with urban action sequences. It follows the plucky Jill throughout the adventure in which she, together with Frankie, not only uncovers a police detective's obsessive stalking of the morally compromised Vicky Lynn but also reveals the true murderer—the nebbish doorman. It is only in the very final scene, a return to the nightspot in which Vicky makes her seemingly triumphant debut, that "Somewhere over the Rainbow" materializes in the diegesis. Paralleling Vicky's ultimately unsuccessful emergence into the gesellschaft of the city, Jill's music is played by the band as she and Frankie dance cheek to cheek. However, after so many encounters with compromised legal, moral, and social behavior, hers is an undermotivated urban success story. Unlike Rose, for Jill, and by extension all women who would thrive in the city, self-determination has been reduced to companionate marriage. This theme reverberates in all of the movies in this era that used "Street Scene" iconically to represent the city. Whether motivated by threats of physical violence, as in *The Dark Corner* or *Kiss of Death*, the battle against ethnic and religious discrimination as in *Gentleman's Agreement*, or economic need, as in *My Friend Irma*, the women in these films all eventually accommodate themselves to life in the city through the pursuit of marital ties, usually after they have exhausted other options.

## A Tale of the Suburbs, 1953

> The "failure" of stereo suggests that the combination of widescreen images and multitrack stereo proved to be too much of a revolution in the mid-1950s. Widescreen cinema and stereophonic sound, as idealistic phenomena, conceived by the film industry to provide a perfect illusion of reality, proved to offer an excess of spectacle that could survive only in the most artificially theatrical of venues—in high-priced, reserved seat, first-run theaters that, like the legitimate theater they sought to emulate, adopted theatrical schedules, featuring matinees in the afternoon, one show in the evening, and three shows on weekends and holidays.
>
> —John Belton, *Widescreen Cinema*

Although *How to Marry a Millionaire* was only Twentieth Century-Fox's second release in CinemaScope, in its use of sound to imagine the metropolis it already demonstrated these signs of excess. The CinemaScope concert prelude sought to produce something other than "a perfect illusion of reality." In doing so, it revealed a fissure in cinema as neurological modernity, a rupture produced by the distance between the material circumstances of life in the city and the artificiality of its cinematic reproduction. Both in terms of exhibition practices and as a form of aesthetic address directed at contemporary audiences, the opening of *Millionaire* represents the city through fantasy as a patently nonexistent space, or a nonplace.[17] "Street Scene" as Hollywood's sensorium no longer realistically represents the city to itself.

In the *Millionaire* clips that appeared in the CinemaScope demonstration reel (of April 1953) and in the print advertising for the film's release, it was the CinemaScope process itself that drew top billing, even over the stars, Marilyn Monroe, Lauren Bacall, and Betty Grable. Twentieth Century-Fox had pinned its very survival on the success of *Millionaire* and CinemaScope. The film industry's timing and interest in promoting wide-screen processes has been extensively analyzed in relation to changing social and economic conditions in the United States after World War II (Belton, *Widescreen;* Hincha; Cohen). In particular, CinemaScope can be seen as Hollywood's most successful response to two related forces: a broad demographic shift that moved film spectators out of metropolitan areas and the associated changes in the way American consumers used their diminished leisure time—the explosion of both the suburbs and television (Jackson). Although the introduction of wide-screen cinema drew spectators back to theaters to an extent, John Belton has dubbed magnetic sound a "frozen revolution," and the economics of wide-screen stereo suggest a simple and compelling explanation for its lack of acceptance (Belton, "1950s Magnetic Sound"; Erffmeyer). Along with a proprietary screen, CinemaScope's largest installation costs were for multichannel sound, and theater owners resisted the expenditure. After March 1954, stereo was no longer required in the CinemaScope exhibitor's installation package, and the studios released films in both magnetic stereo and optical mono prints. However, stereo was neither indispensable nor even highly endorsed by technicians and critics (Crowther, "Sound and [or] Fury"; Ryder). Nor, if we consider trends in spectatorship through the mid-1950s, was it particularly sought by audiences. A measure of American culture's lack of concern for this improvement in "realistic" audio reproduction can be surmised in dominant exhibition practices of the mid-1950s. Even leaving aside the exponential growth of television viewing, by June 1956, for the first time, more people heard films at suburban drive-in theaters through monophonic speakers in the car window than went to traditional "hard-top" theaters (Gomery 91–92). Available almost exclusively in first-run urban movie palaces throughout the 1950s and '60s, film exhibition with stereophonic sound quickly became associated with urban opulence rather than the emergent postwar practices that, following the dominant social trends, had moved the population and entertainment activity away from the city toward the suburbs and the tract-house living room.[18] The experience of film spectatorship and its relationship to metropolitanism had substantially changed in the twenty years since "Street Scene" first appeared.

Wide-screen processes of the early 1950s sought to lure postwar audiences back to the theater to fulfill a long-standing promise. By almost entirely filling the audience's peripheral vision, and engulfing them in multichannel, stereophonic sound, wide-screen processes promised total immersion of the senses—a complete transportation of the audience out of their Real and into the narrative of the film itself. Replacing the convention of a single monophonic soundtrack broadcast from behind a sound-permeable screen, CinemaScope in exhibition used, at minimum, four channels: three loudspeakers across the centerline of the screen at the left, right, and middle, as well as a supplemental surround channel with speakers located at the rear corners of the theater (Sponable). This arrangement of speakers (in conjunction with stereo recording practices) produced a stereo soundstage: a three-dimensionality of sounds across the proscenium arch, in which sounds could emanate from any position in relation to the image on the screen, even extending back into the image and past the edges of the screen. The surround channel was intended to bring the sound additionally out into the theater itself, beyond the plane of the image. The goal was to exceed the boundaries of film as a representational form, and make the jump toward a reproductive one: to make the experiences seem more "real."

Despite the potential of these technological innovations to more completely and powerfully realize the aesthetic project of the sensorium, this presentation of "Street Scene" represented a departure from previous uses of the music. Rather than engage the film viewer more fully in the modernity of the city through the equation of a musical vernacular modernism with metropolitanism, this incarnation of "Street Scene" contradicted these previous associations in two ways. First, *Millionaire* introduced the newly devised CinemaScope stereo sound system within the context of a distinctly unreal space—an ethereal concert without a visible audience or any specific temporal or spatial connections to the city. The sequence is presented utilizing the most conventional of Hollywood codes: a visual continuity created through musical themes "conversing" via the technique of shot/reverse shot, and an "intelligible" stable point of enunciation for the audio that throughout remains fixed six feet behind the conductor. The directionality observed by critic Crowther was only in the service of the narrative commands. In its aesthetic mode of presentation this concert sequence functioned as a regressive element, reassuring audiences of sound's conventional functions rather than heralding any new meanings in film audio practice. What did distinguish this concert scene was the immersive quality of the audio experience—and that was only available to urban film spectators in those first-run movie houses. If the goal in previous uses of "Street Scene" was to cement the associations of this music with an almost documentary-style presentation of street life, the grandeur of this sonic space, though compatible with exhibition sites such as the Loew's State theater on Broadway, was incompatible, even discordant, with that previous cityscape. Rather than pull the film spectators into a re-creation of their own "real" metropolitan world, this "Street Scene" acoustically drew them off to a fantastic Olympian pleasure palace.

Second, following the pattern of every movie that had used "Street Scene" for more than twenty years, *How to Marry a Millionaire* employed a visual montage of Manhattan as its establishing sequence. However, this time there were two differences. First, though "Street Scene" is amply present in the concert prelude, it is Lionel Newman's song "New York" that supplies the instrumental theme for the main titles. The song itself, sung by a chorus, backs the Manhattan montage. In place of the established semiotic equivalence of a musical vernacular modernism married to a gritty, realistic visual sequence of everyday urban life, this tune is closer to the spiritual ancestor of an upbeat "I 'heart' NY" commercial. Second, although this is easily the most splendiferous portrayal of Manhattan of the bunch—displaying the visual power of CinemaScope to full advantage—this Manhattan is strikingly distant, sanitized, and unpeopled. It is only in the final phrase of the song, when we arrive outside the apartment building where our heroines live, that we finally see some signs of urban life at a human scale—and then only a few strolling pedestrians on Sutton Place. Gone is the sensuous and tactile city at the scale of individuals. Instead, these fantasies of spatial excess reveal a radical shift in the conception of urban modernity with which "Street Scene" had been associated, a shift that is confirmed in the gender politics of the narrative itself.

Ironically, even though the medium had increased its capacity to represent urban life in ever more realistic detail, the city of this narrative is organized as mere fantasy space. "Street Scene" in its reference to urban modernity is no longer the unalienated public sphere of *Street Scene* or the paranoiac world of *I Wake Up Screaming*. Instead, the "Street Scene" of *Millionaire* inadvertently reveals the modernity of the city as an absence, a lack. Paradoxically, in this instance of cinematic stereo lies the kernel of a revolutionary shift away from a modernity associated with the American city. There are few scenes in *How to Marry a Millionaire* in which the three female leads appear out and about in the city. Instead, they lounge in insular indoor splendor, comically plotting seduction and dreaming of wealth, even while the eventual romantic pairings are obviously predetermined from the start. There is no intimation whatsoever here of the city as a site of possibility for women, or even one in which their actions carry substantial personal consequences. Though CinemaScope heralded a new type of film spectatorship, its function was as a compensatory realm, a replacement for the loss of a real urban public sphere. Rather than representing urban modernity to itself, or negotiating its tensions, *Millionaire* should be considered against the reality of the post–World War II city that was succumbing to financial and racial turmoil, white flight, and loss of employment opportunities, especially for women. *Millionaire,* as a fantasy space, is really crypto-suburban. The new modernity was really to be found elsewhere.

## Notes

1. *How to Marry a Millionaire* opened simultaneously in two Broadway theaters (Loew's State and Brandt's Globe) on November 10, 1953. See display ad 38, *New York Times,* November 10, 1953, 39.

2. For a discussion of early cinema's link to preexisting musical forms of entertainment, see Altman, "Introduction: Sound/History" 114–17. In fact, the ties between live and filmed entertainment were still quite tangible and present at the time of this premiere. The Roxy Theatre had only just eliminated live stage shows, and fired twenty-six union musicians. On September 18, 1953, a strike by Local 802, American Federation of Musicians, AFL, protesting this contract violation, was resolved, and the premiere of the first CinemaScope feature, *The Robe,* went forward as planned ("Roxy Dispute Is Settled," *New York Times,* September 19, 1953, 17).

3. Gary Marmorstein claims that "Street Scene" (221 and 222– 23) was initially slated as the title music for *Laura* (Otto Preminger, USA, 1944). The liner notes to the soundtrack CD for *The Greatest Story Ever Told* (George Stevens, USA, 1965) claim that "Street Scene" served as a temporary track, but was discarded by Newman for his newly composed music.

4. For a list of examples of scores taken from previously composed film scores, see Brown 343. See also this chapter's epigraph, a one-line memo dated September 17, 1948, to Newman in which Darryl Zanuck, apparently promoting a piece of music he admired, advocated the continued recycling of "Street Scene" (quoted in Behlmer).

5. The foundational literature in this branch of the sociology of the city includes Max Weber, "The Nature of the City," and Georg Simmel, "The Metropolis and Mental Life," both in *Classic Essays on the Culture of Cities,* edited by Richard Sennett. For a specifically Americanist perspective, see also Lewis Mumford, *Technics and Civilization* and *The Culture of Cities;* and Richard Sennett, *The Fall of Public Man.*

6. See Kracauer, *Theory of Film;* and Benjamin, "The Work of Art in the Age of Mechanical Reproduction," in *Illuminations,* 217–51.

7. See Rice, *Minority Report* 276, for a discussion of Rice's early Hollywood experiences as a member of Samuel Goldwyn's scenario department, and a brief description of Vidor's "complete sympathy with the spirit of the play."

8. In the original 1929 Broadway production (which Rice directed himself), sound effects were supplied as follows:

> A recording instrument set up in an open window on Times Square had provided a disk that reproduced the hum of city traffic, punctured by rumbling trucks, shrieking brakes, honking horns, even a distant steamship whistle. Two record players, started a minute apart, produced an overlapping effect that was always varied. The records kept going throughout the entire play, sometimes at a roar, sometimes at a murmur, as the mood demanded. One man was kept busy just supervising them. I do not know how many hundred disks were worn out in the course of the play's run. (Rice, *Minority Report* 252)

9. In "Scoring the Film," Max Steiner provides a compelling firsthand account of the interrelationship of developing film music scoring practices and sound technologies in the 1930s.

10. Leonard Bernstein's 1955 article "Why Don't You Run Upstairs and Write a Nice Gershwin Tune?" argues that a nuanced analytic approach to Gershwin's music is essential, especially because it seems to flow so effortlessly and seems to have required little effort on the composer's part—a mythology that effectively ties musical style to the ideology of vernacular modernism.

11. Music was a crucial element in the fusion of African American forms with a white cultural establishment that Ann Douglas identifies in *Terrible Honesty*.

12. Subsequently, Robbins released even more arrangements. In 1942, it was published for dance band with singer under the title "Sentimental Rhapsody" as part of the "Robbins Popular Standard Dance Hits" series. Then it was republished for solo piano (Albert Sirmay, arranger) in 1950, another concert band arrangement (Bennet) in 1952, as part of a medley of popular piano tunes in 1953, for school orchestra (Alfred Rickey, arranger) in 1955, and twice in 1958: as part of a medley for brass choir (which also included M. Rozsa's "Hail Nero" Triumphal march) and for organ (Bennet, arranger).

13. Although Leizden was the Goldman band's principal arranger, I have been unable to confirm through programs that this arrangement was actually performed by that band. I wish to thank Professor Myron Welch, curator of the Goldman Band Collection at the University of Iowa, for his assistance.

14. As recently as 1979, Morton Gould and the London Symphony Orchestra included it on a demonstration disc titled *Exploring the Digital Frontier*.

15. This renewed interest in "Street Scene" coincided with Alfred Newman's appointment as music director of Twentieth Century-Fox in 1940 with the long-standing support of Darryl Zanuck. However, I have been unable to trace the passage of intellectual property rights from studio to studio, from Goldwyn's own production company, United Artists as the original distributor of *Street Scene*, MGM as the apparent owner of the publishing copyrights, and finally to Twentieth Century-Fox.

16. For the film spectator, the figure of the psychopathic killer would have been very visible in American urban mass culture at this moment, due to what Estelle B. Freedman has described as a media- and medical establishment–fed public hysteria. Lisa Maria Hogeland also identifies this phenomenon in hard-boiled novels of that time.

17. Michel Foucault's idea of heterotopia as a distinctive type of urban space in "Of Other Spaces" is particularly relevant here.

18. Of particular interest here is the comparison with the postwar practices of listening to the stereo at home. Similar to the present study, Keir Keightley's "'Turn It Down!' She Shrieked: Gender, Domestic Space and High Fidelity, 1948–1959" analyzes gender politics in the context of audio practices.

# 7

# Film and the Wagnerian Aspiration

## Thoughts on Sound Design and the History of the Senses

**JAMES LASTRA**

Rightly or wrongly, Francis Ford Coppola's *Apocalypse Now* (USA, 1979) has become the film that stands for the emergence of modern, immersive sound design. Partly because it is the first film where a technician, Walter Murch, is credited as "sound designer" (understood as analogous to the role played by the cinematographer with regard to the image), and partly because of its intense, thematic reflection on the role of technological media in reshaping sensory experience, it seems to insist on that designation. Like the earlier Coppola-Murch collaboration, *The Conversation* (USA, 1974), *Apocalypse Now* both embodies and reflects upon new possibilities and dilemmas produced by the emergence of sensory technologies. In the case of *The Conversation,* the mystery plot foregrounds a range of epistemological questions raised by sophisticated audio surveillance technologies: How trustworthy are these electronic senses? Is the meaning of hearing transformed by its electronic mediation? Are sensory technologies more or, perhaps, less objective? And how do artificial senses affect our notions of the human? *Apocalypse Now* poses those questions in a world-historical context, ultimately asking whether our historical experience of artificially created sensory environments has had an effect on how humans (and especially Americans) understand their own power, knowledge, and place in the world.

Around the time of *Apocalypse Now*'s release, both film audiences and filmmakers found themselves confronting this issue, and each sought to accommodate the new possibilities offered by stereo surround-sound design to existing cinematic norms. They also understood them in the context of the norms of the broader audio culture of the era. For instance, by the time of *Apocalypse Now,* pop and rock music recording had assumed the vanguard of the craft (a fact acknowledged by *Apocalypse Now* diegetically and in technique) and surely served as a crucial frame of reference for audiences. However, where most owed the bulk of their experience of complex audio environments to the Beatles, Pink Floyd, Phil Spector, and the like, the particular demands of filmmaking required more than a simple application of pop music norms, which, after all, were usually unaccompanied by images, and more often than not unencumbered by storytelling.

What they did share with the movies was the ability to create rich, imaginary worlds of a density and perfection unmatched by "mundane" forms of experience. The emergence of sound design as a practice and stereo surround sound as a technology in the late 1970s signaled a reawakening and reinvigoration of cinema's recurring aspiration to provide a total artwork of full sensory immersion. Central to this aspiration was a reexamination of the relationship between the audio-visual constructions of the cinema and the human sensorium. The new ability to create complex and three-dimensional sonic environments to accompany the wide-screen image prompted filmmakers to ask how the new sensory technologies could be used and whether they had redefined the relationship between perception and the human understanding of sensory experience and cognition. As in earlier encounters with photography, phonography, cinema, synchronized sound, stereo, and 3-D film, the emergence of sound design produced a flurry of commentary trying to define the nature of the new form.[1] One of the most common refrains of the era stressed the new technology's ability to immerse or surround the auditor in a more replete and realistic sound environment, and to provide a greater capacity for point-of-view or character subjectivity. Soundtracks promised to bring the auditor more fully *into* the film whether by constructing a total environment that included the auditor who would perceive the film as if it were his or her own world or by asking that auditor to identify with the perceptions of another. That is, it reopened the question of whether the cinema was a reproduction of real perceptions or rather a more obviously conventional system of rhetorically structured experiences designed to produce a particular narrative effect. Decades of Hollywood history had decided firmly in favor of the latter option, but the former breathed new life into this era as a form of audio "spectacle" joined the arsenal of film's formal attractions. It is perhaps fitting that Coppola is so closely associated with modern sound design given that not since Wagner has an artist had such lofty aspirations for his work and its capacity to serve as an alternate reality.[2]

As in earlier eras, the process of coming to terms with new sensory technologies worried the boundaries between human and machine, and between perception and representation. Overlaying these issues is a series of larger concerns that locate the emergence of sound design within what we might call a history of the senses—a *longue durée* covering the modern era's attempts to grapple with technologically mediated forms of experience and to divine what they mean for human history. The question is how such a history might relate to the specific situation of the Hollywood film.

Phenomena such as the emergence of sound design can be approached in a variety of ways. Technological and economic factors surely exerted a determining pressure, and the stylistic norms that emerged surely derived from existing techniques in both the film and the music recording worlds. Although David Bordwell makes a convincing case that stylistic developments need to be explained by their most proximate causes, proximate causes do not necessarily explain the *meaning* of a technical or stylistic shift. I would like to suggest that, while keeping Bordwell's caveats in mind, a history of the senses might provide a feeling for what impact surround-sound design may have had within and beyond the cinema (*On the History of Film Style* 141–42).

Bordwell is clearly correct that a history of the senses cannot mean that the basic operation of the sense organs changes radically with changes in society or culture. All the same, this does not mean that what changes are merely "habits and skills," as he would have it (142).[3] Surely, these do change, but what changes more profoundly are the basic categories of sound (say, sound versus "noise"), the kinds of sounds (mechanical versus organic), and the structured situations and environments in which sound is encountered (concert hall, insulated home, or pop music recording). In effect, what changes are the structures of listening typical of an era, an institution, or a form of sound practice.

What I propose to sketch here is what a history of sound design would look like if it were to consider both the immediate formal, stylistic context—its meaning within film history—and how it might relate to a larger history that understands sound design as encompassing such variables as architectural acoustics, sound proofing, performance norms, and the like—anything designed to "shape" a sound experience in a particular way. My goal is to suggest one approach to a history of sensory experience (both discursive and as embodied in particular regularized practices) that respects both the demands of empirical research and the more speculative but deeply embedded tradition of debate about the implications of particular forms of sensory manipulation. Such an approach would seek to respect what was actually *done* with sensory technologies, but also understand what effects commentators believed them to be having.

A growing and impeccably researched body of literature amply demonstrates not only that the history of controlling (or "designing") sound reaches back centuries but more pointedly that the nineteenth century saw a dramatic escalation in concerns about such controls and in techniques for achieving them. As John Picker shows, a veritable furor arose in 1850s London over the explosion of urban noise, specifically that attributed to hurdy-gurdies. Associated primarily with Thomas Carlyle and Charles Babbage, an intense multiyear campaign was waged to control what some called music but what many more considered "noise." In pushing for its suppression, social elites sought to construct a predictable and managed sound environment with little or no noise and copious sonic privacy (whether achieved by legislation or insulation). In short, they tried to design their own soundscape (Picker 41–81).[4]

Likewise, Emily Thompson's impressive history of architectural acoustics demonstrates how the Victorian obsession with managing noise and managing the soundscape developed in the United States into seminal research in both sound insulation and concert hall architecture. Concomitant with the development of a more scientific approach to controlling the sound environment, Thompson argues, came a far-reaching redefinition of the nature of sound (particularly musical sound), which was decreasingly understood as a unique phenomenal event in a particular space and increasingly as a "signal"—a conceptual entity defined by a particular *use* of sound and defined by specific institutional or aesthetic standards or both (*Soundscape of Modernity* 228–93, esp. 235–66). Perhaps most familiarly, she reminds us, sound technologies such as the telephone and the phonograph were widely perceived to have altered the relationship between sound and space. More subtly, those devices encouraged engineers and consumers alike to adopt a redefined understanding of sound itself, as "voice" became identified more and more with the electrical "signal" traveling over a wire—a kind of quantifiable unit of pure intelligibility (235–36). In every case (even those, like the hurdy-gurdy, that are remote from the cinema) it is clear that from the 1850s onward, the broader culture came to realize that technology (including architecture, insulation, and the like) could alter the nature of one's sensory world and could do so with pleasing (or displeasing) effects.

Ultimately, what Thompson, Picker, and scholars such as Jonathan Sterne and Carolyn Marvin make clear is that it is impossible to speak of "sound" or "hearing" in a pure state. Both are necessarily defined in specific cultural and social terms. Perhaps the clearest evidence of this is the shifting boundary between acceptable and unacceptable sounds, that is, between "sound" and "noise." Whether driven by concerns about public health, as was argued in London, or by changes in architecture or technology, the redefinition of the term *sound* within a particular cultural or social realm causes related shifts in habits, expectations, and sound forms. As beliefs about what an experience "should" sound like change, so too a complex web of practices changes.[5]

Admittedly, this cursory sketch of the kinds of issues that might be associated with a broader conception of sound design is not meant to be exhaustive; it is simply meant to point to a history of shaping the sound environment that goes beyond the cinema but nevertheless impinges on it. *Apocalypse Now*, in fact, asks us to reflect on the impact of sound design on cinema, and sensory experience more generally, by directing us toward at least one key nineteenth-century theorist of the senses, Richard Wagner.

## Sound Design, 1979, Part 1

*Apocalypse Now* is not only a landmark of modern sound design but also a sustained reflection on the meaning of technology, spectacle, and the politics of sensory experience. It is a flawed but audacious work—a film that risks participating in a mode of sensory manipulation that it places at modernity's heart of darkness, a film that even more than the novel upon which it is based becomes obsessed with voices—with sound. And just as the novel allegorizes the phonographic moment—remember, Kurtz is repeatedly figured as a disembodied voice (Kreilkamp)—Willard notes that it is Kurtz's *voice* that "put the hooks in me." One scene from *Apocalypse Now,* in particular, put its hooks in me: the notorious air-cavalry attack on a Vietnamese village accompanied by Wagner's "Ride of the Valkyries."

Truly, there is no better term to describe this sequence than *Wagnerian,* and not only in the banal sense of its musical content but more profoundly because it embodies, thematizes, and reflects upon the cinema's deepest Wagnerian aspirations—aspirations that are brought to the foreground by modern forms of sound design. Like the Wagnerian musical drama, born of a desire to provide a total and totalizing sensory experience, cinema has, from its earliest days, harbored something like the same wish, offering phantasmagoric worlds of multisensory stimulation, albeit in a more quotidian form—giving us this day our daily *gesamtkunstwerk.* By bringing this impulse to the surface in dramatic form, *Apocalypse Now* encourages us to reflect on the interpenetration of the apparatuses of human sensation and of cinema—to ask how the advent of sensory technologies has altered our understanding of human experience. The cinema (and the cinematic history) engaged by *Apocalypse Now* is not a purely visual one; rather, it is a resolutely *audio*-visual one, a cinema born not out of painting or theater but out of the spirit of publicly performed music, where drama and audience exist in a shared space united by a sonic envelope. *This* cinema's nearest relative is, perhaps, the Wagnerian musical drama, not because it derives from it but because both harbor the wish for a reintegration of sensory experience with the whole of human being.

Rather than Valkyries arriving to collect heroes for Valhalla, however, this musical drama gives us wrathful, prosthetic gods raining fiery death upon anonymous mortals inhabiting not only a different world but what seems like a different century. And if this war takes on the character of total sensory immersion for the participants in the drama, it does so for the spectator as well. It is a thrilling and disturbing scene, conveying the terrifying sublimity of a godlike power over life and death, and our own guilty implication in an aesthetic regime that renders war the ultimate *gesamtkunstwerk*. One could hardly find a more apt embodiment of Walter Benjamin's 1936 claim that "mankind, which in Homer's time was an object of contemplation for the Olympian gods, now is one for itself. Its self-alienation has reached such a degree that it can experience its own destruction as an aesthetic pleasure of the first order" (*Illuminations* 242).

*Apocalypse Now* threatens to aestheticize violence in just this way, but simultaneously reminds us that, historically speaking, this aesthetics is, in fact, an anesthetics, where the senses no longer serve to connect sensation to memory and meaning but instead ward off experience, hoping to protect the organism from trauma.[6] I remember thrilling with an opening-night audience to the terrifying sublimity of this scene, finding the experience exhilarating even as I believed I deplored the violence it depicted. In subsequent weeks, as "Charlie don't surf" and "I love the smell of napalm in the morning" became the mindless (and completely unironic) catchphrases of thoughtless teens across the United States, I began to question my own complicity in a technological apparatus that could aestheticize the destruction of others so persuasively. My own sense of being overwhelmed by the audio-visual pyrotechnics of the film and of being thoroughly situated within its structures of subjective engagement had almost completely overcome my critical relation to the events depicted, and rendered me numb to the meaning of the violence. At the same time, it offered me an almost omnipotent sensory experience of power, whose terrifying appeal I could not discount, and raised in me the question of how, against my better judgment, this could be so. Over the years, I have come to feel that the question the film poses most brilliantly is precisely the one generated by the audio-visual argument of this scene: what are the ultimate consequences of a regime in which sensory technologies have become the central component of the collective apparatus of perception? What is finally at stake in this story is the fate of the human sensorium—indeed, humanity—in its encounter with technological modernity, and whether an alien and alienating technology can avoid its destiny as an agent of anaesthesia, to become a part of a new sensory regime. This led me back, via the allusion to Wagner, to the nineteenth century and how it grappled in a more sustained and public way with the impact of technology on the history of the senses.

## Sound Design, 1856

When he offered the observation that mankind had come to the point where it could enjoy its own annihilation, Walter Benjamin expressed his fear that as a result of the ever increasing perceptual and psychic shocks characteristic of modern life, the human perceptual-cognitive apparatus would cease to be a conduit for experience, and become instead a buffer against it, thereby anesthetizing the subject to the events around him. Perceptions would be shut off from memory, and, consequently, humans would begin to respond to stimuli without thinking in order to insulate themselves from the traumas of history as much as those of everyday life. Anesthesia, paradoxically, became part of our perceptual equipment. As I write this, I find I can easily ignore the sounds of (apparently endless) elevator repairs that bothered me immensely when they began next door—a fairly familiar and benign form of urban anesthesia. However, such an anesthetized individual, Benjamin worried, was also the principal audience for the fascist spectacularization of politics, which could only culminate, he rightly reasoned, in war. This is why he felt that *aesthetics* (in its etymological sense as the science of perception) had become the ground of political critique and why an analysis of the dominant structures of perception seemed crucial. The factory worker and the soldier were on the front lines of this emerging perceptual "crisis," and each became emblematic of the damage done by the new perceptual regime by what he called a mimetic reception (or innervation) of technology.

According to Benjamin, workers respond to the factory system by mimicking the machines that determine their work, "learn[ing] to coordinate their own 'movements to the uniform and unceasing motion of the machine'" (*Baudelaire* 132). As one commentator puts it, "The factory system, injuring every one of the human senses, paralyzes the imagination of the worker" (Buck-Morss 17). On this view, his work is "sealed off from experience"; memory is replaced by conditioned response, learning by "drill," skill by repetition: "practice counts for nothing." Just as the worker numbs himself to the deafening sounds, noisome smells, and enervating sights of the factory, in order to protect himself from their assaults, his repetitive tasks are walled off from memory so that concepts such as craftsmanship lose meaning. One does not become better and better at assembly-line work regardless of years of labor. Every movement on the line is as disconnected from every other as one roll of the dice from the next.

In a parallel fashion, the soldier numbs himself to the physical and psychic assaults of the battlefield. His anesthesia has a clinical name—shell shock—which Benjamin argues had become a common state of affairs. Being "cheated out of experience" (*Baudelaire* 137) has become the general condition, as the perceptual-cognitive system is marshaled to parry technological stimuli in order to protect the body and the psyche. In response to such defenses against an increasingly assaultive environment, more and more powerful shocks were needed to break through this shell, and the widening spiral of shock-defense–greater shock led to the proliferation of spectacular effects for their own sake.

Granting that this account has some validity, what does this ultimately have to do with cinema? A great deal, I would argue. As Ben Singer has argued, cinema (and by extension phonography) easily found its place within the culture of neurasthenia—a kind of nervous exhaustion—that dozens of writers claimed defined the era of their invention. This, they tell us repeatedly, was an era of "shattered" nerves, or "nervous breakdown," of people "going to pieces."[7] The supposed cures for this state (whether prescribed by one's physician or by oneself) were intoxication and anesthesia—preferably opium, but cocaine, alcohol, ether, morphine, and a variety of other substances would do just as well. At the cultural level, it is argued, the numbing intoxication took the form of the phantasmagoria, which "made a narcotic of reality itself" (Buck-Morss 22). These technologically produced forms of sensory experience (dioramas, magic lantern shows, World's Fairs, and so on) produced sensations that were real enough, but their function was wholly compensatory. Their goal was to flood the senses with controlled and collective experiences of intoxication, experiences that were to provide a pleasing but socially false antidote to the seamy underside of modernity. In our present example, the shattered bodies of Vietnam are the dark underside to the phantasmagoria of prosthetic divinity. More generally, the nineteenth-century technologies of the senses (photography, phonography, telephony, cinema, and the like) produced a phantom experience of collective sensory omnipotence. In a manner of speaking, these prosthetic senses became the common uniform.

This idea became a common trope of late-nineteenth-century literature with J.-K. Huysman's des Esseintes and Oscar Wilde's Dorian Gray serving as models for a new type of individual. These characters regard their present as debased and hopelessly compromised, and strive to create more perfect, artfully crafted worlds of sensory ecstasy to compensate for a world no longer adequate to human aspirations, desires, or even capacity to experience. Responding to a felt separation of the senses into shattered pieces of a forgotten whole, each constructs idealized fantasy worlds catering to every sense individually. That these desires have not abated should be clear from the fact that one of Huysman's most compelling yet absurd inventions—des Esseintes's "symphony" of aromas stored in vials—has appeared lately in commercial form as the CD-like "Scent Stories." According to the advertisements, you pop a disc in the "player" and a complete olfactory narrative is at your command.[8]

With or without this historical background, Coppola's choice of Wagner here is hardly innocent. As Theodor Adorno argued more than sixty years ago, Wagner is perhaps best understood as the composer of intoxication—of the phantasmagoria. Adorno argues that poetry, music, and theater combine to create an "intoxicating brew." Wagnerian drama floods the senses and fuses them in a "consoling phantasmagoria" in a "permanent invitation to intoxication, as a form of oceanic regression" (*In Search of Wagner* 87–100). In an argument that harks back to Marx's discussion of the commodity and looks forward to his analysis of the culture industry, Adorno argues that Wagner's music replicates the phantasmagoria of the commodity. In its structure, it predicts and anticipates both advertising and the pseudoindividuation of mass culture. In lieu of an internal musical logic, Wagner's music is unified by a surface coherence of style, resulting in a kind of pseudototalization. On this account, the musical drama responds to the shattering of the human senses into discrete and disconnected zones, by providing an arena within which they appear to reconnect (101–9). As such, it takes its place within a broader explosion of phantasmagoria of many kinds, each corresponding to the individual senses, which no longer cohere in a total system. These are privatized fantasy worlds designed to insulate the sensibilities of the new ruling class—upholstery and fabrics for touch, perfumes for the nose, spices for the tongue, photos and recorded sound for the eye and ear. Their function is to provide what reality no longer can—they are designed to manipulate the perceptual-cognitive system by total control of environmental stimuli, to alter consciousness like a drug.

Intoxication and phantasmagoria—two words that describe *Apocalypse Now* to a tee. It is a story told by a perpetually drunken narrator, about soldiers who see through binoculars, speak through microphones, hear through radios and tape recorders, eat steak and drink beer on the battlefield, and take drugs to systematically derange their senses. The soldiers deliberately and in full consciousness stage their own attack as opera. "We'll come in low out of the sun . . . ," for example. Their goal is to make the Vietcong experience their own deaths as theater of the highest order—a tragedy of epic scale, with the air cavalry as displeased gods who control the entire experience of death just as surely as any Zeus. They move among a series of artificial environments ranging from the mildly transformative effects of popular music and waterskiing to the completely bizarre worlds of the Playboy show and the Do Lung bridge, where the last United States outpost becomes part Gates of Hell and part carnival.

But the film does not treat the creation of artificial environments—or the notion of war as *gesamtkunstwerk*—as restricted to the soldiers. Instead, the film knowingly manipulates *our* relation to the war and to the depicted events by manipulating our point of audition. We prepare for the attack located securely with the soldiers in the helicopter—we hear and see as they do. As the attack begins, we find ourselves with the villagers. At other points, however, we are surely listening to a recording of the opera. Indeed, at every moment when recorded music appears, we are reminded that the entertainment industry (pop music, opera LPs, television news, the movies) produces consoling or intoxicating environments for us too, an other world of carefully calibrated and better-than-perfect performances.

But *Apocalypse Now* seems to be suggesting something more specific, more telling, about sound recordings in its narrative and in its techniques. Just as the film begins by juxtaposing the world of the battle with the world of the intoxicated narrator, joining them by way of a pop song (a "studio" rather than "live" recording, at that), it also suggests that recorded music speaks to the issues raised by *Apocalypse Now* in a double register. It represents the tail end of a historical inquiry into the nature of sound recording as a practical activity that has often outpaced theory in its insights, on the one hand, and a reflection on the role that cinema (in its various audio-visual guises) has played in elaborating the significance of technologically mediated sensory experiences, on the other. Along the way it sounds the death knell for any remaining sense of an innocent perception we might still harbor.

## Sound Design, 1915, Part 1

With the advent of Thomas Edison's series of "Tone Tests" in 1915, a certain basic logic of sound design reached its purest embodiment. Arising out of a long tradition of discussion and debate (made evident in instrument design, scientific research, aesthetic practice, and vernacular epistemology), the Tone Tests promoted the idea that representational technologies are literally mechanical senses, and representation therefore a matter of mechanical seeing and hearing. Simply a more "perfect" ear, the phonograph should be able to reproduce a recorded performance so that it would be indistinguishable from the original. Understood in this manner (and it does make a certain intuitive sense), the goal of recording was the absolute duplication of a prior event. Development of both devices and practices (as well as aesthetics) proceeded from this assumption.

For the Tone Tests, performers under Edison contract would tour with a phonograph and a set of recordings, sing or play beside them on a darkened stage, and challenge the audience to discriminate between the human and mechanical voices. Contemporary reports indicate that audiences were satisfied with the illusion. As improbable as this may seem to contemporary ears (and it has been reasonably well established that the performers routinely sang so as to sound like the machines), Edison clearly believed that absolutely faithful duplication was the obvious and paradigmatic representational goal, and he enshrined it as his chief marketing strategy.[9] In fact, Edison insisted that his "Diamond Disc" series be sold not as records but as "Re-creations" (Frow 236–42).[10] Regardless of whether any recording actually achieved the ideal, the currency of such an explicit standard helped to shape the popular and professional understanding of phonography's *raison d'être*, and the iron law of sound design—absolute sensory duplication. In other words, the recording was to be an unmediated experience of perception that completely effaced representation.

As effective a practical heuristic as this was, it effected a forced resolution to the senses-technology split by means of a precipitous and misleading anthropomorphization of processes that could just as often be profoundly inhuman. This extreme humanizing of technology dialectically overturned itself by producing its own opposite—the mechanization of the human. Nowhere was this more evident than in what we might call the prehistory of sound design, where devices that were literally constructed of dissected human ears suddenly seemed to demonstrate the *insufficiency* of human hearing and the superiority of the machine. What is more, the profoundly inhuman capacities of the machine soon came to serve as an ideal for the human senses themselves (Lastra, *Sound Technology* 46–47; Daston and Gallison).

In such an environment, the simple equation of machine with sense organ and mechanical inscription with human perception could not long endure. In fact, the emergence of sound design as a concept and a practice required a threefold renunciation of the anthropomorphic position, involving, first, dividing apparently singular devices (that is, the mechanical senses) into multiple technologies—of detection, depiction, and reproduction, for example; second, dividing the process of recording (acts of mechanical perception) into coordinated zones of manipulation—including performance, inscription, and mimetic construction (here, mixing and editing); and third, abandoning the idea that the original object or sound bore any essential and determinate relationship to the finished representation (Lastra, *Sound Technology* 61–91; Maynard). In the end, the idea of recording as mechanical perception was replaced by a mimesis (in the Aristotelian sense of the term) defined by plausibility and internal consistency rather than by slavish replication of preexisting, and notionally autonomous, "originals."

Edison's recording *practices* from the same period suggest a very different understanding of phonographic representation. However much he may have promoted the slogan "Comparison with the Living Artist Reveals No Difference" (Harvith and Harvith 12; Frow 236), he clearly assumed that live performance and phonography were different enterprises altogether, involving techniques and standards appropriate to each medium. The nearly deaf Edison insisted that all his phonographic talent satisfy his own personal auditory standards, which were attuned not to musical norms but to the demands of the machine, saying, "I am like a phonograph. My ears, being a little deaf, seem to catch all the useless noises more readily than the musical tones, just as the phonograph exaggerates all the faults of a singer" (quoted in Morris 263).

His sense of aesthetics was grounded less in music than it was in mechanics. Edison "considered dramatic personality intrusive on discs and developed a stringent, mechanical perfection aesthetic for recordings that included purity of tone, extreme clarity of enunciation, and the abolition of extraneous noises, which, he conceded, would not be objectionable in the concert hall or opera house" (Harvith and Harvith 13–14).[11] His sense of artistry was determined by the quality of the phonogram, and not at all by the performance that produced it. When finally convinced to hire top-notch singers for his Diamond Disc series, Edison forced even the most famous to adjust to the demands of the device and thereby register with greater clarity on the disc (8).

As testimony from Anna Case and others makes clear, under Edison's relentless pressure, singers came to understand that sound recording was a process that could involve more than passively duplicating what was effectively a concert hall performance. Performers and technicians alike learned that it might even be advantageous to change aspects of a particular musician's performance style in order to take advantage of the machine's peculiarities. In Benjamin's terms, these might be considered instances of the innervation of technology—performing a kind of constitutive self-alienation before the device, incorporating its faculties into a kind of second nature.

Through repeated exposure to the most mundane and practical aspects of musical recording, both performers and engineers came to believe that sonic representation was *not* simply a matter of precisely transcribing completely prior and autonomous events, and to concede that a performance, for instance, might deviate from its customary *presentational* norms in order to achieve a particular *representational* effect, like intelligibility, regardless of its effect on the character of the "prophonographic" event.[12] In essence, what a phonograph *detected* was not necessarily what it *depicted* (Maynard 263–66).

This was no minor realization, but one as fundamentally transformative of phonography as the process of diegetic, continuity editing had been for cinema. In effect, Edison and his associates discovered that the specific acoustic qualities of the prophonographic performance were fundamentally irrelevant to the phonographic process, as long as the final cylinder or disk "reproduced" (in fact, "produced" would be a more accurate term) a satisfactory performance.[13] Gradually, they came to realize as well that, beyond performance, the processes of inscription could similarly *create* sonic effects, and even reproduction or playback could enter into the complex calculus of sound representation. It is hardly an exaggeration to say that abandoning the idea of an "original sound" was the founding gesture of all sound design—something that might have been realized with the advent of modern architectural acoustics but needed to be relearned again and again over the twentieth century.

By emphasizing the sonic character of the final product over all other considerations, Edison gradually weaned his performers and technical staff from the prejudice that inscription was the only properly phonographic act, instilling in its place a flexible and essentially "diegetic" understanding of sound representation that created internally coherent and plausible but spatiotemporally nonliteral musical worlds. In short, they had come to realize that rather than canning perceptions for a listening subject, they were in effect, creating the perceiving subject—the auditor—as a precipitate of their acts of sonic construction.

## Sound Design, 1958

Wagner returns. John Culshaw pushed the idea of sound design forward by severing the finished representation from any real or realizable "original" or "theatrical" performance, nowhere more dramatically than in his 1958 recording of Wagner's *Das Rheingold,* which convinced thousands of listeners of the importance and aesthetic possibilities of stereo production. Emphatically insisting that records are listened to in the home and not in a concert hall, he created a sonic world more "perfect" or more fully realized than could exist on stage. Though often recognized for his attempts, in the *Rheingold* recording, to re-create the dramatic use of stage movement in live performance, he also went far beyond the possibilities of the stage. He suddenly and dramatically granted particular characters an impossible and superhuman spatial ubiquity, and incorporated the thunder and myriad other sound effects demanded by the score (Gelatt 317–18). Culshaw's 1968 *Elektra* was even more daring (and more controversial), because it relinquished the theater completely, powerfully reimagining operatic space and sound in terms appropriate only to the record, and wholly impossible on the stage. As Culshaw wrote in a spirited and insightful defense of his work, "For most listeners, a recording is not a souvenir to remind them of an evening at the Met, nor does it need to bear any essential relationship to anything that ever happened in any opera house anywhere" ("The Record Producer Strikes Back" 68).

Pop music engineers were to go further still, giving records a distinctive audio signature or feel that, though obsessively unified, more often than not would have been impossible to achieve in performance. That is, music producers worked in a fashion that anticipated Walter Murch's comment that the idea of the sound designer was to have him serve as "somebody who took on the responsibility of 'auralizing' the sound of the film and making definitive, creative decisions about it. Someone the director can talk to about the total sound of the film the way he talks to the cameraman about the look of the film" (Kenny, "Walter Murch" 20).

More important for my account is that practitioners had discovered that the founding gesture of sound design, as it were, is both the complete severing of sensory experience from representation and a compulsive linking of the two in an indissoluble unity that appeared to efface representation, and by effacing it restore reality and to offer it to us in its fullness. It aspired, in effect, to the status of total artwork, which would redeem a world no longer capable of being experienced in itself. Though born of, and in part complicit with, a crisis in perception, it harbored within itself the utopian desire to restore to the senses the ability to experience.

## Sound Design, 1979, Part 2

*Apocalypse Now* ratifies a concern with the history of the senses in two ways: the practical, formal way in which it recapitulates and integrates the insights of vernacular and scientific inquiries into sensory technologies, and the way that it engages the political and historical issue of anesthesia. Murch's approach to *Apocalypse Now* is a distinctive mix of the traditionally conventional and the radically new. For the most part, Murch grounds the sounds of the film in real, usually visible sources. He keeps narratively important dialogue intelligible and keeps his acoustic space as continuous and ordered as that in the image. He edits seamlessly and shifts between omniscient and restricted forms of narration in a clear and systematic manner, making allowances for the greater emphasis on character subjectivity that Hollywood's assimilation of some art cinema norms had made allowable. That is, he eschews the idea of sensory duplication and instead takes advantage of the fully rhetorical, fully diegetic form of sound representation made commonplace by decades of previous filmmaking—for the most part.

What he adds, I believe, is a new emphasis on the possibility for what might clumsily be called "audio spectacle," or "audio sensationalism." In instances such as the hallucinatory opening, the attack, the Playboy show, and the Do Lung bridge, the sound design clearly exceeds its narrative demands and develops into something different, something that presents an almost purely sensory "argument." What these scenes try to convey in addition to their narrative information is something akin to a sensual, experiential equivalent or correlative to the depicted events. This is not to say that such correlatives are always in agreement with the narrative, although often they are. At Do Lung, for example, there is a fleeting moment when the camera tracks with Willard as he moves right to cross the bridge and the lighted guy wires take the form of a carousel canopy while the synthesized effects and music briefly become calliope-like. It is a hallucinatory, delirious moment that emerges and vanishes like a wave of terror. It is an audiovisual account of the scene far more concise and visceral than the subsequent search for the commanding officer. In the case of the Wagner-fueled attack it strikes me that the sensory argument is more equivocal, more interesting, and more revealing.

Put briefly, *Apocalypse Now* tries to show that the impulse toward prosthetic sensory experience is primarily an impulse to create a substitute, compensatory world, where fullness and apparent perfection strive to replace the world of actual experience with a better, more consoling, and more oblivious one. Here, to sever sound representation from a documentary function (remember the film's facile mockery of the documentary news crew) is not a simple technical decision designed to allow for better sounding recordings, but a technique of anesthesia, of refusing the meaning of what one experiences. To feel oneself the hero in one's own personal movie (skiing, surfing, killing, dying), it seems to say, is vastly preferable to acknowledging one's complicity in the messy worlds of politics, war, and the media that enable them. In *Apocalypse Now*'s account, the brutalization of the senses demands a protective shell, a screen of obliviousness that allows distance from the acts in which we participate. Allow experience to penetrate too deeply, and you become Kurtz.

## Sound Design, 1915, Part 2

So, in the end, what does *Apocalypse Now* have to say about the anesthetic impulse I decry? Does it set itself apart from the too common genre of the pointlessly dense, superabundant, overwhelmingly loud soundtrack that anesthetizes in the name of aesthetics, and finally celebrates the eclipse of perception? Does the film as a whole doom us to a senselessness like the one it suggests with the death of Clean, where a disembodied voice speaks mechanically to deaf ears, where two deaths speak to each other? I wish I could say for sure, but let me offer an old observation in the guise of a conclusion. According to an argument proposed by Siegfried Kracauer, scoring a film with well-known music, whose signification is given or clichéd, is "apt to produce a blinding effect," removing from the spectator's consciousness "all visual data which do not bear directly on that [meaning]" (*Theory of Film* 141). The "Ride of the Valkyries" threatens to blind us in just this manner. In fact, the soldiers *are* blinded, as are the legions of spectators who *simply* thrill to the vision of death from above.

On the other hand, the very sound that anesthetizes may also enable us to see, to cure a blindness, as it were. It may, in fact, enable us to see with sharpened historical clarity, and with a collective vision. It addresses us not only as a filmgoing collective but also through the experience of filmgoing as the collective subjects of history. *I* used *Apocalypse Now* to direct us to the 1915 of Edison's Tone Tests that foregrounded the essence of sound design. The film *itself* leads us to another 1915—the 1915 of D. W. Griffith's *Birth of a Nation,* whose chilling use of the same musical selection in a strikingly parallel sequence forces the historically aware spectator to see with horrifying clarity the suddenly visible, transhistorical patterns of U.S. racism and imperialism—patterns as apparently out of place as a cavalry in Vietnam, but just as disturbingly present. That is, *Apocalypse Now* revisits a key moment in the history of film style in order to reawaken it as well as to criticize. The scene alerts us to the history of form, and to how form can be understood as an empty vessel, but it simultaneously *uses* form to point to what is not entirely contained by that term. In short, it argues by way of sensation—the fell of the scene—that affect, emotion, and sensation have a history and, beyond that, a meaning as well.

## Notes

1. See Lastra's *Sound Technology and the American Cinema* for a fuller discussion of these matters.

2. Coppola discusses his wildly ambitious plan to develop a site-specific, multiday film of Goethe's *Elective Affinities* in Riviere.

3. Bordwell cites Michael Baxandall's wonderful *Painting and Experience in Fifteenth Century Italy* as a good example of a history of vision that does not seek to posit a single "mode of perception" for an era but focuses on "habits and skills." I agree that Baxandall's approach is the better one, but nothing in Benjamin or Adorno's invocation of sensory shifts (as will be explored shortly) requires such a homogeneous formulation.

4. The term *soundscape* is usually attributed to R. Murray Schafer's *The Tuning of the World*. Schafer defines the soundscape as "any portion of the sound environment regarded as a field of study" (274), and such study as located "in the middle ground between science, society and the arts" (4).

5. Recently, three works have tackled this issue in the specific realm of musical performance: Mark Katz, *Capturing Sound*; Robert Phillip, *Performing Music in the Age of Recording*; and Colin Symes, *Setting the Record Straight*.

6. Though clearly present in Benjamin's "Artwork" essay, the theme of anesthesia is most forcefully developed in Susan Buck-Morss, "Aesthetics and Anaesthetics: Walter Benjamin's Artwork Essay Reconsidered." I am indebted to this analysis, which I believe diagnoses the dynamics of shock and response accurately. See, originally, Benjamin, *Charles Baudelaire* 133, on the "mimetic shock absorber." Miriam Hansen's "Benjamin and Cinema: Not a One-Way Street" offers the richest and most detailed account of these issues.

7. Ben Singer's "Modernity, Hyperstimulus, and the Rise of Popular Sensationalism" places this in the context of film history. See also, Buck-Morss 18–24.

8. http://www.scentstories.com/.

9. Anna Case's memories of this campaign are included in John Harvith and Susan Edwards Harvith, *Edison, Musicians, and the Phonograph*, 44. An account of a related test involving Case can be found in Oliver Read and Walter L. Welch, *From Tin Foil to Stereo*, 202–3. See especially, Emily Thompson, "Machines, Music, and the Quest for Fidelity" and Lastra, *Sound Technology and the American Cinema* 85–86.

10. For accounts of earlier versions of this marketing idea, see *Collier's* (October 1908) and *Munsey's* (December 1914), which include advertisements.

11. Vachel Lindsay notes the performance style dictated by the machine when he laments the necessity for dialogue to be "elaborately enunciated in unnatural tones with a stiff interval between question and answer" (223 and, more generally, 221–24).

12. I mean the term *prophonographic* to recall Etienne Souriau's *profilmic*. Here I want to designate a realm encompassing all manipulations of a sound that occur anterior to the processes of technological inscription.

13. Harvith and Harvith note, for example, that Edison evaluated a test recording of Caruso's as follows: "This Tune Will Do For The Disc, but not so loud and distorted as this. Take it low and sweet. The Phonograph Is Not An Opera House" (13; emphasis in the original).

# PART III

# Sound and Genre

# 8

## Asynchronous Documentary
### *Buñuel's* Land without Bread

**BARRY MAUER**

### Structural Tensions

A peculiar relationship between sound and image underlies the structural tensions of *Las Hurdes* (Land without Bread) (Luis Buñuel, Spain, 1933). When Buñuel shot the film in 1933, he was unable to record synchronous sound due to a lack of funds and because travel conditions made transporting sound equipment to Las Hurdes—the setting for the film—prohibitive. Buñuel accompanied screenings of the silent film with recordings of Brahms's Fourth Symphony and with his own narration. E. Rubinstein comments on the effects of such live sound:

> It is . . . implicit in the . . . genre "travelogue" that voice and images may be subject to an ontological disparity normally supposed foreign to cinema; the voice need not even be recorded on the sound-track, need not be a synchronized element of that total recorded entity that is, in our usual sense of things, a film—the voice in practice as in theory could alter with every showing of the film. . . . [T]he narration of *Land without Bread* could be, and originally was, mechanically as well as emotionally disengaged from the images. (8–9)

This disparity between image and sound remains the basis of Buñuel's approach in subsequent versions of the film, which included recorded sound. In 1937 Buñuel acquired funding to add sound to the picture, yet he retained the asynchronous quality that we expect from live film sound. According to Michel Chion, sound and image transform one another in the spectator's perception. This transformation occurs as a result of an "audio-visual contract," wherein "the two perceptions mutually influence each other . . . lending each other their respective properties by contamination and projection" (*Audio-Vision* 9). Chion uses the term *synchresis* to refer to "the spontaneous and irresistible weld produced between a particular auditory phenomenon and visual phenomenon when they occur at the same time" (63).

Each "track" of Buñuel's film—the visuals, the music, the narration, and the text—does more than merely avoid synchresis; each represents a different register of discourse. Rubinstein, paraphrasing Pascal Bonitzer, describes "*Land without Bread* as a field of battle in which the subversive (as embodied in the images) is pitted against the institutional (as articulated in the voice)" (12). Bonitzer writes, "The voice-over is assumed to know: that is the essence of its power. The commentary is cold, but the visuals howl. . . . What is calculated in *Las Hurdes* is a radical testing of the mastership of commentary, of the essential imperialism and colonialism of documentary" (31).

Bonitzer and Rubinstein locate the "imperialism and colonialism" of documentary in voice-over narration: "What speaks is the *anonymity* of 'public service' . . . of information in general," writes Bonitzer (26). Because the voice of narration passes as the authoritative voice of information, it can appear neutral or objective without being either. In general, documentary narration denies the subject's power to speak and weakens the spectator's desire to interpret reality independently. Buñuel's film responds to the authoritarian qualities of documentary narration by politicizing narration, stressing its discontinuity with the images.

Tom Conley, elaborating on this argument, claims that Buñuel's narrator represses the excesses of the images, often doubling the spectator's desire to avoid seeing unsettling things:

> The voice-off . . . tends to freeze the images by directing the viewer's eyes to only several of many elements (generally the human as opposed to the natural or organic or seasonal ones) in frame. Betraying, trivializing, or, better, repressing many of the visuals, it marks a difference of consciousness. When the voice reflects the view of a focused, "Western" or industrial view of continuity, history, culture, humankind, or missionary reason, the visuals provide a rich flow of images exceeding—in pleasure, disgust, wonder, Eros, marvel—what the voice . . . cannot express about them. (179)

Conley asks whether the visuals possess enough power to stand up to the narration. Yet photography, it appears, is usually no match against the voice. Janet Malcolm argues that photography, unaccompanied by narration, lacks the decisive power to convey truth because of its ambiguity: "[The camera] can reflect only the usually ambiguous, and sometimes outright deceitful, surface of reality" (77). Thus, a narrator may take advantage of photographic ambiguities to manipulate an audience.

Buñuel's film demonstrates how such manipulation occurs. The narrator states a few obvious truths about the images—by means of the "doubling" that Conley notes—using a visual-verbal form of tautology; for instance, the narrator speaks of "barefoot children," and we see an image of children with bare feet. But the narrator also recounts events outside of the frame that are impossible to verify; for example, the narrator claims there are caves near Las Batuecas monastery with ancient depictions of men, gods, and bees, but we never see them on-screen. Conley finds other gaps between what the narrator says and what we see: "When disaster or plight is reported (in the British accent of the colonizer), oblivious to what is being said of them, children smile at the camera" (179). The narrator says that the "idiots"—seen in shots 188–201—are "probably dangerous." Rather than appearing dangerous, these "idiots" appear comical (resembling the Three Stooges).

The off-screen voice is a common device of the nonfiction filmmaker. Mary Ann Doane makes the following connection between this voice and its power: "Two kinds of 'voices without bodies' immediately suggest themselves—one theological, the other scientific (two poles which, it might be added, are not ideologically unrelated): 1) the voice of God incarnated in the Word, and 2) the artificial voice of a computer" ("Voice in the Cinema" 174–75). Buñuel's narrator speaks in even tones, using terms that are familiar to audiences; meanwhile, the film's tableaus and montage—even the music, which expresses a passion absent from the narration—work to discredit the narrator. By pitting the tracks of image and narration against each other, *Land without Bread* reveals the manipulations involved in film construction and trains spectators to adopt a skeptical attitude toward the images and the voice of narration.

## Forces of Redemption

The voice of narration persuades because it can easily fake an ethos, that of the trustworthy expert. The mantle of expertise may be gained by citing a few "facts" not available to the audience. For a speaker to gain the mantle of trustworthiness, the audience must believe that the speaker has the interests of others in mind and that the speaker does not appear to be seeking personal gain. Thus, imperious documentaries justify their authority in part by claiming redemptive power, presenting their crews as forces for the relief of suffering. For example, the television documentaries of the 1980s about the Ethiopian famine intended for the audience to believe that filmmaking itself was a form of intervention that relieved suffering and that the viewer could join the team by contributing money. Films about the miseries of others typically pose as a kind of charity, as though the fact of watching others suffer serves as a benevolent sacrifice. The narrator may encourage such identification with the documentarians, and the narrator's voice may acquire authority by the same means.

*Land without Bread* frustrates the desire to believe film works as a form of benevolent intervention. Buñuel's film, in fact, produced the opposite effect; it was at best a demonstration of documentary's impotence to relieve suffering, and at worst a form of malevolence.

Though Buñuel's film crew appears on-screen three times during *Land without Bread* (we never see the faces of the crew, only their hands), their interventions never help the Hurdanos; rather, Buñuel and his crew transform the Hurdanos into spectacle. When the cameraman intervenes in a scene about a dying girl (shots 99–102), he merely examines the girl; in fact, he displays her throat for *our* examination. The narrator says of the sick child: "We could do nothing for her. Three days later we heard she had died." Of this event, Conley writes

> The moral implications of the shooting sequence are obvious: the recording crew did no more than film a calamitous social condition as if it were tourism. It invested money in cinema rather than welfare. Even worse, the viewers are rendered culpable for witnessing what they should have remedied, not filmed. The double bind of the human predicament is seen as an esthetic spectacle. Once a relation of voyeurism is established in the relation of the film to the spectator, the unassailable distance held between viewers and subjects disallows any relation of enraged empathy, in this instance, that would have marked the collective perception of the child in its plight. (182–83)

Likewise, the crew member holding a book about mosquitoes for the camera demonstrates the cause of malaria to the film spectator but not to the Hurdanos, as Ken Kelman points out (124), and does nothing to address the disease (shots 179–83). The film crew appears again during a close-up of a child's casket, as a crew member's hands remove a blanket to reveal the child's corpse for the camera. According to Agustín Sánchez Vidal, a Buñuel scholar who grew up near Las Hurdes, the casket seen in the film is actually a trough used for making bread (unpublished letter). The image of the bread trough suggests that while bread is the stuff of life, the breadless trough in Las Hurdes has become the vessel of death. The bread trough poses a contradiction to the title of the film and the voice of the narrator who claims that bread is unknown to the Hurdanos; why would the Hurdanos have a bread trough if they have never made bread?

The film raises moral questions that implicate filmmakers and spectators alike. Identification with the film crew is possible only if spectators are willing to accept their impotence and that of the film crew. The presence of the film crew in the visual field of the film marks their failure as redeemers. Not only did the production of *Land without Bread* fail to help the Hurdanos, it seems to have produced no beneficial effects for the Hurdanos after its release. Sánchez Vidal writes, "The Buñuel movie was intensely manipulative and in many ways hurt the Hurdanos, whom he converted into strange bugs, and to whom everybody approached to see if it were like a zoo" (unpublished letter).

## Reflexive Documentary

Though *Land without Bread* fails as a work of redemption—and we should acknowledge that Buñuel may not have intended it as such—it does produce at least one beneficial effect: Buñuel's film opens audiences to reflection. Just as the razor that slices the eye in his *Un chien andalou* (France, 1928) opens our vision to the workings of the inner world of the mind, so the sight of the conditions in Las Hurdes opens our vision to the outside world. It is primarily a work of psychological inquiry; it explores the conflicts and drives that motivate the viewer's perceptions of external reality.

*Land without Bread* is, as Ken Kelman writes, "a work of high *imagination*" (122). Buñuel elaborates on the distinction between his film and others:

> To my mind there exist two different kinds of documentary films: one which could be called *descriptive* in which the material is limited to the transcription of a natural or social phenomenon. For example: industrial manufacture, the construction of a road, etc. Another type, much less frequent, is one which, while both descriptive and objective, tries to interpret reality. Such a documental film is much more complete because, besides illustrating, it is moving. . . . Thus besides the *descriptive* documentary film, there is the *psychological* one. (from Bunuel's first [and unpublished] autobiography, quoted in Higginbotham, *Buñuel* 54)

*Land without Bread* addresses psychological issues by means of conflicting textual tracks. The film is episodic, covering various aspects of Hurdano life: school, the countryside, the economy, the cultivation of food, disease, deformity, and death. In each episode, the textual tracks overlap, producing a metafilm that traces the psychological processes of encoding and decoding the film.

Jay Ruby explores the idea of *Land without Bread* as a reflexive work in his essay "The Image Mirrored: Reflexivity and the Documentary Film." According to Ruby:

> To be reflexive is to structure a product in such a way that the audience assumes that the producer, the process of making, and the product are a coherent whole. Not only is an audience made aware of these relationships, but it is made to realize the necessity of that knowledge. To be more formal about it, I would argue that being reflexive means that the producer deliberately and intentionally reveals to his audience the underlying epistemological assumptions that caused him to formulate a set of questions in a particular way, to seek answers to those questions in a particular way, and finally to present his findings in a particular way. (65)

Although Buñuel does not present an explicitly reflexive narration to accompany *Land without Bread*'s images, the disparity he creates between image and sound produce a type of implicit reflexivity. Ruby points out that *Land without Bread*, because of its implicit reflexivity, has caused confusion among critics. He cites Basil Wright's review of the film in which "Wright assumed that the narration and music score were errors and not a deliberate attempt on Buñuel's part to be ironic. 'Unfortunately, someone (presumably not Buñuel) has added to the film a wearisome American commentary, plus the better part of a Brahms symphony. As a result, picture and sound never coalesce, and it is only the starkness of the presented facts which counts.'" Ruby responds: "It is sufficient for our purposes to realize that it apparently never occurred to Wright that some audiences might regard the juxtaposition of music, sound, and images as ironic, perhaps even as a parody of travelogues and information films" (68). Though Ruby's point about the film's reflexivity is well taken, his argument raises a series of unanswered questions. *Land without Bread* may indeed be reflexive, but what are the effects of this reflexivity? For whom does the film become reflexive? In which contexts does it become reflexive?

## Historical Contexts

With available historical information, we can reach some conclusions about how *Land without Bread* functioned in the past. Buñuel was an active member of the surrealist movement led by André Breton and had made two films prior to *Land without Bread*: *Un chien andalou* and *L'âge d'or* (France, 1930). These earlier films were screened in Paris and attracted the attention of art audiences and a more general public. Both films were intended as provocations, and indeed they did provoke violent reactions, especially *L'âge d'or*, which drew the wrath of militant right-wing Catholics who objected in particular to its depictions of Jesus attending an orgy. *Land without Bread* retains the quality of surrealist "shock" apparent in Buñuel's first two films, yet Buñuel's social concerns are made explicit in *Land without Bread* in a way that was not apparent in his earlier films.

The inscription—yet another semiotic "track"—that begins the film (seen in shot 2) describes the project as a study of "human geography." In his autobiography, Buñuel elaborates on this point by citing the work of anthropologist Maurice Legendre as his inspiration for the film. Legendre's work *Las Jurdes: Étude de geographie humaine* is a doctoral thesis published in 1928; it was the product of Legendre's research in Las Hurdes over a twenty-year period. Human geography is a form of anthropology that views cultures and environments as interdependent entities. It is also a redemptive project since it searches for ways to relieve human suffering; human geography discerns the environment's role in determining the boundaries of the subjects' fates and seeks to make the most of available options for improvement.

Buñuel and Legendre agree in their depiction of the Hurdanos' condition, which Legendre called "antagonism between physical features of geography and the population." Legendre also describes the economic and social weaknesses of the Hurdanos in a chapter titled "The Forces of Misery." Legendre acknowledges the difficulty of social and economic development in Las Hurdes due to the Hurdanos' physical isolation. He writes, "The mountains are petrified waves that the good news can't cross. . . . [T]here is no reason for the passerby or tourist to be there and no reason to stay" (xiv). Buñuel's film follows Legendre's perspective of the Hurdanos practically to the letter.[1] The political struggles in Spain that served as the immediate context for *Land without Bread* have faded into history. Why then does the film still attract so much critical attention? What relevance does it hold for us today?

## Cinema as Truth; Cinema as Sacrifice

*Land without Bread* raises two related issues that make it relevant for audiences today: the status of documentary as "truth" and the function of cinema as ritual sacrifice. *Land without Bread* addresses the issue of documentary as "truth" by breaking from the form routinely used by other documentaries. Robert Flaherty's *Nanook of the North* firmly established the "naturalistic" conventions that Buñuel wanted to undermine. Buñuel recognized the dangerous power of such films to create an impression of reality that audiences might accept passively; audiences may believe that documentary films arise organically from the subjects themselves. *Land without Bread* demonstrates that the filmmaker, not the subject, constructs the impression of reality.

*Nanook of the North* is a work of salvage ethnography, an attempt to preserve a disappearing culture. Flaherty presents the Eskimos as though they are still in their "golden age," before European intervention radically changed their way of life. He re-creates Eskimo life in order to achieve this effect. In the hunting scenes, Nanook uses a spear, although the Eskimos had used guns since the turn of the century. Flaherty avoids images that might diminish the Eskimos' nobility. He also manipulates events for the process of filming. For example, three days of production time were required to record the scene in which Nanook built an igloo for his family. An intertitle informs us that the igloo construction represents an hour of narrative time. The screen time for this scene is less than five minutes.

Though Flaherty's film is silent, it deploys strategies that undermine film's innate capacity to produce delirium, a capacity that the surrealists were quick to exploit. Flaherty employs elements of continuity editing and scripted intertitles to "anchor" the meaning of his images. By contrast, surrealists such as Benjamin Fondane were arguing for filmic discontinuity in order to produce disorienting effects. Paul Hammond summarizes Fondane's 1929 argument that "silent film fortifies everything with mystery because by doing without the spoken word it dispenses with the logic that supports it. The dream-like nature of film is automatically asserted. He approves of the role which misunderstanding plays in the cinematic experience. Talking pictures will make this all the more difficult. To counter this he proposes the dissonant use of sound and speech" (13).

*Land without Bread's* asynchronous structure undermines the audience's easy acceptance of documentary conventions established by films such as *Nanook of the North.*[2] Buñuel's film attacks the illusion of reality in documentary, paradoxically, by using the extremes of realist aesthetics. Audience expectations, derived from other films in the tradition, are evoked: a capsule of knowledge about a region and people, a summation of existing problems, and a homologous narration. Although Buñuel maintained standard elements of documentary throughout *Land without Bread,* he created disquieting deviations from the very beginning. The title shot imposes the written words "Land without Bread" against a backdrop of clouds. The "land" is in the air. This contradiction between image and text—a variation on the asynchronous voice-over narration—interrupts realist continuity.

Whereas most documentary films try to present reality as transparent, *Land without Bread* poses the following question: to what extent do we perceive the world through frames created for us by others? Buñuel's film addresses this question by demonstrating how cinema functions as sacrifice. A close examination of the scene showing the death of a goat (shots 108–16) exemplifies his strategy. "One eats goat meat only when one of the animals is killed accidentally. This happens sometimes when the hills are steep and there are loose stones on the footpath," says the narrator during two shots in which a goat plunges off a mountain (shots 115–16). Logic dictates that it is unlikely for the filmmaker to have had two cameras in position to capture the event of a goat's fall if it had been purely accidental. Buñuel, through the editing, creates the impression that two cameras filmed the goat's death plunge—the image of the goat's fall is seen from two angles—yet Buñuel had only one camera. Buñuel shot the goat—both with a gun and a camera—dragged the corpse up the rock, and pushed it down, filming it again from above. A close look at the first shot reveals a puff of gun smoke bursting from the corner of the image. Vivian Sobchack writes:

> In this sequence, simply more blatant than others which function similarly, we are confronted with a lie, with a manipulation of reality which we can see is a manipulation *for the film.* We are led not only to mistrust the narrator and regard him as unreliable and unethical (he has, after all, lied to us), but also to mistrust the real-

ity and spontaneity of the images we see and the way in which they are offered to us for viewing. Indeed, we need to consider the fact that the "real" images of the Hurdanos are relatively uncontextualized but for their relation to the narrator and what he chooses to tell us about them. ("Synthetic Vision" 74)

The goat's death matches a description of Judeo-Christian sacrifice in Henri Hubert and Marcel Mauss's *Sacrifice:* "The goat of Azazel, on the day of Atonement, was . . . thrown down from the top of a rock (Talm. Babl., *Yoma,* Mishnah, 67A)" (quoted in Hubert and Mauss 131–32). Hubert and Mauss explain the goat of Azazel's death as an expiatory sacrifice, designed to rid the self or community of impurity and sin. "The sacrificer remains protected: the gods take the victim instead of him. *The victim redeems him*" (98). By leaving the audience to understand that *he* sacrificed the goat, Buñuel forces aesthetic aspects to recede and moral concerns to advance. Buñuel committed this violent act so that it could be filmed. Why? Since *Land without Bread* is not about the redemption of the Hurdanos (as already discussed), what sort of redemption is Buñuel calling for here?

The discussion of ritual sacrifice in Georges Bataille's College of Sociology essays may provide some indication. Bataille notes that sacrifice puts the participants' integrity at stake and disrupts the normal flow of existence. In this case, the goat's death suddenly calls into question our identities as witnesses of a spectacle, as well as the filmmaker's role as social agent (the Hurdanos play no role in this scene, so we should not mistake it as being any kind of commentary about them). The sacrifice is a transgression against identity boundaries and a catalyst for their transformation. The goat's death is a sacrifice made on the audience's behalf.

The narration does not independently produce the "meaning" of sacrifice manifest in the scene of the goat's death. We read this scene dialectically in order to understand it as sacrifice. First, the audience interprets the event as "an accident." Next the audience observes the images, realizing that this event is not an accident; the goat's death serves as spectacle for the camera. Finally, the audience concludes that the narrator is not trustworthy.

Sobchack sees the dialectical processes of *Land without Bread* as a ticket to the spectator's freedom. "Our liberty to see is confirmed as we recognize the very impossibility of freedom and clear vision. Indeed, it is this process of questioning freedom and vision that *Las Hurdes* sets in motion through its dialectical structure and method" ("Synthetic Vision" 81). As *Land without Bread* draws the spectator's attention to sacrifice through its dialectical relation of sound and image, it increases the film's investment (as well as the spectator's) in reflexivity; it commits us to examine our responses to film. More than anything, it demonstrates that film is itself a luxury, a peculiar waste of resources expended in spite of, even spiting, the people it represents.

## Notes

1. The Spanish government banned *Land without Bread* soon after it was released in 1933. Five hundred years of monarchy had ended by 1931, but the ruling elite still controlled enormous wealth, whereas the vast majority lived in poverty. The country was bitterly divided into hostile factions. The largely ineffective Republican-Socialist coalition under Manuel Azana proposed reforms in education and land redistribution in an effort to placate angry peasants, and introduced legislation aimed at the Catholic Church, which owned approximately two-thirds of the land (Besas 13–14), but the government failed to deliver the promised land reforms. Workers and peasants revolted, burning several churches through much of the country.

The wealthy elite demanded that the government defend their interests. They formed a common front of landowners, church patricians, and military leaders. This front sponsored fascist gangs and later supported the military uprising led by Franco. The Spanish civil war of 1936–1939 resulted in the deaths of more than a million people and the long, tremendously repressive regime of Franco (Higginbotham, *Spanish Films under Franco* 72).

During the early 1930s, Buñuel, living in France, followed the changes occurring in Spain. His friend Ramon Acin, an anarchist, won the lottery after Buñuel lent him money to buy a ticket. Acin delivered a portion of the winnings to Buñuel, enabling him to return to Spain and make his film (Hopewell 17). Buñuel did not go to sites of recent conflicts, but instead went to Las Hurdes, an isolated region.

*Land without Bread* established a context for understanding the political upheavals in Spain without referring to them directly. Buñuel's portrayal of the Hurdanos' condition was an embarrassment for Spain's politicians, who maintained the fiction that the nation was progressing into the twentieth century. Buñuel focused on the Hurdano "society" that had degenerated over a five hundred–year period, composed of the inbred, illiterate offspring of escaped convicts and Jews fleeing the Inquisition (R. Conrad 9). Buñuel was fully aware of the history of the Hurdanos, but he evidently did not deem it necessary to mention it in his film.

The liberal Zamora administration banned *Land without Bread* and asked other countries not to show it. The right-wing administration of Lerroux and Gil Robles, which followed that of Zamora, saw the film as a threat and labeled Buñuel as a "criminal" (Agustín Sánchez Vidal, personal letter). The Spanish secret police kept a file on Buñuel, including a notice that he be taken dead or alive for his involvement with the Communist Party—which was minimal—and for his creation of *Land without Bread*. Randall Conrad comments:

> [The film's] negation of the "progressive" options in bourgeois ideology, including liberal politics, make it a radical work. In fact, this radicalism perversely made its appearance at the most inconvenient moment in Spanish politics. The year-old republic, internally torn, blocked in its efforts to deliver promised reforms and defending itself from right-wing bids for power, had failed on many fronts but could claim one outstanding liberal reform, the opening of many secular schools in backwards

parts of the country. Yet here was a film that not only put Spain in an unattractive light generally but attacked the new government's proudest achievement. (9)

When Buñuel made *Land without Bread* in 1932, he intentionally undermined the government's reformist image. Then, in an apparent switch in 1937, Buñuel supported the leftist Popular Front governing Spain. He made an effort to transform *Land without Bread* into a propaganda tool for the embattled republic when the film was released in Paris in 1937. Randall Conrad cites a postscript that Buñuel added to the film in 1937, "insisting that the poverty of Las Hurdes is not irremediable, that the Spanish people had begun to unite to improve their conditions until Franco's bid for power, and that the present anti-fascist fight is a continuation of the struggle to end the poverty documented in the film" (R. Conrad 11). This postscript contradicts the fundamental messages of the film, but at least it attests to Buñuel's pragmatism during the fight against fascism.

2. Ironically, Flaherty liked *Land without Bread,* though he opposed Buñuel's candidacy for a post of documentary team leader in the United States (Agustín Sánchez Vidal, personal letter).

# 9

# "We'll Make a Paderewski of You Yet!"

## *Acoustic Reflections in* The 5,000 Fingers of Dr. T

### NANCY NEWMAN

[A] Ten little dancing Maidens, dancing oh so fine.
Ten happy little Fingers and they're mine all mine.
[B] They're mine, they're mine, now isn't that just fine?
Not three, not five, not seven and not nine,
But ten all dancing straight in line!
[A1] And all of them are mine, mine, mine, yes, they are mine, all mine.

—Lyrics, "Ten Happy Fingers"

## "Practice until You Are Perfect!"

Maidens that are fingers, fingers dancing at the piano, the piano encompassing all that is music. In the condensations and displacements of dreams, one object stands in for another, is exchangeable and interchangeable. In the fantasy film musical that is *The 5,000 Fingers of Dr. T* (Roy Rowland, USA, 1953) body parts, even more than whole bodies, are the instruments that make music. But whose body parts, whose fingers? Even the film's title refuses to respect the usual physical boundaries, much as the later Dr. Mabuse's eyes were not affixed to any single body. In Rowland's musical, the five thousand fingers "possessed" by the maniacal piano teacher, Dr. Terwilliker, actually belong to the five hundred boys he has imprisoned. The film's climax takes place when the boys are forced, like individuals possessed, to demonstrate Dr. T's pedagogy, the "Happy Fingers Method." Seated at a phantasmagoric, two-tier keyboard, they are poised to perform the song that glories in fingers that are, with terrifying ambiguity, "mine, all mine."

There are few films in the history of cinema that portray the individual's search for musical identity as wittily and artfully as *The 5,000 Fingers of Dr. T.* Yet critical and audience reaction at the time of the film's release was hardly enthusiastic, two of several factors that relegated the film to an obscure corner of the Columbia

Studio film library for nearly four decades. A "strange and confused fabrication of the dreams of a 10–year-old boy," wrote Bosley Crowther in the *New York Times* (June 20, 1953, 8). Alluding to a standard criticism of the story's fabricator, Theodor "Dr. Seuss" Geisel, Crowther concluded that there was as little to recommend the film as there was in the comic books that it resembled.[1] Box office returns were disappointing as well. The film grossed less than one-sixth what it cost to make, frustrating producer Stanley Kramer and driving another nail into his coffin at Columbia.[2]

Despite its unsuccessful initial reception, *The 5,000 Fingers* gradually managed to develop something of a cult following. Occasional screenings in revival houses, design schools, and on television kept it in modest circulation until Sony purchased Columbia and proceeded to convert the latter's film archive to videotape (Thomas). Released on VHS in 1991, *The 5,000 Fingers* quickly became a staple of the children's section of video stores, a phenomenon attributable to a number of developments, including renewed interest in Dr. Seuss among baby-boomers. Theodor Geisel's whimsical influence permeates the film. In addition to contributing the story concept and cowriting the screenplay (with Allan Scott), Geisel sketched the sets and wrote the lyrics for composer Frederick Hollander's tunes. A DVD became available in 2001, restoring the extravagant Technicolor effects and extending the film's fan base to yet another generation of people and machines.

In addition to its association with Dr. Seuss, the persistent appeal of *The 5,000 Fingers* can be attributed to its clever treatment of the emotional trials associated with piano lessons. The story focuses on a boy, Bartholomew Collins (Tommy Rettig),[3] who finds practicing particularly disagreeable. Falling asleep at the keyboard one day, Bart dreams that his teacher, Dr. Terwilliker (played by the inimitable Hans Conried), has hypnotized and kidnapped his widowed mother. The doctor plans to use them both in the grand opening recital of his Happy Fingers Institute, the initial salvo in his quest for world domination through piano. The institute is half well-appointed castle and half prison, replete with cells for the boys Dr. Terwilliker has entrapped.[4] After a series of perilous adventures in its depths, young Bart is able to foil Dr. T's megalomaniacal plot with the aid of the family plumber and a little bottle of "Music Fix." When Bart awakens all is well. His practice session concluded, he is allowed to take his dog and baseball mitt outside to play. His mother accepts a ride downtown from the handsome plumber, their romantic theme music swelling in the background.

The Oedipal dimension of *The 5,000 Fingers* was immediately apparent to critics such as Crowther, who characterized Bart's situation as a "psychic rebellion." But Crowther was disturbed by the film's blatant Freudian imagery, particularly the "beanie with a slightly flaccid hand" worn by the film's hero and the other boys of the Terwilliker institute. It must have been particularly irksome that the offending headgear, symbol of the "rebellious clan," was available for purchase at Macy's department store, along with souvenir shirts and nonpiano musical instruments, as part of Columbia's promotion of the film (see advertisement, *New York Times*, June 19, 1953, 7). If it caught on, the beanie could become a badge trumpeting "the rights of small fry," advertising a film that seemed "utterly to condemn study of the piano" (*Christian Science Monitor*, quoted in Spoto 150). Even Stanley Kramer's biographer, Donald Spoto, called the movie "an anticultural diatribe."

Spoto's judgement seems to disregard the musical's actual musical elements, which include a lengthy balletic sequence choreographed by classically trained Eugene Loring. How else could he deem "anticultural" a film that earned Hollander an Academy Award nomination for best music and best scoring of a musical picture? As this essay argues, whatever the vicissitudes of the film's exhibition and reception, the various elements of *The 5,000 Fingers*' soundtrack—score, source music, effects, and dialogue—coalesce to offer a refreshing and novel perspective on the complex and personal nature of aesthetic experience.

Two diverse themes, the Oedipal and the musical, are clearly operative in *The 5,000 Fingers*. But whereas the story appears to tell us that Bart's options are between making music and not making music—between piano and pastimes such as fishing or baseball—the soundtrack "says" something more complicated to listeners. Through a progression of production numbers and auditory events, the audience experiences, with Bart, the pleasures and terrors of the sonic realm. These spectacular moments display the delights of an imagination allowed free range, extending beyond the Hollywood musical's conventional opposition between classical and popular music into new creative territory that draws without inhibition from both. As the Oedipal drama unfolds, and with his mother's voice as guide, Bart demonstrates not only a capacity for taking control of the auditory environment but also for developing his unique "voice." Instead of abandoning musical culture, he will ultimately develop the inner resources to wield Dr. T's baton with fingers that are, without doubt, his own.

## Magic in the Mother's Voice: "You Really Are Missing the Beat!"

In its fusion of Freudian and musical themes into one ideological operation, *The 5,000 Fingers* harks back to one of the first sound pictures, *The Jazz Singer* (Alan Crosland, USA, 1927). From the beginning, Hollywood musicals located the struggle over musical identity within the family or the potential family. The central drama of *The Jazz Singer* is an Oedipal struggle. Al Jolson's title character must reconcile his musical career choices with his mother and father, both of whom prefer that he honor his heritage and become a cantor. The picture concludes with the death of the father, Jolson taking his place at the synagogue. *The 5,000 Fingers* similarly locates the quest for musical identity within the nuclear family, requiring the cooperation, encouragement, and acquiescence, if need be, of Mother and Father. In order for Bart's musical development to proceed, he must overcome his dependence on his mother's musical judgment and abilities, and use the skills he has thus acquired not only to find, but also to shape, a father who is a suitable role model. It is only through the successful completion of this Oedipal quest that harmony, both musical and matrimonial, can be restored to the Collins household.

The structure of another landmark film, *The Wizard of Oz* (Victor Fleming, USA, 1939), resembles that of *The 5,000 Fingers*. Both pictures deploy a narrative frame that contrasts the main character's circumscribed quotidian existence with a vibrant dreamworld constituting the bulk of the pictures' running time. In the "realistic" framing story that opens *The 5,000 Fingers*, we learn that Bart has a tendency to nap when he should be practicing. He is abruptly awakened by his piano teacher from a dream in which he is chased by darkly clad grown-ups brandishing colorful butterfly nets in a balletic pantomime accompanied by full orchestra. After this foreshadowing of the dreamworld, the two have a contentious exchange over Bart's lack of preparation for Dr. Terwilliker's student recital the following month. Dr. T is fussy, rigid, and old-fashioned; his attitude toward Bart is one of superiority. His association with European classical music is established through the grandiose claim, "We'll make a Paderewski of you yet!" a name that Bart does not understand until Dr. T Americanizes the pronunciation, "PaderOOski." From an adult perspective, Dr. T is associated with the piano virtuoso tradition inaugurated by Liszt and perpetuated in the United States by Rachmaninoff and Rubinstein. From a child's perspective, Dr. T is obsessed with the piano.

Bart's waking life is situated in a modest but comfortable midcentury middle-class single-family home. The piano room adjoins the kitchen, where Mother (Mary Healy) is baking, within earshot of Bart's efforts, and the plumber, Mr. Zabladowski (Peter Lind Hayes), happens to be fixing the pipes.[5] Through an expedient direct address by Bart we learn that his problem is twofold: he hates his piano lessons, and he lacks a father.[6] The death of the latter (which contemporary audiences would presumably have attributed to World War II) has reactivated Bart's Oedipal complex, forcing him to seek a new parent from the available males in his life. The father's absence has also become synonymous with a lack of musical direction, as Bart has linked his mother's desire that he play the piano with her lack of a mate. At the same time, Bart fears that her interest in the piano is merely the result of Dr. T's manipulations. This anxiety takes the form of her capture and internment through Dr. T's Caligari-like hypnotic powers, underscored by a Theremin, in his dreamworld.

Mrs. Collins's attempt to offer Bart musical guidance takes the obvious form of providing piano lessons, but we soon discover that her influence is much more profound. She is able to transfer the spirit of Dr. Terwilliker's pedagogy to Bart's fingers through her voice. After he makes several halfhearted attempts to play the first section of Dr. T's didactic etude, "Ten Happy Fingers" (marked A in the epigraph), Mrs. Collins approaches the piano to sing the melody. Suddenly, Bart proves an accomplished accompanist, able to play not only the tune's first section but the entire song with an elaborate harmonization. The effect of the mother's presence is magical on his performance, and she assures him that he has improved. When she leaves the room, Bart reprises the entire song. He is now capable of singing and playing simultaneously, and even adds a touch of resentment to the execution.[7]

This number, a momentary departure from the general realism of the narrative, seems calculated to portray a psychological truth, that Bart's musicality is brought forth and sustained through the mother's voice. The "maternal voice" has been described by Kaja Silverman as providing an acoustic mirror for the child, analogous to the visual reflection that initiates the child's self-identification during the "mirror stage" proposed by Lacan. Just as the child discovers its own unity out of a jumble of indistinguishable body parts through its gradual differentiation from the mother in the Lacanian mirror stage, the child, according to Silverman, "could be said to hear itself initially through [the maternal] voice—to first 'recognize' itself in the vocal 'mirror' supplied by the mother" (*Acoustic Mirror* 80). The mother's voice, as Silverman takes great care to point out, emanates from a position within culture and the symbolic order of which she, too, is a subject. However, from the retrospective position of the child's entry into language, the maternal voice appears to have constituted a "sonorous envelope" for the individual, extending back even to the womb (72). The image of the mother's voice reaching the child prior to the distinction between external and internal, between self and other, is both powerful and pervasive. Film theorist Guy Rosolato, for example, has argued that "this primordial listening experience is the prototype for all subsequent auditory pleasure, especially the pleasure that derives from music" (paraphrased in Silverman 84–85).

We might observe that the psychological truth analyzed so cogently by Silverman is also historically contingent. In *Discourse Networks*, Friedrich Kittler argues that the cultural fantasy that posits the mother's mouth as the originary site of discursive production is an effect produced beginning around the year 1800 by the combined apparatuses of a new type of pedagogy, romanticism, and the emergence of the nuclear family as the locus of subjectivity (25–69). The new pedagogy made mothers responsible for primary instruction; most important, mothers taught young children to speak and read by carefully and patiently demonstrating the oral qualities of the letters of the alphabet. Children observed and imitated their mothers, internalizing the voice that first enables them to enter into discourse. The harsh and disciplined rote learning of the past was replaced, beginning in the nineteenth century, with an emphasis on nurture and play in education. A disdain for rote learning continues to characterize pedagogical reform. The almost magical effectiveness of an understanding, maternal voice in an educational setting is portrayed in *The 5,000 Fingers* through the juxtaposition of Dr. T's strict admonition to "practice, practice, practice!" and the mother's coaxing, mellifluous tones in the scene just described.

If Mrs. Collins seems to have no musical ambitions for herself, that is not too surprising. In the discursive network analyzed by Kittler, Woman's "function consists in getting people—that is, men, to speak" (25). This formulation might be extended by adding that mothers enable men—that is, boys—to make music. Yet Bart's dreamworld shows him to be profoundly ambivalent about his mother's musical direction. If the sonorous envelope of the maternal voice is capable of conjuring images of plentitude and jouissance, it can also produce fears of entrapment and impotence. According to Silverman, the individual's defensive mechanisms require that whatever is incompatible with the phallic function—including subordination to the mother, an admission of incompetence—be displaced onto the female (*Acoustic Mirror* 86). Bart's negative feelings are projected onto Dr. T and manifested when the latter banishes Mrs. Collins to her cage. Forcibly removed to her "lock-me-tight" in the inner recesses of Dr. T's lair, she is visibly relegated to the interior of a luxurious but inescapable sonorous envelope. Mrs. Collins's containment in the Happy Fingers Institute reduces the mother to the helplessness of a child.[8]

## The Man of My Dreams; or, "You're Working for Me Now!"

Once the possibility of Bart's musicality is established through his mother's presence, it must be secured with a father, a father whose musical style evokes the mother's voice in all its warmth and immediacy. Initially, the plumber appears an unlikely candidate for this role. When Bart appeals to Mr. Zabladowski to tell Mom what he really thinks of Dr. T and piano lessons, the plumber refuses to reveal his opinions. He repeatedly claims that he knows nothing about music. It is therefore up to Bart to elicit Mr. Zabladowski's natural talent and to show his mother how musically compatible the plumber really is.

In Bart's dream, Mr. Zabladowski has been hired to install sinks in the Happy Fingers Institute in anticipation of the grand opening concert the following day. Dr. T's intention is to marry Bart's mother immediately after this event. Alarmed at the idea that Dr. T could become his father, Bart tries desperately to enlist Mr. Zabladowski's help in rescuing Mrs. Collins. The plumber, however, displays a marked lack of interest in assuming the roles of husband and parent, responsibilities that he must fulfill, given the premises of the movie. Mr. Zabladowski's failings are those of the ordinary American man—an ambivalent relationship to money, an obstinate determination to do his job, and an insistence that nothing more than his job be asked of him. In short, he lacks not only authority but imagination. In Lacanian terms, it is up to the young Bart to turn this penis bearer into a phallus wielder.

As Bart's fantasy progresses, we see that Mr. Zabladowski is indeed musical. Despite his protests and unwillingness to compete with Dr. T over musical expertise, he is a competent and charming singer and dancer. At one point, the two men have a wordless battle in the form of a hypnotic dance. The orchestration elaborates a short musical idea first heard as underscoring when Bart was chased by the guards of the institute. For the dance-duel, this motive is developed melodically and rhythmically into a dizzying array of exotic, mainly Latin-based, musical styles. When the two men, panting, reach a stalemate, Dr. T asks, "Where did you study?" Mr. Zabladowski shrugs and replies, "I just picked it up," nonchalantly dismissing any interest in, or need for, formal training.

Although the plumber is reluctant to challenge Dr. T either musically or sexually, with Bart's coaxing he proves capable at both. It is thus crucial that Bart's mother, despite her enchantment by Dr. T, is a witness to Mr. Zabladowski's emerging talent. After the men's musical duel, Mrs. Collins joins them in a song-and-dance trio that ostensibly affirms their mutual interests in the institute. More important, "Get Together Weather" allows Mr. Zabladowski to display his musical versatility. Having shown ease with social dances such as the tango and rumba, he proves equally comfortable with the trio's humorous mix of Irish step and Anglo-American square dancing. At the scene's conclusion, Dr. T realizes that Mrs. Collins may indeed be inclined to "get together" with the plumber. "You're beginning to build up an immunity to my little hypnotic trances," the doctor remarks while securing her cell.

The psychological dimension of this sequence of numbers helps distinguish *The 5,000 Fingers* from that subgenre of musicals in which a child plays Cupid to the grown-ups (Altman, *American Film Musical* 104). What is significant in Bart's guileless staging of Mr. Zabladowski's musical ability for Mrs. Collins is that this staging is, fundamentally, the representation of his own desire. In this phase of his development, Bart's desire appears coterminous with that of his mother; because of her emphasis on the piano, Bart has assumed that she wants a musical partner. The mental construct through which Bart has linked his lessons to the mother's marital options seems to instantiate Lacan's maxim that "the unconscious is the discourse of the Other" ("Agency" 172). Bart's unconscious desires "are those of an already constituted social order" in which his mother participates (Silverman,

*Subject of Semiotics* 166). Furthermore, the mother's desire "awakens in the son the impossible wish to supply her with the missing phallus" (185).[9] As Bart perceives his task, he must present to his mother the musical partner that she lacks.

The transformation of the plumber into a musically capable father is advanced when he is momentarily persuaded that he should care for, and take care of, Bart. With a crooner's baritone, Mr. Zabladowski sings a sentimental, lullaby-like tune, "Dreamstuff," while cuddling with Bart in an overstuffed chair. The arrangement of this Tin Pan Alley song form (AABA) follows a conventional format, with the reprise of the tune's first section (AA) stated by instruments alone. However, the convention is employed here to reveal a further dimension of the musicality Bart has projected onto Mr. Zabladowski. Instead of singing the first section's words, Mr. Zabladowski whistles the tune, demonstrating another of his "natural" talents. Soon Bart begins to hum along. At the B section, Bart joins Mr. Zabladowski with the lyrics and added harmony, making this number the closest thing to a love duet in the picture (cf. Altman, *American Film Musical* 37).

The pair's intimate sharing of this song, and the ease with which Bart joins Mr. Zabladowski, is also a step in Bart's transformation from an inept, unwilling student to a competent musical partner. After the two quarrel, Bart has his solo, "Because We're Kids," a pensive song about justice built around the refrain, "You have no right / You have no right / To push and shove us little kids around." The audio quality for this number is impossibly close-up, as if this boy soprano is right next to the listener. The gentle immediacy of his delivery recalls Mr. Zabladowski's crooning and confirms their compatibility.

Perhaps the most important aspect of Mr. Zabladowski's musical character is that he acquires stylistic and generic associations in Bart's fantasy that contrast sharply with those of Dr. T. Just as Bart partitions his feelings about his mother into positive and negative images, his need for a father is split into good and bad candidates for the position. In Bart's dream, this opposition is expressed musically. Whereas Dr. Terwilliker represents the exclusionary tradition of classical piano and the cool calculations of disciplined practice, Mr. Zabladowski's music exudes spontaneity, warmth, and the expression of individual sentiment. Mr. Zabladowski's musical behavior is characterized by the ease with which he "picks up" a given style. Dr. T's music, on the other hand, is pedantic, extroverted, and designed for effect, as in the Lisztian, multiple-keyboard "Happy Fingers" Rhapsody that Bart performs at the institute. Constructed rather than natural, Dr. T's music constitutes a contrivance rather than an outpouring. As Jane Feuer might say, it doesn't come from the heart, and thus cannot partake of the "folksy" qualities associated with popular music during the classic period of Hollywood musicals (*The Hollywood Musical* 57).

Later we will see how Bart, as he begins to exert control over his sonic environment, surpasses these opposing paternal models to flirt with a musical discourse that is considerably more artsy, modern, and free-spirited. But first it is worth noting that the association of the two adult males with specific musical spheres links *The 5,000 Fingers* to a plot type employed in Hollywood since the advent of talking pictures. Again, *The Jazz Singer* forms a foundational text. The title character's dilemma is that he must choose between two musical worlds: between traditional religion and entertainment. Bart and Mrs. Collins seem to face a similar dilemma, as they must choose between Dr. T's classical piano and the plumber's popular song. In *The Hollywood Musical,* Feuer observes that the opposition of elite and popular art forms was a standard plot device during the studios' heyday (54–60). This opposition took several forms, as in the "opera versus swing" narrative of *Babes in Arms* (Busby Berkeley, USA, 1939), or the "ballet versus tap" story line of *Shall We Dance* (Mark Sandrich, USA, 1937). Regardless of the particulars, however, the Hollywood musical valorized spontaneity, directness, and its own music—popular music—in virtually every instance.

It is not only in Hollywood that one finds immediacy and accessibility idealized at the expense of other kinds of musical behaviors. This is also an unexamined premise of certain recent writings about music, especially those that describe popular and non-Western musics. Christopher Small's *Music of the Common Tongue* is one such text. Small's intention, and it is not an unadmirable one, is to advocate for African American music, or what he terms music of the "vernacular." Unfortunately, however, some of his ideas about that music seem to have been formed primarily as a reaction to what he perceives as lacking in the music with which he grew up, European art music. He confesses that, although he was a latecomer to African American–based popular music, "the music of this tradition fulfilled in me not only an emotional but also an intellectual and social need which European classical music, however much I loved and admired much of it, did not, and if I was honest, never had fulfilled" (3). According to Small, classical music has become elitist and stagnant, whereas vernacular music has remained nonexclusionary and vital. Classical music may persist into the present because of its connections to privilege, but African American music has endured because of its "openness to development, its universal accessibility and the ability of its musicians to evade capture by the 'official' values of the industrial state" (6).

Not surprisingly, many of the qualities that Small finds desirable in "vernacular" music correspond to those that Feuer describes as characteristic of the "swing" side of the "opera versus swing" narrative: spontaneity, directness, informality, and the rejection of professionalism (*The Hollywood Musical* 54). This is not to imply that Small was led to his "conversion" by watching Hollywood musicals; if it seems as though this is the case, that is only because such a conversion is culturally overdetermined. Despite Small's disdain for "the official values of the industrial state," it is precisely those values that produce in the subjects of the modern state the desire for a compensatory aesthetic realm. Jochen Schulte-Sasse has observed that writers in the early eighteenth century already realized that art had the potential to ameliorate, at least temporarily, the ill effects of the Industrial Revolution and the division of labor. Schulte-Sasse summarizes this view in terms of the relative autonomy of artistic endeavors: "Art that is distinguished, indeed liberated, from technological utilization is radically opposed to quantifiable 'wage-labor,' since its functional autonomy guarantees compensatory 'satisfactions'" (38). For Small, the problem with classical music today is that it fails to provide these compensatory satisfactions, and because of that lack he has abandoned it and embraced another kind of music.

Small uses the term *vernacular* to "remind us that the ability to take part in a musical performance is as natural and universal a part of the human endowment as is the ability to take part in a conversation" (7). Again, what Small views as a truth about the human condition—that conversation is natural and universal—is actually a historically produced effect. As Jürgen Habermas has shown, the ideal that anyone, regardless of social status, could take part in a conversation openly and freely was forged in the coffeehouses, salons, and various "societies" of the eighteenth and early nineteenth centuries. The novelty of these institutions was that within them, "the mind was no longer in the service of a patron; 'opinion' became emancipated from the bonds of economic dependence. . . . [The] intent was that in such manner an equality and association among persons of unequal social status might be brought about" (33–34). Gradually, and despite resistance, this ideal became a decisive force in the development of modern, enlightened society. Small fails to acknowledge that his desire for a vernacular music that flows as easily as conversation is predicated on his participation in a culture that has as one of its official, or at least constitutive, values the possibility of free, open conversation between equals.

Accordingly, the conventional Hollywood musical also idealizes a form of music that resembles the seemingly unrestricted outpourings of speech, the popular song. Just as the hallmark of a distinctively American style of choreography is those scenes in which ordinary walking and moving about are imperceptibly transformed into dancing, the epitome of musical presentation in these films occurs when talking seems to slide into song. Fred Astaire was a master of such delivery, of course, but as Jane Feuer has observed, "In musical comedy ordinary speech and music are linked each time a song is born out of a scene of spoken dialogue" (*The Hollywood Musical* 53). We might note that this link is also valorized by Small, who claims that "in many if not most of the world's societies . . . talking, singing and even dancing may flow into one another as elements of daily social intercourse" (52). If Small admires this lack of boundaries between speech and song because it seems to have liberating potential, we should recall that the modern subject's experience of free speech as a revolutionary force is the result of a particular historical situation.

Feuer has shown that the typical Hollywood musical idealizes singing for a conservative, rather than revolutionary, reason. By redefining the entire field of music as popular song, the musical celebrates its own kind of music, which is essentially commercial (*The Hollywood Musical* 51–54). To accomplish this, the plots of these pictures often focus on "putting on a show," or "the business," permitting ample opportunity for self-reflexive gestures about the attractiveness of various accessible, jazz-inflected styles. And even in films where the positive values of "classical" music (such as respectability and longevity) are stressed, whatever style is associated with the "popular" always wins when characters must choose between different kinds of music (56). In the next section, we will see that a small number of musicals, including Gene Kelly's films of the early 1950s, attempted to transcend the conventional binary opposition of elite and popular arts. Like these films, *The 5,000 Fingers* tries to reconcile this opposition by incorporating stunning "set pieces" that draw on a wide range of influences. The startling revelation of an alternative path to music making in the depths of the Happy Fingers Institute proves a key development in Bart's Oedipal journey.

## "I Don't Think the Piano's My Instrument!"

Just as Bart stages the encounter of his mother and chosen father in "Get Together Weather" to satisfy her (presumed) desire, he stages his own musical aspirations by transforming elements of Dr. T's expertise into something distinctive and original. This is manifested in the most astonishing sequence in the film, a six-minute ballet in Dr. T's dungeon. Again, the notion that "the unconscious is the discourse of the Other" is illustrated through a musical number. If the particular form of Bart's desire originated with his mother, the content of that desire is elaborated through his piano teacher, representative (due to the real father's absence) of the Lacanian Name-of-the-Father. In the ballet, Bart assimilates what Dr. T has rejected in waking life—all instruments that are not the piano—into his dreamworld. He thus exhibits the capacity to turn what the "bad" father considers negative into a positive. Several brief shots of Bart watching this fantastic number remind us that he is a character in his own dream, staging his desire to find a place in the musico-symbolic order.

Bart's seemingly fortuitous stumble into that place—he is chased there by Dr. T's guards—is preceded by a scene that requires his manipulation of the sonic environment of the Happy Fingers Institute. In an act of bravery and defiance, Bart steals the key to Dr. T's treasure chest by replicating with his pocketknife the steady beat of the metronome in which it is secured. Notably, there is no difference in timbre between the knife and the metronome: not only does this ensure the doctor's slumber is undisturbed, but it also seamlessly positions the audience as sharing Bart's aural perspective, no matter how fantastic. The timbral consistency of the two objects functions in a manner analogous to the loudness of "point of audition" sound. According to Rick Altman, such sound "carries with it signs of its own fictional audition." Typically, a point of audition is conveyed through manipulation of volume in order to represent the aural perspective of a particular character in the narrative. The effect is to "lur[e] the listener into the diegesis not at the point of enunciation of the sound but at the point of its audition" ("Sound Space" 60). In the scene just described, the operative element is timbre rather than volume. In hearing the sound color of metronome and knife as identical, we (and the sleeping Dr. T) are positioned to hear the world through Bart's ears.

Emboldened, Bart deliberately risks waking the doctor by smashing the glass box that holds Mr. Zabladowski's execution order. He is chased to a part of the dungeon where all the "nonpiano players" are held, the "scratchy violins, screechy piccolos, nauseating trumpets" that Dr. Terwilliker disparages in the framing story. But these performers are no ordinary instrumentalists. From his hiding place, Bart watches while half-clad men in ghoulish green body makeup and tattered tuxedos play section upon section of oversized, impossibly curved, fanciful music-making devices. The instruments are pure Dr. Seuss, the dance an extreme vision of Hollywood's "prop number." More than twenty different types of music makers are seen: trombones with bells shaped like Victrola horns, a treelike headpiece with dangling cowbells, tubas wrapped from knee to shoulder, a gigantic marimba played by five men wearing pastel boxing gloves, dress manikins converted into double basses, radiators, bicycle horns, a tire pump, fireplace bellows, etcetera, etcetera. What kind of music could they possibly make?

The bewildering fact is that they do make music, coming together in a stylistically eclectic manner that is by turns jazzy, late romantic, and European village dance. The allusions in the score range from sailor's hornpipes, brass fanfares, and ethnic polkas to the sophisticated big band arrangements of Duke Ellington and the lush strings of Tchaikovsky. A pianist himself, the classically trained Hollander took full advantage of the tradition of orchestration liberated from the keyboard that had begun with Berlioz, passed through Richard Strauss and Stravinsky, and reached new heights in the jazz-inspired compositions of Milhaud and Gershwin.[10] Hollander's brief description of his work on the film shows that he found inspiration in the manic sounds and wild colors of Ted Geisel's sketches and lyrics, from the "Tick-Tick-Tick . . . Eine-Zwei-Dreie, dideldadeldudel" of assorted timekeepers to the parody of the tonal system itself in the "Dressing Song" (in which "Do-re-mi" becomes "Do-mi-do duds") (Hollaender 391–95).

The centrality of the dungeon ballet to the film is attested by the extraordinary effort that went into its development and retention. Eugene Loring, who is generally regarded as one of the founders of a distinctively American approach to ballet, considered this scene one of his finest achievements (Delamater 104, 233). His participation in *The 5,000 Fingers* was unusually extensive for a choreographer. Although he did not write the actual script, "I was in the story conferences so that I could make the scene take the plot along, instead of just having a number," he recollected (Delamater 233). Loring was especially pleased that Roy Rowland had allowed him to shoot the dungeon ballet, as he had planned not just the movements but how they could best be filmed (229). He was also grateful to Stanley Kramer for not abandoning the number when it became apparent that his complicated plan to use 64 dancers appearing as 104 on different levels with "those crazy instruments" would be extremely expensive to film. After Rowland made Kramer aware of Loring's innovative ideas, the producer raised the extra fifty thousand dollars needed for the number's completion (233).

This scene is not "ballet" in any strict sense, of course. As Delamater has pointed out, the term "is often applied to those rather longer dance numbers—irrespective of the type of dance used in them—not usually introduced by or associated with a specific song" (103). It is worth recounting how this extended notion of "ballet" developed in order to understand the cultural context of the scene in *The 5,000 Fingers*. Peter Wollen has noted that there was something of a fad for classically influenced dance toward the end of the Hollywood studio period. In his study of *Singin' in the Rain,* Wollen outlines a trajectory for ballet from *Oklahoma!* (1943) on Broadway to the film *An American in Paris* (Vincente Minnelli, USA, 1951). Eugene Loring contributed to this trajectory in several important ways. A classically trained dancer, he pioneered a new approach to traditional ballet, specifically a style of choreography "dealing with American subjects and dramatizing the dancing" and incorporating elements of both jazz and modern dance (Wollen, *Singin' in the Rain* 13). Loring's best-known work for the concert hall from this period is *Billy the Kid* (1938), with music by Aaron Copland.

The first stage musical to successfully incorporate the new type of ballet was *Oklahoma!* and the year after it opened more than half the musicals produced on Broadway "had some kind of ballet number" (35). In response, Hollywood devised numerous ways to work ballet into narratives, often as daydream or fantasy. Loring was hired by MGM's Freed Unit as a choreographer for *Yolanda and the Thief* (Vincente Minnelli, USA, 1945), and his innovative approach to depicting dance on-screen influenced other dancer-choreographers, including Gene Kelly. A further impetus to incorporating ballet into film came in 1949, when Michael Powell's *Red Shoes* (UK) offered stunning examples of dance choreographed specifically for the camera's eye and dramatically integrated into the story. The picture's success in the United States demonstrated that audiences were interested in ballet, and Kelly, who persistently sought ways to extend his tap-based character dancing with classical technique, seized on this opportunity. According to Wollen, the seventeen-minute ballet in *An American in Paris* was a direct response to a set piece of the same length in *The Red Shoes* (40). And Kelly was determined to incorporate a ballet into *Singin' in the Rain* (Gene Kelly and Stanley Donen, USA, 1952), even if it meant adding yet another tangent (and a third female partner) to an already convoluted plot.

*The 5,000 Fingers* was made at virtually the same time as Kelly's two pictures, and it is in the context of their eclectic mix of different artistic worlds—traditional ballet, modern, jazz, and tap—while at the same time producing something that is distinctively cinematic that the dungeon ballet originated.[11] What may seem today like a confusing mixture of high-brow pretense and childish silliness, the ballet in Dr. T's lower dungeon has its source in a broader dialectical movement to enliven classical technique through the inclusion of vernacular elements and to complicate popular entertainment through traditional vocabularies. Hollander's instrumental score is similarly eclectic, drawing on the resources of the romantic orchestra, the jazz band, and urban experience (hence the prominence of the accordion). It is episodic and quixotic, with unexpected turns amusingly related to the exhibition of Dr. Seuss's chimerical instruments.

The dungeon scene is, in a sense, the quintessential "dream ballet," a dream within a dream that displays the subject's deepest wishes, his quest to shape a distinctive musical "voice" out of the disparate elements of waking life (Wollen, *Singin' in the Rain* 35). Made audible through the richness and variety of Hollander's composition, Bart's musical imagination partakes of the Seussian qualities of cleverness, artifice, and what might be called a "vernacular modernism," drawn freely from the musics and noises of contemporary life. By integrating whistles, bells, and canvas sneakers into an orchestral ballet, the dungeon scene offers a striking instance of American cinema as an aesthetic praxis that "articulated and mediated the experience of modernity" (Hansen, "Mass Production of the Senses" 60). As Miriam Hansen has observed, "Hollywood did not just circulate images and sounds, it produced and globalized a new sensorium; it constituted, or tried to constitute, new subjectivities and subjects" (70). The fact that the dungeon is a lyric-free zone is especially significant, as it marks a difference between Bart's unfettered imagination and the popular song associated primarily with Mr. Zabladowski.

Bart's sonic authority reaches a new level when he requires the plumber to take an oath to help in the struggle against Dr. T. The shared and elaborate speech act between Bart and Mr. Zabladowski, parodying the Boy Scout oath, both resolves Bart's Oedipal quest and enables Mr. Zabladowski to take action. "Don't you know? This makes you my old man!" Bart tells the plumber. The two seal their bond in blood, by the cutting of fingers. Their wounds will be the proof of their shared adventure—the answer of the Real—when Bart awakens later, safe at home.

Immediately upon taking his vows, "Pop" attempts to rescue Mrs. Collins and vanquish Dr. T's minions. But he and Bart are captured, and in keeping with the narcissism of a child's dream, it is Bart who will be the true hero. He convinces Mr. Zabladowski that they can transform the latter's bottle of "Air Fix," which eliminates odors, to a "Music Fix," which will prevent Dr. T's grand opening recital from proceeding. After several failed attempts to create a sound-catching device from the contents of Bart's pockets, the two realize they need some "sound equipment." Luckily, they notice that the guard snoozing nearby is wearing a hearing aid. "If it brings noises into his ear, why couldn't it bring noises into our bottle?" reasons Bart. The addition of the guard's sound-collection device works so well that the Music Fix becomes explosive; it is both sonic and atomic.

The manufacture of a device that puts ultimate, deadly power on the side of justice seems to simultaneously reassure cold war audiences of the legitimacy of the U.S. pursuit of nuclear weapons and mock the reductiveness of such simplistic ideas. But Bart has other powerful weapons in his arsenal. Much as he had contained his antagonistic feelings toward his mother by caging her, he neutralizes Dr. T by reducing the latter to a figure of ridicule. This is played out in the other showstopper of the film, the campy "Dressing Song." Like the dungeon ballet, the "Dressing Song" is a homosocial piece that revels in the male figure. This time, however, the figure is comic. Stripped down to a T-shirt and boxer shorts, Dr. T sings, "Come on and dress me, dress me, dress me in my do-mi-do duds" to a fawning, liveried male chorus line. This "Mephisto in velvet gown," who uses the language of solfège to prattle, "I want my undulating undies with the maribou frills," is destined to lose his dictatorial authority over the institute (Hollaender 393).

The psychoanalytic principle at play in this number seems to be Lacan's thesis that "the superego is the imperative of jouissance—Enjoy!" ("On Jouissance" 3). The "Dressing Song" is the silly mirror image of "Ten Happy Fingers," a voyeuristic but delightful indulgence that contrasts wildly with the simple, didactic tune of Dr. T's serious public persona. For much of the film, "Ten Happy Fingers"—Dr. T's leitmotif—resounds through the underscoring, permeating the institute, conflating diegetic and nondiegetic space. Like the song "Brazil" in Terry Gilliam's film of the same name (1985), its persistent rhythm "serves as a support for the totalitarian order" (Zizek, *Looking Awry* 128). As a product of Bart's imagination, the "Dressing Song" allows the boy to challenge Dr. T's fascistic tendencies by revealing the piano teacher's propensity for "idiotic enjoyment." The pleasure of Dr. T's dressing room is the non-sense of the totalitarian order, of the authority figure in his most ridiculous moment.

The composer, Frederick Hollander, recognized the terrifying aspects of Dr. Terwilliker, describing him as a "schreckliche tyrannklavier" (frightening piano tyrant) and associating him with Hitler in the nickname "Etüdenadolf."[12] But Hollander teasingly admonished his readers that punishment is to be expected when young pianists refuse the classical approach to music. "So geht's, wenn man seinen Clementi nicht kann" (That's how it is, when one can't play one's Clementi, 394).

A long-standing fascination with the relationship between terror and the imagination is seen in the following anecdote from Dr. Seuss's biography. Late in his career, Ted Geisel was asked if Goethe's "Erlkönig" had influenced his 1937 children's book, *And to Think That I Saw It on Mulberry Street*. Although the question surprised him, he acknowledged having memorized the poem in German—the language of his childhood—during high school (Morgan and Morgan 277). A ballad set most famously by Schubert, "Erlkönig" is an Oedipal tale of two males' competition over a young boy. It is also a meditation on the ambiguous nature of terror. The boy is feverish, frightened, and seriously ill, and the listener is never certain whether the malevolent figure he describes to his concerned, frantic father is real or imagined. The story ends with the boy's death. As we will see, Ted Geisel's association of the piano with terror went well beyond average experience.

The climax of *The 5,000 Fingers* occurs when the five hundred boys, seated at the two-story keyboard, lift their hands to play "Ten Happy Fingers," under Dr. T's baton. His count-off, "uh one and uh two . . . ," is interrupted by the repetition and alteration of his voice through postproduction effects. It is as if the "Music Fix" has seized Dr. T's voice, without which he can neither coordinate the boys' performance nor order his guards to seize Bart. Dr. T is rendered powerless as the institute resounds with clanks, pops, and other cartoon effects. The villains scatter when they realize the Music Fix is atomic, and, finally, Dr. T begs Bart for mercy and agrees to free everyone: mother, father, all the boys. But rather than running from the keyboard, Bart picks up the baton Dr. T has dropped in defeat. Stepping into the doctor's place, Bart conducts the boys in "the most beautiful song ever written": "Chopsticks." This is no mere child's play, however. What we hear is a circuslike fantasy keyboard, shrill treble with dissonant clusters in the bass, a tour de force of musical leadership. Soon there are gurgles, screams, and fireworks as the Music Fix explodes in a predictably phallic column of smoke, waking Bart from his dream. Maybe the piano wasn't Bart's instrument, but he does have something musical to say, after all.

## Coda

Shortly before Ted Geisel interested Kramer in his "vicious satire" of piano teachers, he had developed the character of Gerald McBoing-Boing, a little boy whose "speech" consists of sound effects rather than words (Morgan 129–33). Gerald's story was elaborated in two works, including a picture book telling Gerald's back-story and a cartoon, *Gerald McBoing-Boing* (Robert Cannon, USA, 1950), which won an Academy Award for Best Animation in 1951. What is pertinent about Gerald's story here is the preposterous dissociation between the human body and the sounds it produces. Instead of ordinary speech, Gerald's utterances consist of the imitation of nonlinguistic sounds: train whistles, explosions, sirens, and so forth. Incredibly, he manages to turn this peculiarity to an advantage through employment at a radio station. However, that cannot change Gerald's isolation, his fundamental difference from everyone else in his world.[13]

As the chaotic scenarios of the *Cat in the Hat* series have come to epitomize, Dr. Seuss was not particularly mindful of ordinary boundaries. In addition to the crazy settings and whimsical drawings, the genius of those early readers is their delight in the sounds of language. First published in the mid-1950s, they were controversial, for many years, for privileging rhyme over reason. Yet their lasting power, like Gerald's and Bart's, is their expression of the individual's attempt to wrest meaning from sound, to obtain mastery over self and environment, to become whole from disparate parts.

The intensity of Geisel's relationship to sound is everywhere apparent in his work. It is therefore not surprising that he had an early experience that linked the body, the family, music, and language in a tight nexus. Long before he was forced to take piano lessons himself, Geisel associated music with separation and death. Like many German American families, the Geisels often made music together at home. Father was a baritone, Mother accompanied on the piano, and they spent much time in the music room. But shortly before he was four, Ted's eighteen-month-old sister died of pneumonia. It was a difficult experience for the family, and somehow the young Geisel conflated his sister's casket with a cabinet of Caruso records stored in the music room. "No matter how thrilled I was later by my father's voice and my mother's accompaniment, I always saw Henrietta in her casket in the place where the [music] cabinet was" (Morgan and Morgan 9). Perhaps without realizing it, Geisel revealed something of the play of symbols in his imagination when he described Bart's dilemma in a memo to Kramer: "The kid, psychologically, is in a box" (quoted in Jenkins, "No Matter How Small" 201).

Instruments, voices, the anxiety of loss, separation, and death: what began with *The 5,000 Fingers* as Geisel's attempt to get even with the piano teacher of his teenage years turned into something beyond his control (Spoto 150). He was unhappy with the film's outcome but never explained why. "As to who was most responsible for this debaculous fiasco, I will have nothing more to say until all the participants have passed away, including myself," he declared (Morgan 138).

That moment is now behind us, and the answer is still murky. As this essay has shown, the film merged Freudian and musical narratives in its witty and artful portrayal of the intensely personal nature of musical experience. The result was hardly recognizable as a conventional Hollywood musical. *The 5,000 Fingers* was ahead of its time in many ways, and more than merits rehearing. Just as the film's sets, costumes, and dance hail a viewer in love with the image, the use of music, language, and effects hails a listener in love with sound.

## Notes

1. The film was released several years before Theodor Seuss Geisel initiated the *Cat in the Hat* series for which he is best known today. Beginning in 1937, he had published elaborate children's books that balanced text and illustration in fairy-tale settings but served no obvious moral or didactic purpose, a radical concept in children's literature (Morgan and Morgan 81–84).

2. The film cost $1.6 million to produce, but grossed only $250,000. Kramer describes his difficulties with Columbia studio head Harry Cohn in his autobiography. According to child actor Tom Rettig, Cohn decided to withdraw the film's promotional budget soon after the picture opened. Following a highly publicized New York premiere, what was supposed to have been a nationwide tour of the cast was abruptly canceled by Columbia. "The story that went around was Stanley Kramer and Harry Cohn were having their famous fight, and Cohn pulled all support from every project that was Kramer's at that point" (Rettig quoted in Chusid).

3. Rettig is best remembered as Jeff in the television series *Lassie* (1954–57).

4. Ted Geisel's sketches were the basis for Rudolph Sternad's set designs, merging absurdist cartooning with German expressionism, all spectacularly realized. The keyboard featured in Bart's dream, for example, was one of the largest sets ever built at Columbia, filling two soundstages. On this and the lighting innovations of Frank (Franz) Planer, see "Facts about 'The 5000 Fingers of Dr. T,'" *American Cinematographer* (January 1953).

5. The eroticism of the plumber's concealment in the kitchen would have resonated for an audience knowledgeable of Healy and Hayes's marriage a few years before. The couple were well known as a comedy team on radio (Kramer 107).

6. Apparently, this surprisingly postmodern technique was employed in a last-minute attempt to reconfigure the picture, which was considered too long for audiences ("Bill's Tribute").

7. Although Tom Rettig appears to sing on-screen, his songs were overdubbed by boy soprano Tony Butala.

8. See Silverman's critique of Kristeva's concept of the "chora," which the former views as reducing the mother to the status of infancy (*The Acoustic Mirror* 101–40). Interestingly, the cells in which the boys are held in Dr. T's institute have the suggestive ovoid shape often seen in Dr. Seuss's illustrations. See, for example, *The 500 Hats of Bartholomew Cubbins,* a book whose title resonates with the film in both number and name of the main character.

9. For an insightful discussion of the mother's dual position in the imaginary and the symbolic order (and Lacan's equivocations on woman's relationship to the phallus), see Silverman, *Subject of Semiotics* 185–93.

10. Interestingly, Hollander's career as a film composer began when a young actress asked him to provide the piano accompaniment at her audition for *Der blaue Engel* (The Blue Angel) (Joseph von Sternberg, Germany, 1930). The part went to another actress, Marlene Dietrich, but Hollander was asked to stay on as musical director, resulting in the tune for which he is best known, "Ich bin von Kopf bis Fuss" (Falling in Love Again).

11. The films share other similarities that could be explored elsewhere. For example, both *The 5,000 Fingers* and *Singin' in the Rain* presented audiences with reassuring images of "sound-collection devices" in the right hands (that is, Music Fixes and microphones) at a time of rapid innovation in such technologies.

12. Henry Jenkins has argued that Ted Geisel characterized Dr. T as "the reincarnation of der Führer" in his initial conception of the story, and numerous visual and narrative associations with Hitler are apparent in the finished film ("No Matter How Small" 200–203).

13. See sound designer Walter Murch's recollection that he was nicknamed McBoing-Boing as a child, after the cartoon was released (Ondaatje 6).

# 10

# Paul Sharits's Cinematics of Sound

## MELISSA RAGONA

The material performativity of film—its ability to recode its own record-ing process—was at the center of Paul Sharits's work, as well as that of other experi-mental filmmakers during the heyday of 1960s and '70s structuralist filmmaking. Sharits, along with his contemporaries Tony Conrad (*The Flicker*, USA, 1965), Hollis Frampton (*Zorn's Lemma*, USA, 1970; *Critical Mass*, USA, 1971; *Mindfall*, USA, 1977–80), and Michael Snow (*Wavelength*, Canada, 1967), used sound, in particular, to generate new translations of the cinematic frame. Posing a chal-lenge to the epistemology of silence set up by filmmakers such as Stan Brakhage and Andrew Noren—which underlined the anticommercial force of silent films; critiqued notions of synchronous, "realist" sound; and emphasized the purity of visual forms—Sharits and his peers used the gritty materiality of film to amplify the incongruous as well as contiguous relations between sound and image.[1]

Sharits's theory of sound and hearing, which he referred to as "cinematics," was deeply informed by both the integrated serialism that began its genealogy in film through Austrian filmmaker Peter Kubelka as well as his own work in the textuality of Fluxus-based practices. But whereas language and text are central to the way Sharits rethought the silent image-based models of American avant-garde film, he used these systems not as linguistically determined forms but rather as open systems that had more of a relationship to conceptual mathematics than the prevailing psychoanalytic and semiotic readings of film during this period. This essay is an attempt to rethink the research that has been done around structural film in terms of filmic self-reference and apparatus theory, with its focus on the primarily visual effects of such a cinema. I propose that a rethinking of this work through sound will shed new light on how Sharits was attempting to radically alter the temporal and, thus, epistemological definitions of film. By examining part of Sharits's curriculum for cinematics—which calls for a close study of the relations between mathematics and language through Wittgenstein, as well as sound and composition via Iannis Xenakis—I reveal how Sharits began to sense the limitations of a purely visual single-screen cinema. Toward the latter part of his career, he gradually moved his work into multilevel experiences, allowing him to experiment with audio and visual arrangements on equal terms.

In part, the work of Austrian filmmaker Peter Kubelka informed Sharits's turn toward sound. Kubelka's understanding of film sound was heavily influenced by modern music, especially "Second Viennese School" composers Arnold Schoenberg and Anton Webern. These composers tended to rely on abstract, sometimes-mathematical relationships between notes and to compress gestural changes within a composition. That is, rather than a theme being slowly elaborated, gradually going through its several musical variations, a composer such as Webern (especially) would reduce the theme to its absolute briefest, densest expression. Then, he would subject it to various permutations in as condensed a manner as possible. This resulted in rather jagged, intellectually demanding atonal compositions. Kubelka borrowed heavily from Webern in terms of how he thought about both visual and sound montage in his films.

Kubelka's film *Arnulf Rainer* (Austria, 1958–60) is often cited as the first flicker film (preceding Conrad's and Sharits's work with flicker effects), composed entirely of alternating stretches of black-and-white leader, varying in length from twenty-four seconds to the brief duration of a single frame. This film has also become emblematic of what Kubelka calls the "metric film," whereby every part of a film is precisely measured into equivocal elements that are then set in relation to each other as a compositional whole.[2] Or, as Kubelka states, "every one of the chosen elements meets every other one" (148). Kubelka applied this notion of metrics not only to the visual segments of film but to film sound as well, giving sound equal valence in relation to the visual. Although Kubelka's application of serialism to film inspired the audio experiments of Sharits, Snow, and Conrad, it was engaged in a more modernist vocabulary than his successors.

Tony Conrad makes clear that his soundtrack for *The Flicker* (often compared to Kubelka's *Arnulf Rainer*) differs from Kubelka's in terms of both aesthetic approach as well as technical application: "*Arnulf Rainer* is a completely different kind of work from *The Flicker*. Kubelka's effort is recuperative of a constructivist ethos very much akin to that of Anton Webern, while my film—though unhappily labeled 'structuralist' by P. Adams Sitney—is much more rooted in the surrealist tradition: it has transparency, activation of the unconscious, and effects a radical endorsement of individual subjectivity" (Conrad, *OxFF* n.p.).[3]

Where Kubelka utilized alternating patterns of silence and white noise[4] to create the soundtrack for *Arnulf Rainer,* Conrad was interested in "pulse trains whose frequencies would fall into the perceptual region lying between rhythm and pitch" (n.p.).[5] Likewise, Sharits was interested in sound concepts such as phasing and overtonality—the after- or supraeffects of rhythmic structures.[6] In contrast, Kubelka grounded the sound work for his early films in the "natural rhythms" of organic life (the body, nature). For example, Kubelka (and his critics) often compared the soundtrack from *Arnulf Rainer* to the undulating waves of thunder and lightning: "I can exactly place these events up to 24 times per second. A meeting of light and sound, thunder and lightning" (157). So whereas Kubelka set the stage for experimental film sound design, examining sound seriously as its own object, Conrad's and Sharits's innovations moved sound out of its purely rhythmic goals and into the realm of nonrepresentational, atemporal arrangements.

Still, Sharits's sound experiments must be thought of in relation to Kubelka's pioneering work in terms of serialist-inspired compositional forms (*Adebar,* Austria, 1957; *Arnulf Rainer*), as well as his articulation of "sync events" (*Unsere Afrikareise* [Our Trip to Africa], Austria, 1966) in which one can separate sound and image, and combine them in new ways, to create perceptual "events" that can exist only through composition in film (for example, the sound of a gunshot and the image of woman's hat blowing off). Sharits's involvement in Fluxus during the sixties extended his interests in compositional practices inspired by the integrated serialism of 1950s Darmstadt school composers such as Pierre Boulez, Karlheinz Stockhausen, and Iannis Xenakis. The text-based practices or "event scores" of Fluxus, in which short-form performance instructions were written as scripts or served as conceptual tools for different kinds of activities (musical, performance or action-based, cinematic), had their origins in the aleatory and indeterminate compositional work of John Cage. In turn, Cage's compositional work had an intimate relationship to writing and reading—two practices that became part of his method for measuring durational time in music. Sharits's most emblematic Fluxus works are *Word Movie* (where reading replaces in a sense the experience of viewing) and a series of silent shorts (ranging from ten to thirty seconds each): *Unrolling Event, Wrist Trick, Dots 1 & 2,* and *Sears Catalogue* (all USA, 1966). These Fluxus works were all to some degree filmic scores, already taking film up off the screen and into more performative, event-based experiences.

While involved in Fluxus film events during the mid-1960s, Sharits also constructed multiple-screen installation pieces that he referred to as "locational pieces." His interest in installing film coincides with his interest in the "spatiality of music" ("Hearing" 74). He had already begun to make comparisons between the kinds of aural overtonality achieved in postserial compositions (that is, in the work of Xenakis, La Monte Young, Steve Reich, Conrad) and the complex visual chords, the fades and dissolves of color and light, his flicker films created.

It was not until his work in film installation, however, that he felt he could come close to achieving the "compositional dimensionality" of music and sound. "My early 'flicker' films—wherein clusters of differentiated single frames of solid color can appear to almost blend or, each frame insisting upon its discreteness, can appear to aggressively vibrate—are filled with attempts to allow vision to function in ways usually particular to hearing" (70). Sharits's early work is very much focused on studies of the film frame: they characterize what he terms a period in which he was "tormented by the implications of film as a physical strip" ("Postscript" 2). *Ray Gun Virus* (USA, 1966), *Piece Mandala/End War* (USA, 1966), *Word Movie [Fluxfilm 29]* (USA, 1966), *Razor Blades* (USA, 1968), *N: O: T: H: I: N: G* (USA, 1968), and *T, O, U, C, H, I, N, G* (USA, 1968) all take up the level of the frame through the strategy of flicker. Inspired by Kubelka, Sharits utilizes a "metronomic beat" of pulses to structure these films (Krauss 95). This particular microstudy of the physical nature of film, a kind of "avant apparatus theory," also informs the way he thinks about sound. For example, in *Ray Gun Virus,* he uses the sprockets of the film itself to generate sound. Their metered order mark the film's passage through the projector, as well as the duration of frames, to produce an aggressive, relentless, driving soundtrack. In fact, viewers are instructed to listen to the soundtrack full volume, thus assaulting them at both perceptual as well as visceral levels.

The primacy of the auditory in these works is key to understanding the paradigm shift Sharits makes from a film language inspired by linguistic models of the visual to one shaped by the temporal dimensions of film's recording technology. Rather than analyze predetermined "elements" or independent units of a film as a linguistic model of film analysis might suggest, Sharits encourages us to discover a film's grammar by letting the permutating parts "rub against each other" ("Words" 33). By following his trajectory from *Word Movie* to *Synchronousoundtracks* (USA, 1973–74), it is possible to map how Sharits's notion of cinematics not only attempted to free cinema from anything beyond itself but also provided us with a way to think beyond representational conventions of the soundtrack as accompaniment to the image. Indeed, Sharits suggests that sound could be an analogue to vision, commenting both on an image's function in time as well as its confined position within the frame.

"Cinematics," as a term, first appeared in Sharits's "Words Per Page" (*Afterimage,* 1972), where he suggested that *cinematics* replace *cinema.* He was especially interested in critiquing what he called the more "muddled" metaphoric terms such as *language* or *grammar* of film ("Words" 32). For Sharits, cinematics was a "structural-informational system," a very specific and special kind of conceptual art, growing out of the minimalist sensibility that preceded it in painting, sculpture, and music (33). Cinematics shares minimalist concerns with objecthood in which a work exhibits particular identifiable elements or preoccupations such as the intensification of material, equal division of parts, use of serial or non-hierarchical systems of image and sound organization, and chance, random, or numerical ordering. Sharits speaks of a system of "self-reference" as exemplified in the works of Frank Stella and Jasper Johns, in which the former points to the edge or frame of painting, the latter to the iconic, reflexive asymmetry between picture and painting. Both Stan Brakhage, with his emphasis on the disjunctive, distractive glitches (blurs, splices, flares, frame lines, flash frames), and Andy Warhol, with his focus on the durational material process of recording-projecting, make the viewer conscious by calling attention to film grain, scratches, and dirt particles, and thus, to the sense of the flow of the celluloid strip. Similar to what Stella and Johns suggested for the temporal position of painting, Brakhage and Warhol suggest for the "shape" and "edge" of film. Warhol, for instance, in his early "static" films disregarded the normative idea that a film is composed of parts and that its duration or time scale is the sum of those heterogeneous parts. In step with Warhol, Sharits made it apparent that the "edge" of film "can be generated by, rather than arbitrarily contain the internal structure of film" (38).

Theories of structuralist film arose out of the vocabulary of this self-reflexive, material-centered work and found its most forceful articulation in the theoretical work of film historian P. Adams Sitney, who identifies the key aspects of structuralist film as fixed camera or frame position, flicker effect, loop printing, and rephotography off the screen (348). Propagated largely by the work of Christian Metz, semiotic models of film captured the imagination of filmmakers and critics during the 1970s. These models were repugnant to Sitney but embraced (to a certain extent) by Sharits as important elements of a cinematics curriculum. Although Sharits saw parallels between film's discrete structural elements and the structural elements of language, such as morphemes and phonemes, he also felt that cinema, as language, had been greatly reduced by narrow applications of linguistics to predetermined units of film's structure. Questions such as "What constitutes a shot?" or "What codes are implied in the zoom function?" were not deep cinematic inquiries for Sharits. Rather, the grain, the frame and its duration, the shutter and its rotation, and other infrastructural units of information lay at the heart of film's signification and meaning ("Cinema as Cognition" 78). It was at the level of the phonemic or micromaterial at which larger decisions about filmic language would be made. This material, he argued, could not be accessed by looking at the system one part at a time as in montage theory, but by allowing the microscopic cinematic elements to "come up against each other," or, as I mentioned earlier, to *rub* against each other in time. The macroscopic fabric of film would thus emerge: "Flaws reveal the fabric and 'cinematics' the art of the cinema's fabric" ("Words" 33).

For Sharits, structural linguistics, as a science of language, was more exact-ing than film language or grammar, but it remained a limited model for a truly transformative cinematics. Phenomenology, psychophysiology, cybernetics, mu-sic theory, and mathematics were equally as important (if not more so) as the study of structural linguistics. Sharits was especially interested in the philosophy of mathematics and its relation to language, in particular through the work of Wittgenstein and postserial composer Iannis Xenakis. After he exhausted the mandala form in flicker films (1965–68), which reflected his interests in "time and space symmetries" as well as the chance operations of "nature," Sharits turned to the study of abstract algebra and classical and postserialist music. It was during this time that he began looking to the "actual materials and processes of [his] medium"—the flat screen and frame, the three-dimensional symmetry of the projection beam (its conic shape), the grain of emulsion and film as a "time-ribbon." He realized that the mandala films were in a sense psychophilosophical "traps" that implied a false sense of the "nature of reality" and were imbedded with clichés of "dynamic equilibrium," "order," "symmetries" ("-UR(i)N(ul)LS" 13). By taking the science of linguistics and thinking of it in terms of its relation to mathematics via Wittgenstein, and, later Xenakis, Sharits was able to develop a theory of sound and image that would express his interest in studying the microstructures of grain, phoneme, and scratch, as well as the macrostructures of wave and projection in film. The relationship between chance and structure is a key trope in Sharits's work, which can be traced from his experiments with changing the internal structures of words at the level of the signifier, as in, for ex-ample, *Word Movie*. The latter commented on the arbitrary relationship of mean-ing and sound, as well as the slippage between representation and performance. In his locational pieces, such as *Synchronousoundtracks*, Sharits experimented with temporality and sound by allowing identical soundtracks to phase in and out of synch with one another. Sharits's concern in most of his work during this period (from 1968 to the late 1970s) was how *not* to reconcile the uncontrollable elements of invention with the rigor of material structure. While Steve Reich's phase-shifting experiments served as early models for Sharits's filmic sound proj-ects, Ludwig Wittgenstein's and Xenakis's approaches—the former's concerns with language and mathematics, the latter's focus on mathematical approaches to composing—pushed his already phenomenologically minded projects into more concise statements about the relationships between stasis and movement, sound and image, as well as light and color.

Particularly informative is Wittgenstein's rethinking of mathematical and logical propositions via language: "In real life, mathematical proposition is never what we want. Rather, we make use of mathematical propositions only in inferences from propositions that do not belong to mathematics" (232). For instance, in *Remarks on the Foundations of Mathematics* (1956)—one of the bibliographic entries in Sharits's "Cinematics Model for Film Studies in Higher Education"—Wittgenstein argues that the proof 16 to the 15th power "does not simply consist in multiplying 16 by itself fifteen times and getting this result—the proof must show that I take the number as a factor 15 times." For Wittgenstein, the interesting question is: "For what purpose do I use what has struck one? Is it notation? Thus I write 'a²' instead of 'a x a.'" Wittgenstein encourages us to ask the same kind of question in relation to language: "What do we actually use this word or this proposition for?" (180) In *Word Movie*, Sharits plays with this contradiction in terms of the paradigmatic structure of the alphabet as it relates to the seemingly separate ontological realm of abstract, isolated letters: *L* is capable of generating *lobe, mole, hole, close, splice, manual, cancel, slit,* and *snail*—we cannot form any kind of relational proposition here. A kind of cognitive vertigo takes over, giving the illusion of vertical movement (of letters) when, in fact, our left-to-right, horizontal, reading mode is left intact, though barely operational at such ruthless speed. Text does not lend itself so much to reading as it does to counting. Words or word fragments refer to the limits of the frame (and page) and, in turn, comment on the limits of viewing when thought about within the framework of reading.

Words act as a nontemporal container of the film's length, especially when they are read in asynchronous, repetitive patterns that do not seem to end at any particular time. This is word processing, a letter version of Tony Conrad's *Cycles of 3's and 7's* (USA, 1976), in which the harmonic intervals that would ordinarily be performed by musical instruments are represented through the computation of their arithmetic relationships or frequency ratios on a calculator. Both films refer more to the events of mechanical reproduction and recording than the events of language or music. The recording technologies of film are further accentuated by Sharits's use of the flicker effect. As Rosalind Krauss has argued, "the optical information on the screen becomes the visual correlative of the mechanical gearing of lens and shutter" (96). Moreover, this pulselike effect in *Word Movie* parallels the pulsing voices and makes the words more congruent to the perimeter of the film's shape and less loyal to symbolic meaning.

Xenakis, unlike Stockhausen or Boulez (who held on to the tenets of serialism well beyond Xenakis's limit), further complicated Sharits's approach to sound-image relationships through his claim that music is not like a language: "Music is much purer, much closer to the categories of the mind" (Matossian 89). I would like to suggest that a relationship exists between Sharits's interest in phenomenological investigations of time through Husserl and his study of the work of Xenakis's stochastic compositional methods—especially in relation to Sharits's *S:TREAM:S:S: ECTION:S:ECTION:S:S:ECTIONED* (USA, 1968–70). *S:TREAM* is a conceptual lap dissolve from "water currents" to "film strip current" ("Postscript" 6). Husserl's notion of an "internal time consciousness" identifies the intentional character of the consciousness of time: "A longitudinal intentionality goes through the flux [of consciousness], which in the course of the flux is in continuous unity of coincidence with itself" (107). As Sharits notes in "A Cinematics Model for Film Studies in Higher Education," Husserl "brackets" or "suspends" one's objective sense of the temporal so that one can actually register "perception of one's perceiving during the act of perception" (63). This was one of his goals in *S:TREAM*: the desire to "experientially affect viewing's stream of response, creating mental ripples" or to momentarily suspend perception so that the act of perceiving would become transparent. In other words, the viewer-listener would come "crashing downward into [an] abyss of self-consciousness" ("-UR(i)N(ul)LS" 19). Sharits also acknowledged the critiques of Husserl's work, stating, "Many of his models of perception have been superseded by the psychophysiological research of the last several decades" ("Cinematics" 63). He suggests that an advanced course in phenomenology might be "team-taught with a psychophysiologist, using instruments which would measure alterations in brainwaves, blood pressure, galvanic skin responsiveness, etc. during the act of experiencing various film and video works" ("Cinematics" 63).

Similarly, Xenakis, in his stochastic compositions, was interested in characterizing physical reality in a nondeterministic way by giving probabilities of occurrences, and thus rethinking notational time in terms of a statistical mean of events and movements (Matossian 94–95). Stochastic means conjecture, chance, probability, but also translates from the Greek as *target* and *reflect*—the twin systems of indexical and phenomenological investigations. Xenakis's scores, influenced by his earlier training as an architect under Le Corbusier, were visual models that he animated by manipulating different components of sounds into a complex audio template—mirroring the aggregate movement of particles, whether they be gas molecules, cicadas buzzing in different pulses, a swarm of bees droning in flight, or clouds scurrying across the sky (90). Xenakis's graphic representation of this concept is realized in one of his early architectural projects under Le Corbusier, the *Couvent de St Marie de la Tourette* (1954–57). Xenakis used Le Corbusier's *Modulor* (a system whereby the human scale itself would be a constant unit of all the dimensions of a building)—to produce a detailed polyrhythmic study with light and shade as the dynamic range. Later, in his own music, Xenakis incorporated the idea of several simultaneous layers of durations, each layer individually proportioned in a complex rhythmic polyphony. Likewise, Sharits in *S:TREAM* uses the notion of what he refers to as an "authentic heterodyne, a completely natural chord of sight, sound, and 'thought'" ("-UR(i)N(ul)LS" 15).[7] Both Xenakis and Sharits used systems of reproduction and transmission, with patterns of frequency and duration as compositional elements. Their goals were not to visualize sound (Xenakis found this abhorrent, Sharits found it boring), but rather to discover an auditory picture or audio frame of a certain material proportion.

Through Xenakis (and Wittgenstein), Sharits learned *how to do things with numbers*—an approach to the indexical structures of organic material. In *S:TREAM,* Sharits did not want the sound to be synchronous or to refer directly to "liquid" or "film" but to be a sound "stream," a flow of words nonconnective to the visual currents (15). By recording a woman's voice repeating the word *exochorion* (the outer membrane of a reptilian or mammalian embryo), Sharits builds multiple tracks of voices until they are overlaid or superimposed on one another to create new aural vocabularies. The high frequency of the letters *S* or *X* generate a separation of "S/X" sound-rhythm-lines from the meaning-carrying-consonants and vowels of the words themselves, suggesting a sort of "white noise." Similar to Xenakis's formal study of white noise, especially in pieces such as *Diamorphoses* (1957), Sharits was interested in the process of sound densification. Whereas John Cage had made an important move in replacing pitch with duration, which freed composition from its temporal definitions in terms of musical succession, Xenakis was interested in the larger question: what would remain of music if it were considered outside the dimension of time? Densification focuses on intervals of intensity between sonic events—it increases as complex sonorities move closer to *noise.* Sharits expressed what he had hoped to achieve in terms of sound in *S:TREAM* as "wordsaslinesofparticlesseparated" (16). These six words are fused as one word in his essay on *S:TREAM* and echo how he had described *Word Movie* as fifty words visually (and aurally) "repeated" in varying sequential and positional relationships so that the individual words optically-conceptually fuse into one 3 3/4–minute-long word. In *S:TREAM* words become synonymous with the particles of wave theory; they express an overtonality that is more about temporal simultaneity than language play. Sharits's work on *S:TREAM* is informed by Xenakis's work with aggregate movement. Indeed, one can locate similar approaches to the notion of a sound continuum between Sharits's *S:TREAM* and Xenakis's *Concret PH* (1958).

In *Concret PH,* Xenakis recorded crackling embers from which he extracted very brief (one-second) sound elements. Then he assembled them in huge quantities, varying their density each time. This work can be compared to his instrumental preoccupations concerning "clouds of sound" during the same period and is especially important to think about in relation to Sharits's work because it refers to a parallel visual form: *PH* stands for the hyperbolic paraboloids *(paraboloïdes hyperboliques)* that characterize the Philips Pavilion, which he conceived while working as an engineer and architect for Le Corbusier in 1958. As Sharits says of *S:TREAM,* the "relationship" in simultaneous occurrence and in overlapping structure (or wave-form) congruencies occurs not in the work but in perception itself: a subtle, continuous passage of both auditory and visual events.

What Sharits learned from Xenakis was that the spatializing effects of sound could also help us think about the spatialized nature of the visual world—something we are discouraged from doing because of how most narrative visual representation seeks to localize perception. The kind of compositional dimensionality Sharits was searching for was also limited by the single-screen format of film. Sharits began working with multiple-screen installations such as *Soundstrip/Filmstrip* (USA, 1971), *Synchronousoundtracks, Damaged Film Loop* (USA, 1973–74), and *Shutter Interface* (USA, 1975) in order to "approach the complexities of music's spatial dimension." And though he sometimes uses the language of conventional musical models to describe their effect—like "one screen could state a theme and another could answer it, elaborate upon it; the other screens could respond to this dialogue, vary it, analyze it, recapitulate it, etc." or by referring to his drawings for the films as scores—Sharits kept the notion of recording (its durational force and technological precision) as his main compositional mode ("Hearing" 74). In *S:TREAM*, the image of scratched emulsion, peeling back from the filmstrip, marks the recording process of the film, as well as the soundtrack—indeed, the final erasure of most of the film emulsion in *S:TREAM* accentuates the presence of the optical soundtrack—whose exposed invisibility gives it a voluminous presence. Likewise, in *Synchronousoundtracks* (the title itself is written as a loop) one hears the sound of sprockets passing over a projector sound head, with a frequency whose oscillation stands in direct relation to the sprocket-hole images seen on the screen. The dominant impression, then, is of a synchronicity whose terms are articulated with a definition that derives from a complex phasing, itself the product of a generative technique of recording and rerecording. "The loop itself is composed of analytic variation upon the testing of primary recording materials" (Krauss 101).

Sharits's career is marked by a material performance that rubs the indexical of recording technologies up against the contingency of film. By introducing the mathematical aesthetics of postserialist composition into film work, Sharits was redefining and expanding the phenomenological project of conceptual filmmaking to include not only a gestalt of vision but also a theory of sound and hearing that would think as much about granular permutations as it would about the larger gestures made between ways of cinematic seeing and hearing. Sharits's interest in Wittgenstein's approach to language, in which language is figured as a complex game, is intimately tied to one of the main concerns of structuralist film with "materials as language and with language as material" (James, *Allegories of Cinema* 241). But as Wittgenstein reminds us—like language's relationship to mathematical propositions—only the inferences we make from these propositions are useful, not the math itself. Similarly, the materiality of language is useful only in terms of how we might make inferences from it, not the actual grammar or mark making that language performs. By pointing to Kubelka (as a way to find a beginning point for Sharits's work), I am certainly pointing to a history of avant-garde filmmaking that has been largely influenced by experiments in music and sound, rather than the predictable, often disputed "elements" of innovation in the visual aspects of structural film aimed mainly at the level of the frame.

Understanding Sharits as a filmmaker deeply involved in postserial music and composition, and even more important in the cinematic events of Fluxus, allows us to see how his contributions to a phenomenology of film were deeply influenced by sound's resistance to (and preoccupation with) "objecthood." In Sharits's theory of cinematics, objecthood is achieved by "an intensification of materials," both sonic and visual, rather than a definition or representation of actual materials ("Words" 30). From *Word Movie*, Sharits was able to think about how "speech is patterned on an optical soundtrack of a film" but how this patterning could then move off the track and create overtonal dimensions or sound objects that could not be readily seen. The auditory level of film inspired him to rethink Saussure: "The signifier, being auditory, is unfolded solely in time from which it gets the following characteristics: (a) it represents a span, and (b) the span is measurable in a single dimension; it is a line" (32). The span or the line in Sharits's work took after La Monte Young's "Composition 60 #10," which instructed: "Draw a straight line and follow it." The event of this Fluxus line led Sharits from the single to the multiple, from silence to sound, from cinema to phenomenological object. Sharits perceived the multidimensional possibilities of film through sound—a conceptual move that irrevocably changed the way we think about film projection, presence, and performativity.

## Notes

1. As I have noted elsewhere, Brakhage's relationship to sound and silence was complex—he thought deeply about both. Though a major part of his filmmaking was concerned with what the eye could perceive, he was also interested in modernist music, paying close attention to the works of Olivier Messiaen, Pierre Boulez, Henri Pousseur, and Karlheinz Stockhausen, among others. See Ragona, "Hidden Noise: Strategies of Sound Montage in the Films of Hollis Frampton."

2. Kubelka's theory of the metric film is, in large part, a revision of Sergei Eisenstein's. Whereas Eisenstein located cinematic meaning (or articulation) between shots, Kubelka— in step with Dziga Vertov—argues that it exists primarily between frames. That is, the frame-to-frame relationship is the microlevel at which cinema creates the illusion of movement, and so it is this relationship that is the most crucial site for a filmmaker's intervention.

3. Kubelka was also influenced by the notion of the sublime in abstract expressionist painting, especially the work of his friend Arnulf Rainer after whom he titled his film of the same name. But Rainer is also often cited as a surrealist painter. So I am not so sure the distinction that Conrad makes here between abstract expressionism and surrealism is helpful. I would argue that one of the deciding factors that distinguishes Conrad's

work in sound from Kubelka's is Conrad's work with the Theatre of Eternal Music in the 1960s—what Conrad later deemed "New Minimalism" in music. It is more the distinction between abstract expressionism and minimalism (both in painting and music)—and serialism and psychedelic rock and roll (Murray the K's light shows with strobe effects).

4. White noise is analogous to white light, which contains every possible color. White noise is normally described as a relative power density in volts squared per Hertz.

5. Conrad also claims that in *The Flicker* he was interested in exploring "harmonic expression using a sensory mode other than sound," so that the space between rhythm and pitch was really the space of duration created by flicker frequencies "heterodyned, or rather multiplexed together" (Duguid).

6. The two models that inform Sharits's studies in sound in the 1970s are the phase projects of Steve Reich and the explorations of overtones in the work of La Monte Young. In Reich's *It's Gonna Rain* (1965) and *Come Out* (1966), he took single phrases that he recorded on two channels, first in unison, and then with channel two incrementally moving slightly ahead. As the phrase begins to shift, a gradually increasing reverberation is heard that slowly passes into a sort of canon or round. For an in-depth discussion of this process, see Roger Sutherland's *New Perspectives in Music* (Sun Tavern Fields, 1994) and Steve Reich's *Writings on Music (1964–2000)* (Oxford University Press, 2004). La Monte Young, inspired by the drone used in classical Indian music, began to use the overtone as an important element of his Dream Music in the 1960s. An overtone is derived from the summation of the "fundamental" (a part that is essential to the operation of a sine wave)—Young experimented with this in his durational sine tone installations, most notably in those he installed in his Dream House as early as 1969 in Munich, Germany, and later in its site in New York (1979–85).

7. In electronics, a heterodyne is a synthesis of the alternating currents of two different frequencies used in radio or television.

# 11

# "Every Beautiful Sound Also Creates an Equally Beautiful Picture"

## Color Music and Walt Disney's Fantasia

**CLARK FARMER**

One of my earliest memories concerns a problem that seems fit only for children. For the first time that I was aware of, my hometown's annual Fourth of July fireworks display was going to be accompanied by music. Not with a live orchestra, mind you, but with a prerecorded program on a local AM station. After my family took its accustomed spot on a grassy hill, I clutched my inseparable transistor radio, eagerly waiting. This was in the days before MTV, but like many restless children forced to learn an instrument, I was a miniature Wagner who could not listen to music without invoking images and drama. When it finally got dark, the tinny strains of "Fanfare for the Common Man" rang in my ear, and colors exploded against the sky. After years of being consigned merely to *listen* to music, I finally got to *see* it.

But as the program went on, waves of doubt lapped at my tiny soul. Was this really a carefully coordinated spectacle, or was the music just meant as background noise? And how could I know the difference? Fireworks inherently present a problem of synchronization. I knew from my scientist father that sound travels more slowly than light—which explained why the lights bloomed silently, followed by a commensurate explosion—and that the farther away you are, the longer the lag between what you see and what you hear. Thus, with real-life fireworks you rarely get that satisfying, simultaneous burst of light and noise—and music— that you get in the movies, that "synchronization of the senses" that Eisenstein proposed. But laying aside that question, what would it mean to "coordinate" abstract patterns of colored light with music? I had some sense that loud music should mean big fireworks, but beyond that, what was I expecting? Over the years as our provincial fireworks display metamorphosed into the Stadium of

Fire, certain literalisms crept into the program. During the national anthem, we would see "the rockets' red glare" and "the bombs bursting in air," and of course an arrangement of red, white, and blue fireworks when "our flag was still there." But such lyric correspondences don't get you too far.

To a large degree, the efforts to create "visual music" as a distinct and full-fledged art form have been plagued by many of the naïvetés that I had as a child. Nonetheless, they have had a decided effect on film history. The idea of seeing images move with the fluidity, coordination, and expressiveness of music has a long tradition in both film theory and film practice, and the modeling of a visual art on an aural one remains the guiding metaphor for abstract animation. I will focus on one thread of this tradition, namely, the relationship of color and music.

The concept of "color music" remains largely a lost chapter in the history of film sound, the explanation for which can be found in two contemporary trends in color music. First, whereas artists of the past turned to film because of the technical limitations of color organs, today's most serious artists working in this tradition have chosen computers over film due to film's relative technical limitations. The ability of computers to immediately translate sound into image, and vice versa, on the basis of algorithms is especially appealing for those wanting to test the idea of a close analogy between music and color. And the ability of digital technology to immediately respond to user input has led to a new generation of instruments that can "play" color, a situation more analogous to color organs than abstract animation. Hence, in the history of color music, film appears as a transitional medium.

Second, music accompanied by abstract color images has now entered the general culture, so that what began as a rarefied avant-garde experiment ended up a staple of mass culture. When we encounter color music today, it is most likely to be at rock concerts and dance clubs, at planetarium laser shows, on the "visualizers" of computer audio players, and, yes, fireworks displays. At some point, color music bypassed becoming the art of the future and became a plebeian amusement.

But instead of bemoaning this as another example of high art being co-opted by market forces, it is worth reconsidering the shift in color music from the era of the color organ to that of the laser-light stadium display (via abstract animation) as an example of vernacular modernism, namely, mass-produced phenomena that respond to "new modes of organizing vision and sensory perception, a new relationship with 'things'" (Hansen, "The Mass Production of the Senses" 60). The same public that might scoff at paintings that consist only of swaths of color actively embraces abstract displays of color accompanied by popular music. A modern public has come to see color as a material in and of itself, rather than merely a property of objects.

The transition of color music from the avant-garde to mass culture is captured in one particular example of abstract filmmaking—the Toccata and Fugue in D Minor from Walt Disney's *Fantasia* (USA, 1940). On the one hand, this sequence draws on the traditions of color music and musical painting, and seeks to justify itself via a traditional series of analogies drawn between color and music. On the other hand, the film in practice avoids any strict analogy between color and music, a looseness that opens the door to the forms of color music we are most familiar with today.

## Color Music and Musical Painting

Broad analogies between color and music have existed since classical antiquity, when such familiar metaphors as tone color (timbre) and color harmony were first coined (Gage, *Color and Culture* 227). But it was not until the eighteenth century that "color music" was first conceived as a distinct art form. The following passage by physicist Albert A. Michelson in his 1903 book *Light Waves and Their Uses* is a useful representative description of the concept:

> I venture to predict that in the not very distant future there may be a color art analogous to the art of sound—a *color music,* in which the performer, seated before a literally chromatic scale, can play the colors of the spectrum in any succession or combination, flashing on a screen all possible gradations of color, simultaneously or in any desired succession, producing at will the most delicate and subtle modulations of light and color, or the most gorgeous and startling contrasts and chords! It seems to me that we have here at least as great a possibility of rendering all the fancies, moods, and emotions of the human mind as in the older arts. (quoted in Birren 165)

The first attempt to create an instrument that would "play" colors came with eighteenth-century French Jesuit and mathematician Louis-Bertrand Castel, who referred to his invention as an "ocular harpsichord" *(clavecin oculaire).* Setting an unhappy precedent for the history of the color organ, Castel's color music was realized more in theory than practice. From his initial theorizing in the 1720s up to his death in 1757, there is no evidence that Castel's various contraptions ever got beyond the prototype stage. The various models seem to have been modified harpsichords, some outfitted with a system of mirrors and levers that allowed candlelight to shine through squares of colored glass or paper arrayed on a large frame (Gage, *Color and Culture* 233; Peacock 400).

It was more than a century before another attempt was made to create a color organ. The obscure figure D. D. Jameson described in his 1844 pamphlet *Colour-Music* an instrument that would allow light to shine through bottles filled with colored liquid into a dark room lined with tin plates, but there is no evidence that such an instrument was ever built (Gage, *Color and Culture* 235; Peacock 401). Between 1869 and 1873 English physicist Frederick Kastner constructed what he called the Pyrophone, an instrument that used ignited gas to produce both tones and hues (Maur 90; Peacock 401). American Bainbridge Bishop was the first to take advantage of the introduction of electrical lighting in the 1870s, creating instruments that could use either sunlight or an electric light. In 1893 English professor of fine arts Alexander Wallace Rimington patented his Colour-Organ. Using arc lamps, it was the most complex and technically viable instrument to date, and proved quite successful in lectures and public concerts, during which the silent Colour-Organ was supplemented by a pipe organ and symphony orchestra (Peacock 402).

The first half of the twentieth century proved to be the high-water mark for color organs. Alexander Scriabin was the first composer to include notations for color effects in a score, first in his "Tastiera per luce" and more famously in his score for *Prometheus: A Poem of Fire*. Its Moscow premiere in 1911 was plagued with technical difficulties, and so the first performance with projected lights was in 1915 at Carnegie Hall (Peacock 402– 3). A number of greater and lesser names constructed instruments during this time: Australian Alexander Hector's color organ of 1912; American visionary architect Claude Bragdon's color organ, demonstrated in 1915 and 1916; American painter Van Dearing Perrine's several color instruments, built after attending the 1915 production of *Prometheus;* American Mary Hallock-Greenewalt's Nourathar, first demonstrated in 1919; English painter Adrian Klein's 1920 color projector for theatrical lighting; Englishman Leonard Taylor's 1920 color organ; Italian Achille Riccairdo's instrument for the Teatro del Colore in Rome, built between 1920 and 1925; American Richard Lovstrom's color organ patented in the early 1920s; Ludwig Hirschfeld-Mack and Kurt Schwertfeger's Color-Light Plays, first developed at the Bauhaus in 1922, the same period that Raoul Hausmann worked on his Optophone; Hungarian composer Alexander László's Sonochromatoscope, first shown in 1925; Austrian Count Vietinghoff-Scheel's Chromatophon, played at the second Hamburg Congress for Color-Music Research in 1930; American synchromist painter Stanton Macdonald-Wright's light machine, which caused fellow synchromist Morgan Russell to abandon his own plans to build an instrument in 1931; and American Charles Dockum's MobilColor Projector, first constructed during the late 1930s. The most famous color-music practitioner was Danish American Thomas Wilfred, who built his first Clavilux in 1921 and went on to design the Art Institute of Light in 1930 (Gage, *Color and Culture* 245; Kushner 109; Maur 88; Moritz 27; Peacock 404; Wilfred 205).

In some cases, experiments with color organs led directly to an interest in cinema, often because of dissatisfaction with the technical limitations of the instruments. Italian brothers Arnaldo Ginna and Bruno Corra had built a color organ around 1908, but were disappointed with its poor luminosity. According to Corra's 1912 article "Abstract Film—Chromatic Music," this led Ginna to hand paint a number of abstract films in 1910 based around the color-sound analogy (Gage, *Color and Culture* 245; Leslie 280). In 1913 Arnold Schoenberg discussed with Wassily Kandinsky the possibility of filming and hand coloring a performance of his piece *Die Glücklichliche Hand*, which featured a series of changing lights in scene 3 (244). Alexander László had been disappointed with reviews of his concerts that found the output of his Sonochromoscope dull compared to his music, and so he commissioned Oskar Fischinger to create a multiple-projector film and slide show for his 1926 tour (Moritz 12).

But color organs were not the only antecedents of abstract animation. Of equal importance was the tradition of representing music in the static visual arts, a tradition that came to assume a new importance in modernism. As art scholar Karin von Maur succinctly states, "In search of a visual idiom adequate to the new world view, artists turned to music and its independence of material, physical fetters. With music as an ideal, they divorced their imagery from the objective context, thereby liberating colors and forms. Yet in order to avoid falling into mere ornamentation, arbitrariness, or chaos, they knew that the elements of painting must also be composed and ordered in a way similar to music" (8–9).

The role of music in providing a model for abstraction can hardly be understated, as it provided the starting point for the twentieth century's embrace of the nonrepresentational. Practitioners of "musical painting" represent nearly every major art movement of the early twentieth century: Josef Albers, Giacomo Balla, Charles Blanc-Gatti, Georges Braques, Johannes Itten, Wassily Kandinsky, Paul Klee, Gustav Klimt, Frank Kupka, August Macke, Franz Marc, Henri Matisse, Piet Mondrian, Francis Picabia, Morgan Russell, Stanton MacDonald-Wright. As with the color organ, experiments in musical painting sometimes led directly to cinema. French artist Léopold Survage, encouraged by poet Apollinaire, completed a series of abstract watercolors in the hope of filming them to create a new art "of colored rhythm and of rhythmic color" (Leslie 280). Abstract animators Hans Richter and Viking Eggeling came to film after reaching the limits of easel and scroll painting to create "the music of the orchestrated form" (Richter 79). But as I will discuss next, the exact basis for a comparison between the visual arts and music in general—and color and music in particular—varied greatly for each artist.

## Three Models of the Sound-Color Analogy

Following a suggestion by Rudolf Arnheim, we can break down the rationale for making an analogy between color and sound into three categories: physical similarity, synesthesia, and correspondence of expression ("Rationalization of Color" 206–7). Like many false but productive ideas, the belief in a physical similarity between tones and color can be traced back to classical literature. Aristotle hypothesized that colors might in fact be juxtapositions of minute quantities of black and white (think pointillism), the exact color determined by their ratio. It was already known that musical harmony was based on mathematically elegant ratios, as was demonstrated by the Pythagorians' experiments with plucked strings. Aristotle hypothesized that pleasing colors were ones with a "numerically expressible ratio," just as in music (698). Aristotle, however, held this out as a purely speculative account and did not argue for a direct correspondence between certain hues and certain notes. But undergirding this and later speculation was a confidence in an underlying universal harmony behind surface phenomena, perhaps best exemplified by the metaphor of the harmony of the spheres. The main stumbling block was lack of agreement about the order and relation of colors, so that scholars who wished to defend painting in relation to a worldview that privileged mathematical harmony chose to emphasize geometry rather than color as its basis.

Sixteenth-century Milanese painter Giuseppe Arcimboldo was the first to try to determine a note-to-hue correspondence. He constructed a black-to-white value scale of fifteen shades that were meant to correspond to a musical double octave, then attempted to extend this experiment into hue. But Arcimboldo and his contemporaries "could not shake off the Aristotelian view that the hues lay on a linear scale between light and dark" (Gage, *Color and Culture* 231). The real breakthrough came with Sir Isaac Newton's discovery that white light consisted of a consistently ordered spectrum of colors. Newton cast his new discoveries in the mold of the ancients: "May not the harmony and discord of Colours arise from the proportions of the Vibrations propagated through the Fibres of the optick Nerves into the Brain, as the harmony and discords of Sounds arise from the proportions of the Vibrations of the Air? For some Colours, if they be view'd together, are agreeable to one another, as those of Gold and Indigo, and others disagree" (346). Newton's attempt to divide this spectrum into discreet areas was clearly guided by the desire to make them correspond to the seven tones of the musical scale, which explains the otherwise nonsensical inclusion of indigo. Indeed, his color wheel was based on René Descartes's diagram of musical intervals (171).

The father of the color organ, Louis-Bertrand Castel, followed Newton's basic idea of aligning the musical scale with the spectrum. But as various attempts were made over the centuries to construct color organs, it became apparent that there was little agreement as to which notes were supposed to correspond to which hues. Castel's scale ran from blue (C) to violet (B). Eighteenth-century English painter Giles Hussey proposed a scale from red (A) to violet (A-flat), similar to one used by twentieth-century Australian painter Roy De Maistre (235; Johnson

85). Bainbridge Bishop and Alexander Wallace Rimington's scales likewise ran the spectral course from red to violet, but were set to the key of C major (Bishop 10; Gamwell 160). Swiss scholar Hans Kayser and composer Josef Matthias Hauer sought to align the twelve-tone scale with a twelve-part color wheel, but one that did not follow the spectral order (Maur 83). Gertrud Grunow's classes at the Bauhaus not only linked specific colors and notes but assigned corresponding body parts as well (Gamwell 160). Such examples are plentiful, and by no means extinct, as witnessed by the current popularity in Germany of a new "scientific" scale that runs from green (C) to yellow (A-sharp). As such discrepancies suggest, there is no scientific basis for believing in a universal principle of harmony that relates specific notes to specific hues. Most of those working with color music or musical painting ultimately reject the model of physical similarities, though it is often the starting point for their experiments.

Instead of arguing for a relation of color and tones that exists in the external physical world, those advocating a synesthetic model of the color-sound analogy can argue that it is in fact a product of interior mental processes. People suffering from the rare, hereditary condition of synesthesia involuntarily experience two different sensory responses from a single stimulus. The most common form of synesthesia is color hearing, where the subjects report seeing colors when they hear certain sounds. Interestingly, it is not music but spoken language that is the more common trigger for color hearing, a phenomenon referenced in Rimbaud's 1871 poem "Voyelles," which equates each vowel with a color (Gage, *Color and Meaning* 263). When synesthesia is described in relation to music, it is usually in connection to instrumentation rather than individual notes or dynamics, such as describing the sound of a trumpet as red.

The interest in synesthesia peaked at the end of the nineteenth century, spurred by Francis Galton's *Inquiries into Human Faculty and Its Development* (1883). The popularity of the idea makes it difficult to judge which artists and composers were in fact synesthetic and which were merely swept up in the enthusiasm. There is numerous anecdotal evidence of composers speaking about the color of particular keys: Jean Sibelius absentmindedly asking that his stove be painted F major when he meant green (Finlay 196), or Scriabin and Rimsky-Korsakov agreeing that a piece in D major appeared yellow (Gage, *Color and Culture* 243). Painter Wassily Kandinsky freely described colors in terms of sound and vice versa: "Represented in musical terms, light blue resembles the flute, dark blue the 'cello, darker still the wonderful sounds of the double bass; while in a deep, solemn form the sound of blue can be compared to that of the deep notes of the organ" (182).

Proponents of the synesthetic model of the color-sound analogy usually treat these very rare cases as merely extreme examples of a basic human capacity that operates subconsciously in most people. A 1938 study, for example, concluded that 60 percent of its sample experienced "some kind" of color response when music was heard (Birren 163). But similar to the lack of agreement about note-to-hue correspondence among defenders of the physical similarity model, those who actually do experience color hearing rarely agree on which sounds match up with which colors, a fact already noted by Galton.

The relative rarity of actual clinical cases of synesthesia made it an unlikely basis for establishing an artistic practice whose rationale would be understandable to a large audience. But a few decades later, the synesthetic model of the color-sound analogy found a new avenue of legitimation: psychedelic drugs. The fusion of the senses that could be experienced with LSD, mescaline, and other drugs meant that synesthesia was no longer limited to one in every two thousand people, and was now a familiar-enough phenomenon to be taken up in popular as well as experimental film.

Another major influence on color music was modern theosophy, whose model for the sound-color analogy falls somewhere between those of physical resemblance and synesthesia. The majority of the theosophical writings on color stress its connection to geometrical forms and mental states, without any link to sound. A typical example would be Annie Besant and C. W. Leadbeater's 1901 book *Thought Forms,* which proposes that people's thoughts and spirits could be seen by clairvoyants as colored forms, the color representing mood. But this book goes on to argue for musical thought forms, and includes painted examples of what a clairvoyant "saw" during an organ recital of works by Mendelssohn, Gounod, and Wagner. The theosophical belief in the essential oneness of life suggests an actual (meta)physical connection between colors and sounds, as with the physical similarity model, but the ability of only a few to see "auras" suggests the much more subjective model of synesthesia. Theosophy proved enormously influential among both composers and artists, including Scriabin, Anton Webern, Dane Rudhyar, Kandinsky, Jan Toorop, Piet Mondrian, and abstract animators Arnaldo Ginna, Bruno Corra, and Harry Smith.

The theosophical belief that colors could be aligned with particular moods and states of mind also points to the third model of the sound-color analogy, namely, correspondence of expression. Those holding this view point to "perceptually convincing correspondences between colors and sounds based on shared expressive qualities, such as coldness or warmth, violence or gentleness" (Arnheim, "Rationalization of Color" 206). As might be expected, those holding to the physical similarity and synesthesia models will often take shared expressive qualities of sound and color as proof for their positions. Color organist Bainbridge Bishop, for instance, took his subjective response to two different phenomena as proof for his note-to-hue model: "Violet-blue always gives me a sad impression similar to the music played in A minor. This will be observed in viewing distant violet-blue mountains at sunset or twilight. The melancholy effect is strongest when the dominant or subdominant color of the minor key is present, yellow and orange; one of which colors we commonly see at such times above the mountains" (16).

But the idea that a set of colors and a musical passage can both elicit a similar emotional reaction does not require an assumption of a direct physical or mental correspondence. Abstract animator John Whitney, for example, emphasized a general *structural* similarity between color and music, while at the same time stressing the autonomy of the two phenomena: "The fallacy of mechanically translating previously composed music into some visual 'equivalent' is established repeatedly in critical writings on the subject. A more fruitful, less mechanistic approach is possible today, for truly creative possibilities arise when the image structure dictates or 'inspires' sound structure and vice versa, or when they are simultaneous conceptions" (142).

In this model, the emphasis is shifted from finding equivalences between individual elements to understanding the role of such elements within their respective systems. A single note or a single color by itself may not elicit any predictable response; its expressive potential lies, rather, in its place within a whole: "Just as a given shade of blue is perceived as a different color depending on whether it appears next to an orange or a purple, a tone of a given musical pitch has a different dynamic character depending on its place in the tonal structure" (Arnheim, "Perceptual Dynamics" 220).

Such a model presupposes a basic unity of the arts, as well as the idea that the expressive ends of art can be separated from their respective media:

> The character of any perceptual event resides in its dynamics and is all but independent of the particular medium in which it happens to embody itself. This interpretation relies on the concept of isomorphism, which was introduced by gestalt psychology to describe similarity of structure in materially disparate media. Thus a dance and a piece of music accompanying it can be experienced as having a similar structure, even though the dance consists of visual shapes in movement and the music of a sequence of sounds. (222)

But there is a danger of exaggerating the ability of different media to arrive at the same point, as well as the danger of reducing the meaning of art to its effects on an audience. Those holding to this model also have a tendency to stress supposedly hardwired cognitive reactions to color and music, when it is still very much an open question to what degree the expressive qualities of music and colors are based on untutored psychophysical responses, as opposed to learned cultural associations. As experimental filmmaker Malcolm Le Grice argues, "A child of the new millennium may never see a fire and experience cooking only with invisible microwave; its primary experience of red may derive from holding an ice-cold, bright-red can of Coke—where then is the basis of psychological consistencies in colour associations?" (261).

All three models of the sound-color analogy have definite limitations. In practice, they are often taken up as handy metaphors to get a project off the ground, but are not adhered to in any strict fashion. The linking of sound and color in abstract animation will often throw in all three models, while at the same time taking a primarily instinctive approach to creating color music.

## Color Music and Walt Disney's *Fantasia*

The film *Fantasia* was an outgrowth of "The Sorcerer's Apprentice," which was originally intended as an independent short. This best-known segment of *Fantasia* is a culmination of the Disney practice up to that point, combining as it did the recognizable "star" Mickey Mouse with the format of the artistically more ambitious *Silly Symphony* series, produced from 1929 to 1939. As the name *Silly Symphony* suggests, these works were often self-conscious mergings of high and popular culture. Based on Goethe's poem of the same name, Paul Dukas's "Sorcerer's Apprentice" (1897) is a prime example of nineteenth-century program music, a genre that emphasized music's ability to paint a picture and tell a story. Hence, there is an element of redundancy in the actual illustration of the piece with moving pictures, resulting in the extreme form of synchronization aptly dubbed "Mickey Mousing." (For an abstract setting of the same music, see Oskar Fischinger's *Studie Nr. 8* [1931].) But the decision to expand the planned short into a feature-length series opened the door to alternative models of relating music and image.

The first segment of the original road-show release of *Fantasia* is an orchestral arrangement of J. S. Bach's Toccata and Fugue in D Minor. The choice of music is significant because of the central importance of Bach for practitioners of visual music. Bach's almost mathematical forms had already inspired modernist painters who were seeking to move beyond representational art. As painter Paul Klee noted, "In the eighteenth century music already perceived and resolved the paths of abstraction, which were then confounded again by the program music of the nineteenth. Painting is only beginning this task today" (Maur 9). Bach's emphasis on contrapuntal structure rather than melody made it better suited for creating analogies with visible works: "What fascinated composers and painters alike was the tight yet flexible structure of the Bach fugue; comparison with the fugue became a fashionable way of characterizing preoccupation with structure. Robert Delaunay's work was described in these terms by both Marc ('pure sounding fugues') and Klee, who noted in a review of 1912 that one of his window-paintings was 'as far away from a carpet as a Bach fugue'" (Gage, *Color and Culture* 241).

Apparently, abstract animator Oskar Fischinger had first suggested to conductor Leopold Stokowski the idea of an abstract film set to Bach's Toccata and Fugue in a series of letters in 1936, and then Stokowski passed the idea on to Disney in 1937. Fischinger had even made the suggestion of beginning with a full-length shot of Stokowski conducting the orchestra before segueing into the abstract animation. Fischinger was subsequently hired by the Disney Studio in 1938, where he worked for nine months, during which time his abstract films were shown to the staff on a weekly basis (Moritz 84). Unfortunately, his ideas and sketches for the Bach film proved unacceptable to Disney and the senior animator in charge of the project, Cy Young, leading Fischinger to terminate his contract. Only one of Fischinger's sketches, an undulating wave, seems to have made its way into the finished product. Nonetheless, this semiabstract film was clearly inspired by Fischinger's example. (Fischinger began his own Bach film in 1946, the masterpiece *Motion Painting No. 1*, set to the Third Brandenburg Concerto.)

Certainly, the decision to present a form of abstract animation to an unversed audience was risky, a fact borne out when RKO cut this segment for the theatrical-release version of the film. Also removed from the road-show version of *Fantasia* were segments with a narrator introducing each segment. In the case of the Bach piece, his main function is to contextualize abstract animation for an unfamiliar audience:

> Now, there are three kinds of music on this *Fantasia* program. First is the kind of music that tells a definite story. Then there's the kind that, while it has no specific plot, does paint a series of more or less definite pictures. Then there's a third kind, music that exists simply for its own sake. Now, the number that opens our *Fantasia* program, the "Toccata and Fugue," is music of this third kind, what we call absolute music. Even the title has no meaning beyond a description of the form of the music.

Despite this rhetoric, the Bach segment shies away from full-fledged abstraction; some "more or less definite pictures" recur throughout. The Toccata and Fugue film is divided into two parts, live action and animation, following the musical division between toccata and fugue. In the live-action sequence, sections of the orchestra are always present and recognizable, even when semiabstracted in the form of cast shadow. The animated sequence begins with a background of clouds, setting up a whole series of recognizable natural images, such as hilly landscapes, rippling water, geysers, and sunbeams. Some of the images are clearly representations of the orchestra, such as patterns of sawing violin bows, cut in half. One of the main motifs, a series of four parallel lines, is first introduced by seeing animated strings, which although floating free of their instruments are still clearly recognizable because of their attached bridges. The decision to combine abstract and representational animation seems to have been done to soften the blow for audiences not ready for images of the third kind.

The spirit of Bach's music is also violated by the film's general disregard for musical structure. One of the appeals of Bach for practitioners of visual music was the fugue form, which arranges the musical theme into overlapping blocks assigned to different instruments and registers, a structure that easily suggests visual parallels. Fischinger's suggestion to Disney was to have multiple areas of focus on the screen at once, paralleling the experience of the fugue, where a listener can quickly alternate between following a single musical line and stepping back to hear the totality of the interwoven lines. In its completed form, the Bach segment from *Fantasia* presents only one visual idea at a time, a decision again made in consideration of the audience.

A final charge that could be leveled at the film is that the overlay of representational animation, as well as the film's playback system, robs the abstraction of its characteristic spatial character. A central feature of abstract animation is the foregrounding of the screen as a two-dimensional field, which allows for a unique form of cinema spectatorship. Like traditional figure-ground illusions, a square growing larger on the screen can be interpreted either as a shape expanding its dimensions or as an object of fixed size getting closer to the screen. By providing backgrounds and horizon lines, the makers of the Toccata and Fugue section never allow us to experience the abstract images as flat. The abstract images belong to a three-dimensional world governed by the central downward vector of gravity, rather than a potentially free-floating space that creates its own centers of attraction.

The feeling of three-dimensionality is reinforced through the use of Fantasound, an experimental recording and playback system that can be compared to today's surround sound, created for the film and its road-show performances. Sound engineers for Walt Disney Studios embarked on an ambitious plan to reproduce the sound of a full symphony orchestra by means of multiple-channel recording and multiple sound reproducers, amplifiers, and loud speakers to play back the multiple soundtracks (Garity and Hawkins). The engineers focused on two qualities: "the illusion of 'size,' possible to attain by proper use of a multiple-speaker system, and recognizable placement of orchestral colors important to the dramatic presentation of the picture" (Plumb 18). In theory, with such a system it "becomes possible to obtain a virtually auditory perspective effect" (Peck 29). This "three-dimensional" reproduction of sound further anchored the semiabstract images of the Toccata and Fugue, as groups of instruments became linked to various spaces on the screen.

Although the Toccata and Fugue section could be accused of being merely abstraction-lite, it clearly still belongs in the tradition of color music. Both sections, but especially the live-action segment, clearly owe a debt to color-organ experiments, familiar to both Disney and Stokowski. Disney had seen a demonstration of a color organ in 1928, and Stokowski had actually made use of Thomas Wilfred's Clavilux when conducting *Prometheus* as well as Rimsky-Korsakov's *Scheherazade*. (Oskar Fischinger, quite familiar with various color organs from his time in Germany, did not begin building his own color organ, the Lumigraph, until 1950.) There is also a labyrinth connection to abstract painter and color-organ inventor Charles Blanc-Gatti. In 1935 when Walt Disney was visiting Europe, Blanc-Gatti approached him about collaborating on an abstract film, to which Disney demurred. When Blanc-Gatti heard in 1939 that Disney was working on *Fantasia*, he sent Disney a letter about his own just-completed film, *Chromophonie*, which he thought of as a record and expansion of his color-organ experiments. Like the Bach segment in *Fantasia*, *Chromophonie* combines abstract and representational animation—musical instruments and suggestions of nature—a similarity that caused European critics to accuse Disney of stealing from Blanc-Gatti and other musical painters (Allan 110; Leslie 187–88).

The live-action sequence uses projected color light in two ways: lights shining on the instruments and light projected on a screen. The former is introduced even before the narrator gives his introduction, as we see the orchestra arrive and begin to tune their instruments while silhouetted against a luminous blue backdrop. Tightly focused colored lights single out instruments as we hear them, with the strings bathed in a gentle yellow, flutes in pale green, saxophones in lavender, and the brasses in red, except for the tuba that pulsates in violet. During the Toccata and Fugue, this type of projected color continues to be primarily associated with instrumentation, though there is one moment that suggests the note-to-hue literalism of some color organs: the instruments of three french horn players sitting in profile are lit in succession with lights of red, yellow, and green tints as we hear the three final sustained notes of a passage.

The model of physical similarity that this exceptional passage represents has a counterpart later in the film. After the intermission, the narrator introduces us to "the soundtrack," a line that cautiously shifts over from the left of the screen. With optical sound tracks, we *do* have a direct correspondence between a sound and an image, a relation that the narrator suggests might be not simply functional but aesthetic as well: "Now, watching him I discovered that every beautiful sound also creates an equally beautiful picture." We are then shown what certain sounds (specifically musical instruments) look like on a magnified soundtrack. Of course, the lines and waves produced are artists' renditions (complete with color, which would be superfluous on a soundtrack), again a sort of watering down of earlier experiments. For instance, in 1932 Oskar Fischinger created hundreds of scrolls decorated with black-and-white patterns, which he then shot directly onto the soundtrack, basically turning the projector into an electronic player piano (Moritz 43).

The synesthetic model also played a key role in the construction of the Toccata and Fugue. In fact, in preparation for the piece Disney and Stokowski conducted an experiment in which they recorded the visual images and colors suggested to them while listening to the piece: "A loud crescendo appeared to Disney 'like the coming out of a dark tunnel and a big splash of light coming in on you.' One passage suggested orange to Walt, but Stokowski saw purple. A woodwind section gave Disney the image of a hot kettle with spaghetti floating in it, and so on" (Leslie 187).

The narrated introduction to the piece explicitly gives the audience a synthesthetic model for interpreting the image:

> What you will see on the screen is a picture of the various abstract images that might pass through your mind if you sat in a concert hall listening to this music. At first you're more or less conscious of the orchestra, so our picture opens with a series of impressions of the conductor and the players. Then the music begins to suggest other things to your imagination. They might be, oh, just masses of color, or they may be cloud forms, or great landscapes, or vague shadows, or geometrical objects floating in space.

The keying of colors with instruments is not consistent (the violins are yellow in one section, blue in another), but the grouping of color with instrumentation fits the broad outlines of the synesthetic model.

The least-developed model in the Toccata and Fugue is correspondence of expression. At least one commentator understood the relation between color and music in the film to be something akin to a visual aid for following the patterns of the piece:

> In the "Toccata and Fugue," which is perhaps the most successful of the serious pieces, [Mr. Disney] has introduced rhythmical movements and changing forms as well as color for his "translation." The color itself had not been used literally in terms of single notes, but rather to convey larger moods, or qualities of sound. He has managed, besides, to fulfill with the film some of the functions of a printed score—such as ardent listeners follow at concerts—by indicating, in terms of abstractions instead of notes and staves, the main themes and the contrast themes, and a great deal about the general form of the music. Insofar as he succeeds he has made a distinct contribution to popular understanding of the music. (Isaacs 59)

But when carefully examining the color, it is clear that unlike with other abstract color films, the animators here did not seem that interested in establishing structural consonances or dissonances of color that might correspond to the music.

One is struck by how monochrome the sequences in this film actually are. The live-action sequence contains shots completely made of colored bows of cast shadows of the orchestra. Each of the three or four layers is usually just a variation of a single hue, for instance, ranging from a low-value blue on the bottom up to a high-value blue on top. Arguably, there is an interaction of color *between* shots (an idea played with by Paul Sharits in his color flicker films), but here again the film seems to be taking pains to avoid color shifts that would be dynamic, such as crossing between complementaries.

The lack of rigor in exploring these three models of sound-color correspondence makes the Toccata and Fugue the exemplary film for the shift in color music from an experimental art form to a mass-cultural diversion. A light show at a rock concert or on a screen saver does not make demands on its viewers, force them to look for structure, or make them think about the parallels and divergences between seeing and hearing. Even so, such abstract images do invite the spectator-auditor to consider color in the abstract, apart from the world of material objects, in a much less intimidating environment than the museum or salon. Moreover, such images also invite the spectator-auditor to experience a relation between images and sounds that goes beyond the causal or the illustrative, and instead hints at the autonomy of both.

# PART IV

# Film Sound and Cultural Studies

# 12

# "A Question of the Ear"

## *Listening to* Touch of Evil

**TONY GRAJEDA**

A seedy little town at the U.S.-Mexico border is the divided ground upon which conflict arises in *Touch of Evil* (Orson Welles, USA, 1958), a place sustaining the belief that, as one character insists, "All border towns bring out the worst in a country." This geographical space where national boundaries meet stages the fine line of more conceptual borders, as the film establishes a series of binary relations (light and dark, white and brown, law and crime, purity and corruption), only to see those dichotomies disrupted and rendered ambiguous. Indeed, as so much of the scholarly and critical commentary on the film has emphasized, the narrative's structural oppositions are eventually and irredeemably destabilized by the constant crossing of boundaries (both spatial and metaphorical), effectively casting doubt on the very notion of the border itself. But where the literature on *Touch of Evil* has addressed the "problem of race" in the film, such work has often sustained that border defining "American" and "Mexican," failing to interrogate how miscegenation—where identity itself is in question—functions as a driving force of the narrative. Even less has been said about how sound operates in the film and how it too crosses boundaries—not least of which cross those very same racial-ethnic identities—to create, both diegetically and with regard to spectatorship, what might seem to be yet another state of ambiguity.

In what follows I take up the "border" in *Touch of Evil* as a critical metaphor, exploring the ways in which the demarcations of even these rather overlooked but inexorably linked aspects of the film are traversed—emblematic of that space of uncertainty represented by the border town, or *la frontera*. My discussion begins by examining some of the scholarly work on *Touch of Evil*, focusing in particular on those texts that valorize its thematic ambiguity and stylistic disorientation, much of which treats the film as something of a modernist text that resists definitive meaning. Such work, however, has largely excluded the representation of ethnicity from the field of ambiguity, such that the "Mexican" in this discourse remains a fixed identity, in effect foreclosing the film's question of miscegenation. Instead, I argue how the film blurs the border of identity formation not only by foregrounding the hybrid or "half-breed" subject but also by disguising and displacing the conventional treatment of the voice—the very sound that is believed to guarantee a "true" self—thereby audibly "mixing" identity and thus unsettling the otherwise rigid category of ethnicity.

Although *Touch of Evil* provocatively engages the border-crossing trope of identity-formation and subjectivity, it also stages an encounter at the border between the field of vision and that of sound. Such an encounter first takes place thematically and diegetically within the filmic text, as characters struggle to align seeing and hearing (with what they know or do not know), or, even more fundamentally, to simply hear or to be heard. This complex relation between what is seen and what is heard arrayed across the narrative also constitutes—given the film's singular soundscape—a particular phenomenological experience with regard to spectatorship, insofar as the film often blurs the distinction between modes of perception to create at least momentary confusion for the spectator-auditor. This perceptual confusion, as I aim to demonstrate, signals a gesture toward Brechtian aesthetics. Its guarded but unmistakable effect can be heard at those moments when the film quite deliberately foregrounds the technological apparatus of sound, reflexively rendering both visible and audible the machinery of recording, transmitting, and amplifying not just as tools of the trade but as means of production.

Both the border condition of screened subjectivities and that of perception for a listening subject are overdetermined, finally, by the border of history itself, where the film of 1958, of 1976, and that of 1998 each entail a difficulty in situating the text (or texts) that is *Touch of Evil* within a fixed cultural history. It is worth noting that these different versions viewed during different periods gather together to form their own history of reception, including a rather unsettled place in cinema studies. Given that it now occupies multiple temporal framings, thus problematizing its position in a cultural history of representation, one could argue that *Touch of Evil* itself has become a text that embodies *la frontera,* that condition of irreducible ambiguity with regard not only to space within the frame but also to time outside it.[1] Though mindful of the film's celebrated "restoration," which rather subtly but quite dramatically reedited the two earlier versions into a 1998 release apparently more in keeping with the director's expressed intentions, my reading of *Touch of Evil*—informed by a cultural studies approach to film that also attends to a close analysis of the work of sound—depends less on its most recent reworking and more on its already given audibility. That is to say, though

I will briefly consider the famous opening sequence and its remixed soundtrack for how this reconstruction displaced one aesthetic model for another, I am primarily interested in exploring the film's pronounced treatment of making sound and listening essential to its narrative. Indeed, regardless of version or even the wayward instability in conditions of reception, *Touch of Evil*, as I argue, persists as an audiovisual text that asks us to reflect on the faculty of listening itself.

## A Film's Conflicted Reception and the Politics of Difference

The literature on *Touch of Evil* has generally coalesced around the position staked out by film theorist André Bazin who proclaimed that it exemplified a cinema of "ambiguity."[2] Accordingly, a great deal of the commentary on *Touch of Evil* has stressed its illusive qualities, the ways in which it confounds thematic and stylistic certainty, an interpretation perhaps most thoroughly articulated in Terry Comito's essay "Welles's Labyrinths."[3] Comito argues that the film's "expressive content"— the crossing of borders, the world experienced as "a whirling labyrinth"—is "indistinguishable from its visual style," which seeks to disorient the spectator through unpredictable and rather unconventional camera angles, framing and movement, camera work that destabilizes the real as it assaults "our certainties" (12, 15). This "visual instability" is further matched by what Comito calls an "instability of tone" (31), whereby the film manages to elude genre categories, thus inhabiting that "ambiguous border region" somewhere "between melodrama and farce" (30).

The cultural work of the film—its ability to generate disorientation for viewers and critics alike—seems to reach its own border, however, when it comes to the film's representation of ethnicity. Again, commentary from reviews and film scholarship on *Touch of Evil* is instructive here. If the narrative's racial conflict is not lifted onto some allegorical plane entailing the nature of good and evil, it lingers—in many of these readings anyhow—at the level of hackneyed preconceptions. This is most striking when the "Mexican" is figured as Other, functioning "with almost predictable regularity," as William Anthony Nericcio argues, "synonymous with the exotic, the dark, the sordid, the sexual, the decayed" (56). One film scholar, for example, characterizes the Grandi family as "the scurviest group of misfits this side of *Los Olvidados*" (McBride 137), implying, if nothing else, a certain consistency across genres in the representation of the Mexican. Or consider the following commentary by James Naremore, who makes the classic mistake of substituting "window" for "screen" when he claims that the story's setting "is quite true to the essence of bordertowns": "On the streets we see strip joints and prostitution, a few ragged Mexican poor, and a couple of men trundling fantastic pushcarts. . . . The town . . . exists by selling vice to the Yankees, functioning as a kind of subconscious for northerners, a nightworld just outside their own boundaries where they can enjoy themselves even while they imagine the Mexicans are less civilized" (Naremore 158). Perhaps, as Nericcio points out, the usually astute Naremore "should have written, *true to the essence of border-towns as represented in Hollywood movies*" (49; emphasis in the original).

What is curious about these observations, besides this tendency to conflate representation and reality at very specific (and suspect) moments, is that not only does *Touch of Evil*, as Nericcio contends, work "constantly to challenge patented Hollywood stereotypes" (51), but it also rarely lets up in reminding us of its status *as* representation. The film's awareness of its own fictiveness can be heard precisely when it speaks of "Mexico," which always seems to exist in quotation marks. Right from the beginning, Ramon Miguel "Mike" Vargas (Charlton Heston) is at pains to protest that this border town "isn't the real Mexico." Hank Quinlan's first exchange with Vargas is punctuated with: "You don't talk like one, I'll say that for you. A Mexican, I mean."[4] And while there is also a running joke throughout that Quinlan (Welles) does not speak "Mexican," a more earnest struggle over language and translation (which will be examined shortly) is waged time and again, as when Susan (Janet Leigh), desperately trying to speak Spanish, asks, "Qué es eso?" adding hopelessly, "I don't speak any Spanish," or, more unexpectedly, when Vargas, shouting in stilted Spanish demanding the whereabouts of his wife, attacks Risto, who says, "Talk English, can't you?" Finally, upon seeing Vargas's wife, Quinlan quips: "Well, whatta you know! She don't look Mexican either." As Nericcio notes, characters who either "don't look Mexican" or "don't speak Mexican" in a space that "isn't the real Mexico" (51) reveal the ways in which the film seeks to undermine our overreliance on such handy markers of identity as appearance, voice, and language, although we would barely know this from its reception by film scholars, so many of whom largely insist on reading these characterizations quite literally. Even reviews occasioned by the film's 1998 release persist in accepting that the "Mexican" must represent, must signify: "Susan is terrorized by a Mexican gang that wants Vargas to stop investigating its crimes. She's threatened, waylaid, assaulted, doped up. Soon, Quinlan teams up with the Mexican gang lord to smear the Vargases and worse" (Levy n.p.).

If *Touch of Evil* "calls into question," as Comito claims, "any definition whatever, any attempt at coherent representation" (29), why, we might ask, does the critical discourse surrounding the film exclude the representation of ethnicity from the realm of ambiguity? Why must those identities remain knowable, remain—just that—identifiable? Even when the narrative's surface trappings are treated metaphorically, the question of race returns like the proverbial repressed. For instance, though acknowledging that Welles deliberately transferred the setting of Wade and Miller's novel *Badge of Evil* to the Mexican border, Comito insists that the film's "Mexico is not a geographical place Welles represents" but rather "a place of the soul, a nightmare" (11–12). Yet this poetic displacement does not seem very far removed since, as he observes, this "sinister foreign place" into which Vargas and Susan have been drawn, where "it becomes increasingly difficult for them to maintain the boundaries of their normalcy," is precisely where they are "subject to alien incursions" (11, 20). He writes, "What *Touch of Evil* is most immediately 'about'—certainly more than it is about race relations or police corruption—is the confrontation with the labyrinth or vortex that opens before us once we transgress the boundaries of a world in which we are pleased to suppose ourselves to be at home" (12). Comito's certainty about the film's uncertainty, perhaps ironically, is rather partial if not somewhat opportunistic, a reading that does violence to the narrative's more explicit representation of "race relations or police corruption," a move that subsequently allows Comito to insist on the film's "ethical ambiguity to which Welles's crossing of borders has, in spite of his own protestations, led us" (29).[5]

What is perplexing, finally, about the racial unconscious of such textual production is the extent to which so many critical voices fail to listen to what the film itself has to say—rather loudly in fact—about naming, projecting, and fixing identities perceived as "other." The aforementioned "Mexican gang," for example, which one critic typed "as smarmy members of the Mexican underworld" (Karten), are led by "Uncle Joe" Grandi (Akim Tamiroff) who, in a confrontation with Susan, states, "The name ain't Mexican. . . . The Grandi family's living here in Los Robles a long time. Some of us in Mexico, some of us on this side." The arbitrariness of the border is more fully dramatized in the narrative's many interracial relationships: the newlyweds, Mike and Suzy Vargas; the illicit marriage of Marcia Linnekar and Manolo Sanchez, the lead suspect in the murder of Marcia's father, Rudy Linnekar, and his mistress, Zita, the "sizzling stripper."

And "at the heart of the film," as Nericcio argues, is "an unseen, unnamed half-breed" (53) who, Quinlan believes, strangled his wife back when he "was just a rookie cop," adding "that was the last killer that ever got out of my hands." Functioning as a structuring absence, the film's "obsession" with miscegenation stems from the figure of the "half-breed," that liminal subject of undecidability who "arouses Quinlan's hatred." Yet it can also be said that this "mixed" identity arouses Quinlan's desire as well, given his yearning for the presumably Mexican madam Tanya (Marlene Dietrich), whom Stanley Kauffmann calls "a world-weary Mexican tart" (31). Indeed, I would push Nericcio's analysis further to argue that even Quinlan's identity is less stable than it appears, shot through with both fear of and fascination with the other, that is, at least to the degree that Quinlan's personality—his heaving, sweaty "whiteness"—is structured in part through his desire for Tanya, who haunts him with memories from an irretrievable past, a past that is also haunted by his wife's murder. Moreover, as Stephen Heath observed,

Quinlan strangles Grandi in much the same way as the "half-breed" killed his wife, symbolically at least *becoming* the half-breed himself.[6]

My point here is that in *Touch of Evil* identity itself is largely in doubt, a figure of instability often elided by scholarly reception of a film otherwise celebrated as a "cinema of ambiguity." Although identity as such can be said to serve an ideological function—transgressing the ground of history in becoming something of the limit test to this discursive consensus on the film's status as an ambiguous text—it is also important to acknowledge the film's historicity at its moment of production (let alone its subsequent reception). For Kelly Oliver and Benigno Trigo, "*Touch of Evil* is a film that visually and narratively insists on inverting the racial equation predominant during the forties and fifties in the United States that constructed North American whites as superior to so-called colored races" (118). Given the extent to which such concerns have been equally pressing, from 1958 through 1976 and continuing past 1998, the film's substantive resonance with the "racial equation" is as much historical as it is theoretical. Nevertheless, in the midst of so much film scholarship emphasizing the thematic and formal ruptures of *Touch of Evil*, particularly work that valorizes its disorientation and discontinuity, I would stress instead the film's recurring struggle with identity itself, one that paradoxically threads its way through the entire narrative, indeed functioning as a persistent, if uneasy, trope of continuity.

## The Audible "Mixing" of Identity

Whereas visual signifiers in *Touch of Evil* occasionally concede a state of (mis)-recognition for those who do or "don't look Mexican," another layer of ambivalence over identity formation settles around a speaking subject, especially a subject who triggers a sense of anxiety over miscegenation in terms of language for those who do or "don't speak Mexican." More specifically, the film casts doubt on that most fundamental of markers believed to ensure identity: the voice itself. Issuing this immanent "property" of both a subject and the soundtrack, the narrative's central concern with "mixing" can also be heard to take audible form, in particular with what I would call the fugitive role of the voice, one that serves to trouble the notion of a fixed identity. To amplify this point, let us listen to the voice of "Pancho," or should we say *voices,* for over the course of the film he becomes indeed the man of many voices.

We are first introduced to the character whom Susan will dub Pancho (Valentin De Vargas) when he saves her by pulling her back from stepping in front of an onrushing truck.[7] Apparently unable to speak English, he speaks to her in Spanish, which goes untranslated (to an English-speaking audience) but which Susan "reads" as a proposition. Yet soon after this scene we hear him call out to one of the other Grandi boys, Risto, in Spanish-inflected English: "Wait there. Uncle Joe is plenty mad. He wants to talk to you." And in a few short scenes after this, Pancho will speak to Susan over the motel phone in "perfect" English: "Don't you worry, Mrs. Vargas. I'm the one in charge here. Nobody's going to get through to you . . . unless I say so."

Displaying what could be considered a kind of vocal masquerade, "Pancho" performs a number of voices in rather quick succession, including the imitation of a wise-guy sneer, when he says to Susan over the phone, "Where would you like me to take you, doll?" From what sounds like a spoof of the hard-boiled voice familiar to film noir, "Pancho" suddenly feigns a serious demeanor in his temporary role as the Mirador Motel's switchboard operator ("Very good, ma'am"), before switching to what could be his "normal" voice in speaking to fellow gang members ("We relax and have ourselves a ball"). And when Vargas calls asking to speak with his wife, Pancho puts on once again the role of feigning the concerned proprietor by responding, "I'm very sorry, Mr. Vargas. But your wife left definite instructions: she's not to be disturbed." Finally, during the assault on Susan in the motel, and shot in a tight, wide-angle close-up on his face to enhance the effect, he issues a menacing whisper—"Hold her legs."

Traversing this brief sequence, and across the length of the film, the "true" voice of Pancho is difficult if not impossible to detect. Rather, he seems to perform a number of speech acts, enacting a voice that is multiple, mobile, situational. This fugitive voice calls to mind the work of Michel Chion who, in *The Voice in Cinema*, interrogates what he terms *vococentrism*, the standard filmmaking practice in the construction of "every audio mix" in which "the presence of a human voice instantly sets up a hierarchy of perception." For Chion, *"the presence of a human voice structures the sonic space that contains it,"* a condition bearing nearly ontological weight since, as he claims, "Human listening is naturally vococentrist" (5–6; emphasis in the original). Yet if cinema generally exploits this privileging of the voice in its soundtrack work, *Touch of Evil* instead appears to undermine that strict hierarchy from within the very structure that is supposed to guarantee dialogue intelligibility. There are other moments in the film, as will be taken up shortly, when such intelligibility is in doubt, but here it is the differentiation of dialogue within the vococentric model that in effect offers an experience of liminality through sound. Called upon to summon an authentic identity yet heard in its multiplicity, the centrality of the voice itself here is suffused with irreducible difference—the performative movement of fugitive voices that signal the audible mixing of identity.

Although one could point to the playful staging of how Pancho switches voices precisely when he takes up the role of switchboard operator, I would suggest that this switching of voices and identities by way of technological mediation is a recurring motif in *Touch of Evil*. Indeed, the film's narrative often foregrounds the mediation of voice and listening through audio technologies not only in that so much dialogue, as we have seen here, is conducted over the phone, but also, most famously, in the closing surveillance sequence, in which a recording apparatus serves to separate voice from body. Before turning our attention to the narrative function of audio technologies in the film—technologies that could be perceived as threatening to "denaturalize" the place of the voice—it is worth pausing a moment longer to consider some of the ways in which the border, or *la frontera*, is spoken in *Touch of Evil*, for beyond the performative voice, as exemplified by Pancho, there are also many accented voices of rather indeterminate origins.

As a number of scholars have noted, the narrative is inhabited by what James Naremore calls a "mélange of nationalities," embodied by a wide range of actors, from the Waspish American leads and their various disguises (Heston's bourgeois Mexican official, Joseph Cotton's heavy southern drawl) to the multiethnic Valentin De Vargas (Mexican-Italian-American), and such cinematic émigrés as Dietrich (German), Tamiroff (Russian), and Zsa Zsa Gabor (Hungarian). More important, the linguistic markers of numerous accents approximating spoken English or Spanish problematize the film's audible rendering of realism. This "mélange" of accented voices, as Naremore argues, "was certainly not intended to represent 'the real Mexico,'" yet he adds, "nevertheless the crazy, nightmarish distortion of that film was an *interpretation* of a real place, and the various accents are wittily appropriate to the bordertown" (203; emphasis in the original; Leeper 233). Yet one could also argue that rather than reflecting the mélange of a border reality, the film instead signifies an "outside" beyond the frame, the intertextual universe of Hollywood stars no doubt recognizable to audiences at the time, best illustrated by Dietrich's nearly self-referential performance.[8] Such moments of intertextuality mobilize Welles's own border aesthetic, the hybrid crossing of realism and pulp modernism that occasionally punctures the reality principle otherwise carefully crafted.

But it is within the contours of the narrative proper where the struggle over language ensues, a struggle moreover in which one's facility with translation marks out the contested terrain of *la frontera*. Nowhere is this more in evidence than during the scene in Marcia Linnekar's apartment where Quinlan interrogates Manolo Sanchez (Victor Milan). As the questioning proceeds and Sanchez grows increasingly agitated, he begins "code switching" between English and Spanish, only some of which is translated into English by Vargas.[9] "I don't speak Mexican," declares Quinlan, who becomes agitated himself by the spoken Spanish: "Keep it in English, Vargas!" The capacity to pass from one language to another, as Oliver and Trigo argue about *Touch of Evil*, is treated with suspicion. "If another language must be spoken," they write, "it will be spoken with a heavy accent and will be left untranslated to preserve its foreignness" (131). Thus, Sanchez's inability to repress his "mother tongue"—his apparently unavoidable use of heavily accented English—allows Quinlan to accuse Sanchez by announcing, "He speaks a little guilty." Meanwhile, Vargas, who acts as a switch between languages, has, according to Oliver and Trigo, "mastered the art of speaking English in such a way that he can avoid shifts into Spanish." They insist that since Vargas "has managed to erase all traces of his Spanish accent" (131), this successful suppression of his mother tongue serves to align him with Quinlan in terms of such "linguistic values" (133) of class and custom as the public display of politeness and access to and knowledge of the law.

Yet it must also be pointed out that Quinlan's "English-only command" (133), in which he is dependent on Vargas to translate spoken Spanish for him, basically aligns Quinlan with an audience presumably also placed outside the untranslated Spanish, a condition of spectatorship thus occupying a position of alienation (or at least bewilderment) along with Quinlan. This suggests a split along linguistic lines between the semantic and the phatic. As opposed to the semantic voices that are meant to be heard and understood (those spoken in English), the phatic voices, which are heard but not necessarily understood by a presumed audience, are spoken in Spanish that goes untranslated. For example, following the "discovery" of dynamite in the apartment, Sanchez pleads with Vargas in untranslated Spanish, to which Vargas says to Quinlan: "You'll have to stop him yourself." Quinlan answers by stating, "From now on, he can talk Hindu for all the good it'll do him." But Sanchez's untranslated Spanish—the phatic sound that for Oliver and Trigo marks its "foreignness"—is also laden with affect and other points of identification. With regard to a non-Spanish-speaking audience, sympathy is not attached to Quinlan's rage over finding himself outside this discourse community and thus incapable of mastering all communication; rather, we are asked to identify with Sanchez in spite of his accented speech and unmistakable "mother tongue." We have already been positioned to realize that Sanchez could very well be innocent, given an earlier shot in the sequence revealing that the dynamite had been planted.

Here then is one of the key moments in the film when the visual qualifies if not nullifies the oral and verbal, in which the sight of physical proof trumps any and all spoken language. The rhetorical rendering of events—from careful vocal explanations to the discourse of the law—withers once the thing itself appears, the stubborn objecthood that, like Quinlan's cane left at the scene of the crime, cannot be explained away. It is no accident then that the narrative resolution to the uncertainty of language will lead to a mechanical device of recording, in which words are turned into things, providing "concrete" verbal evidence from which one cannot talk his way out. Yet even the taped testimony ensuring that truth will be told will somehow prove elusive. As with the inscribed truth of identity, the film will call into question the ontology of the audible object, crisscrossing the perceptual border of what is seen and what is heard with doubt over the machine assigned to register the real.

## The "Listening Apparatus" and the Sound of Difference

The blurring of borders legislating identities of ethnicity become overdetermined then by the larger role of sound in *Touch of Evil.* By arguing that a crucial aspect of the narrative pivots on the representation of the faculty of hearing, I want to amplify the ways in which the film foregrounds the mediation of listening by technology. How audio technologies are presented serves to activate, intensify, or interrupt the ways we hear and apprehend our surroundings, such that the border between seeing and hearing, which diegetically for the characters becomes unsettled, also collapses at the level of familiar and more accustomed patterns of spectatorship and auditorship. Yet at another level, the border between seeing and hearing has been restructured, not only by the way the film reverses the conventional cinematic relation of image and sound—where the latter is typically subordinated to the former—but also by its attempt to unmask the signs of its fabrication, quite specifically refusing to repress what Mary Ann Doane calls the material heterogeneity of the soundtrack ("Ideology and the Practice of Sound Editing and Mixing").

If we begin by *listening* to *Touch of Evil,* we cannot help but notice the palpable density of voices, where characters frequently and "habitually speak at cross purposes, or all at once" (Comito, "Welles's Labyrinths" 17). Here, Welles's signature technique of overlapping dialogue, in evidence since *Citizen Kane* (USA, 1941), contributes to the overall sense of disorientation that Terry Comito highlighted, a position also advanced by Phyllis Goldfarb in her essay "Orson Welles's Use of Sound." Goldfarb focuses on those confusing moments in the film when sound has been separated from its source—such as the "post-explosion" cacophony of voices—when, in her words, "the visuals and aurals don't fit" (87). By disturbing instead of defining filmic space so that the spectator-auditor loses "all sense of distance and direction," such spatial mismatching and fragmentation "has the effect of partially disorganizing our perceptions," "undermining," as Goldfarb puts it, "aural reality" (88).

These moments of disorientation need to be heard, however, in relation to the many occurrences of aural interpellation, when dialogue intelligibility is secured through what in film sound studies has been called point-of-audition sound (the aural equivalent roughly but sometimes misleadingly to the point-of-view shot). As Rick Altman maintains, "Point-of-audition sound always has the effect of luring the listener into the diegesis not at the point of enunciation of the sound, but at the point of its audition," adding, "We are asked not to hear, but to identify with someone who will hear for us" ("Sound Space" 60–61). This process of placing us within the narrative as an "internal auditor" is most clearly achieved (or at least attempted) through the aural close-up, such as when Suzy listens attentively to a whispered voice through the motel wall, the hushed phone conversation with her "own darling Miguel," or, near the end of the film, when Quinlan murmurs to Menzies, and us, "See? That oil pump . . . pumping up money . . . money."

Although much has been made in film scholarship of Welles's justifiably celebrated practice of overlapping dialogue, far less attention has been paid to why and when this less radical "other" sound arrives—when this thicket of voices has been pruned back to a relatively long line of uninterrupted dialogue—when the principle of montage gives way to something resembling continuity. That is to say, to emphasize the disorienting use of sound (for what Goldfarb calls "a negation of reality") at the expense of these point-of-audition moments fails to do justice to the full range of sound strategies in *Touch of Evil*, an emphasis that effectively submerges the realist aspect of the film's aural aesthetic beneath its modernist tendency toward incommensurability with the real.

In a similar and related way, a complex play of strategies between aural continuity and discontinuity can also be heard, and nowhere more so than in the surveillance sequence that closes the film when Vargas enlists Pete Menzies (Joseph Calleia) to surreptitiously record Quinlan. In what he calls "a bravura display of montage," Terry Comito describes how "three distinct kinds of sound" can be recognized during this sequence: "'live' voices, recorded voices, and voices that are live but heard from a great distance, often doubled by echoes. A given sequence, or even line of dialogue, will characteristically pass through at least two of these registers and often all three. In this kind of sound montage, words become as fragmented as the images, literally disembodied, sundered from their presumed origin" ("Welles's Labyrinths" 18).

What Comito means here by "sound montage" is not exactly clear since, as he points out, even when a line of dialogue passes through different registers, that dialogue, though not seamless, remains fairly complete, continuous, and semantically coherent. Even in its mediated state, such dialogue is not in fact fragmented beyond recognition, for we can still follow its trail through abrupt cuts in volume (matching scale signaling proximity or distance), reverberation, and frequency or fidelity. For example, at one point in the sequence, we hear Quinlan's voice over the speaker, "And then this Mexican comes along, and . . ."—cut to his "live" voice—"look at the spot he puts me in." Or, Quinlan's voice again, this time echoing: "Vargas! Do you hear me? I'm talking to you. Through this . . . this walking microphone that . . ."—cut to his voice over the speaker—"used to work for me!" The sound work here is split between two seemingly opposite yet not inconsistent levels: the fracturing of the dialogue takes place acoustically, yet at the textual level of audibility it remains continuous and "whole."

Considering the technical limits of available audio equipment at the time, the film's sound crew devised a remarkably complex soundscape within the dialogue track alone, creating phenomenological tension between the technological rendering of sound's ambiguity in the acoustic dimension and that very same sound's legibility as narrative information. When asked, in one of the *Cahiers du Cinéma* interviews, why his "montage fragments" seemed so long, Welles stated, "I seek the exact rhythm between one frame and the next. It's a question of the ear. Montage is the moment when film engages the sense of hearing" (Bazin, Bitsch, and Domarchi 202). In this fascinating statement on cutting to sound, Welles suggests a more nuanced account of his sound practice, less a dichotomous relation between narrative continuity and discontinuity than a dialectical one—"a question of the ear" (202).[10] Such a question is raised in the relation not only between shots but also *within* the long take.

Here we might note a disjunction between sound and its source that occurs with the use of off-screen sound, such as the scene in Marcia Linnekar's apartment, where the most brutal moments of Sanchez's interrogation take place outside the frame. This separation between what is seen and what is heard resonates across the film—a separation often generated by the unmistakable and nearly ever present mediation of sound by audio technologies. Although the role and reiteration of radios feature prominently, it is the telephone, as already noted, that is employed extensively in the narrative, not only to conduct a whole series of conversations but, significantly, to execute the terror that Suzy will be subjected to in the Mirador Motel scene, when she is initially menaced by way of phone and loudspeaker. But it is during the surveillance sequence—instantiating microphone, speaker, and tape recorder—where the film most obviously manifests technological mediation itself, as well as where it heightens the racial conflict, a conflict staged not surprisingly at the border.

In *The Acoustic Mirror*, Kaja Silverman offers some brief notations on this scene, underlining how "hearing is played off against the voice, so that the two are defined in terms of antagonism rather than complementarity" (54–55). During the sequence, "Vargas does not speak." As Silverman indicates, "He is 'pure' ear, his own hearing supplemented by the apparatus, and closely identified with it." At the same time, "Quinlan, on the contrary, neither sees nor hears Vargas," conversing with Menzies "until he hears his own voice echoed back to him by the tape recorder." That Quinlan speaks while Vargas listens implies that together they form a single body, one both joined and divided by the apparatus. Further, this body mediated by technology, where the voice is separated from the body, becomes nearly inseparable from the industrialized landscape of its staging. "The camera cross-cuts between the two scenes," as Silverman notes, "so that the space occupied by each functions alternately as 'on' and 'off'" (55). That is to say, the entire sequence enacts the process of recording, with the space itself materializing as a tape recorder—*mise-en-scène* as machine.

Consistent with Welles's Brechtian aesthetics,[11] *Touch of Evil* quite clearly foregrounds the apparatus in order to call attention to its status as medium, the self-reflexive process of revealing its own means of having been constructed. Such an inference could be drawn from the numerous ways in which sound is denaturalized throughout the film—for instance, with the use of the noise of radio static and interference made audible. As such, the film could be heard as counterhegemonic, going against the grain of the dominant Hollywood practice of effacing all traces of its production, a challenge to convention in fact from within the studio system.[12] What also should be recognized is that this demystification of the work of sound technology—a strategy that could more properly be expected from the European avant-garde or American underground cinema—took place within the realm of mass culture, implying contradiction not only within the culture industry but also within the model of Hollywood as the ideologically seamless, well-oiled factory of artificiality and the illusion of reality, a model promulgated by so many film theorists over the years. At any rate, it is quite possible to imagine that Welles was "baring the device" here, calling upon the aural dimension to breach the theatrical space between screen and spectator while drawing our attention to the material work of sound that usually goes unnoticed and unheard.

The attempt here to denaturalize the work of sound as an effect of technological mediation implies a form of aural realism somewhat at odds with both the film's critical reception as a triumph of pulp modernism and its more recent "restoration" that, at least to some extent, rewrote part of the film's aesthetic of sound to create, ironically, a more traditional form of cinematic naturalism. Let us return then to the beginning of *Touch of Evil* and its justly famous three-minute, twenty-second tracking shot. Without fetishizing the "original" soundtrack for this opening scene, it is important to compare how the sound was dramatically altered for the 1998 version, suggesting altogether different aesthetic strategies for the treatment of film sound.

The 1958 studio release overruled Welles's designs for the scene by not only adding superimposed titles but also superimposing Henry Mancini's score, the theme that established what Welles believed was the wrong tone for the film (Murch, "Restoring" 16; Leeper 231). The studio reedited the film while Welles was out of town and added four mostly expository new scenes shot by director Harry Keller (Ondaatje 184); this in particular raised the director's ire and resulted in his legendary fifty-eight-page memo written to Universal Pictures studio chief Edward Muhl in 1957. Welles begins the memo by stating, "I assume that the music now backing the opening sequence of the picture is temporary," which continues:

As the camera moves through the streets of the Mexican border town, the plan was to feature a succession of different and contrasting Latin American musical numbers—the effect, that is, of our passing one cabaret orchestra after another. In honky-tonk districts on the border, loudspeakers are over the entrance of every joint, large or small, each blasting out its own tune by way of a "come-on" or "pitch" for the tourists. The fact that the streets are invariably loud with this music was planned as a basic device throughout the entire picture. (Quoted in Ondaatje 185)

Further into the memo, moving from the above "pitch" for a specific sound design evidently based on an ethnographic account of the real, Welles describes his plan for using "Afro-Cuban rhythm numbers," "traditional Mexican music," and "a great deal of rock 'n' roll," each of which are given particular "roles" at quite precise moments. He writes, "This rock 'n' roll comes from radio loudspeakers, juke boxes and, in particular, the radio in the motel. All of the above music, of course, is 'realistic' in the sense that it is literally playing during the action" (Murch, "Touch of Silence" 86; Brady, 502).

Finally, in reference to the Ritz Hotel scene with Susan and Uncle Joe Grandi, Welles insisted on a very specific set of effects for the treatment of the sound, with the memo breaking into capital letters for added volume:

IT IS VERY IMPORTANT TO NOTE THAT IN THE RECORDING OF ALL THESE NUMBERS, WHICH ARE SUPPOSED TO BE HEARD THROUGH STREET LOUD SPEAKERS, THAT THE EFFECT SHOULD BE JUST THAT, JUST EXACTLY AS BAD AS THAT. The music itself should be skillfully played but it will not be enough, in doing the final sound mixing, to run this track through an echo chamber with a certain amount of filter. To get the effect we're looking for, it is absolutely vital that this music be played back through a cheap horn in the alley outside the sound building. After this is recorded, it can then be loused up even further in the process of re-recording. But a tinny exterior horn is absolutely necessary, and since it does not represent very much in the way of money, I feel justified in insisting upon this, as the result will be really worth it. (Quoted in Murch, "Touch of Silence" 88; emphasis in the original)

Such statements reveal a certain aesthetic of aural realism that Welles sought, in which "realistic" source music "loused up even further"—perhaps for an even greater degree of realism—along with a noticeable amount of ambient sound of the street, the noise of nightlife and so on, would situate an audience more fully within the diegetic world of the film.

The sound team that produced the 1998 version of *Touch of Evil* aimed at fulfilling what Welles called for but was prevented from achieving. Working from a decent negative for the 1958 version found in the Universal vaults and the original monophonic three-track master (a magnetic mix of separate channels for dialogue, music, and sound effects), Bill Varney and Walter Murch were able to remove the music track that contained Mancini's title music in the opening scene, thereby revealing, in Murch's words, a "complex montage of source music" ("Restoring" 16). This "hidden layer of sound effects that had been suppressed during the original mix" was "restored to its original balance in the film," according to Murch, thus "allowing the audience to hear the town, the footsteps of the pedestrians, their voices, the laughter of the crowds, the sirens—even the bleating of a pack of goats stuck in the middle of the road" ("Restoring" 16; "Touch of Silence" 87). The result of this remixed opening is that, as Murch states, "viewers are immediately engaged with the film's storyline and plunged into its particular atmosphere" ("Restoring" 16), an engagement that had been unduly obstructed in the earlier versions by the splashy titles and Mancini's theme music dominated by blaring horns. As with the classical narrative dispensation of interpellation, the film's spectator, now free of any interference run by the textual intrusion of credits or the nondiegetic music from "nowhere," is more easily absorbed into the diegesis. Moreover, as Murch notes elsewhere, the new mix has become the "aural equivalent of the camerawork" (Ondaatje 186). Given the impossible perspective of a two- or three-story crane shot, tracking in and out of traffic and swooping across the border, this rewritten soundscape implies a form of what can only be called impossible listening: an aural perspective that is not "real" but instead can be gained only by spectacular cinematic techniques. In other words, what Murch and the sound team have achieved here in (finally) matching image scale to sound scale is a return to the cinematic perception of an "invisible observer" and its "aural equivalent" (rather than creating a sense of aural realism), an audible effect of contributing to the fantastic illusionism of standard Hollywood practice.

By trading one form of cinematic artificiality for another, the 1998 *Touch of Evil* now more properly re-creates a fictional world of border "realism." Yet what I would suggest is that the "inappropriate" score and superimposed titles of the original release ironically produced the kind of distancing effect that in some ways is more consistent with Welles's otherwise conflicted aesthetic, one that by film's end reminds the spectator-auditor of the constructedness of the (aural) text.

It is interesting to note that the restoration crew did not alter the original sound mix for the closing surveillance sequence, the far less famous other end to the much celebrated opening "tour de force." We might never know to what extent the studio version of 1958 hindered or fulfilled the director's intentions for an ending that for now at least remains a relatively stable text. What does remain, nonetheless, is a text that renders diegetically the apparatus of sound. Though not exactly Brechtian in foregrounding its alienation effect, the surveillance sequence does suggest a third path between the competing interpretations of the film's aesthetic

dimensions, between the pulp modernism of grotesque fiction and a border referentiality committed to the demands of realism. What the film points to here,
at least in so far as sound itself becomes the subject of the film, is the degree to
which technological mediation has become the reality of experiencing sound.
That is to say, the film recognizes as it makes audible sound's modernity, which is
instantiated in the perceptual realization that the ontology of sound is thoroughly
constructed in a process of material production and mediated reception.

What is less clear, perhaps, returning to the closing scene once more, is whether
the film resolves the racial conflict, the staging for which this binding and dividing of a mediated body at the border is played out through the figuration of
ethnicities in the narrative, where power relations defined in part by identity are
reversed or displaced. After all, this final scene, finally, is where an exchange of
identities takes place, in which Quinlan's "famous intuition" (his game leg "talks"
to him) turns out to be "right after all," whereas the once unimpeachably virtuous
Vargas stoops to "creeping about." If Vargas is the one who crosses over by wad-
ing through the water, becoming—dare we say it—the "wetback," it is Quinlan
who does not make it back, the one who in the end goes belly up. If *Touch of Evil*
transmits a kind of flickering of identities—the back and forth, "on and off," between "Mexican" and "American," thereby embodying the space of the border—it
also stages that movement for an audience then and now as well as for a seeing
and listening subject, an experience not necessarily *at* but *of* the border.

## Notes

1. On the question of "textual authenticity" in Welles's work, especially in regards to
*Touch of Evil*, see Naremore, *The Magic World of Orson Welles;* and William Anthony
Nericcio, "Of Mestizos and Half-Breeds: Orson Welles's *Touch of Evil*." Emerging from a
theatrical practice of forever revising the text and embracing the contingency of performance, Welles seemed to relish the editing process itself, even expressing how he wished,
not unlike Griffith, that he could "go to the projection booth and start snipping away"
(Welles and Bogdanovich 49). On the fundamental instability of cinema not just as event
but as text, see Altman, "General Introduction: Cinema as Event" 10–11.

2. See the 1958 interviews with Welles by *Cahiers du Cinéma* in which a discussion on
the director's formal strategies, specifically privileging depth of field shots over radical
montage that contributes to a sense of ambiguity for spectators, leads Bazin to argue
for a thematics of ambiguity, a topic that will be taken up shortly. See Bazin, Bitsch, and
Domarchi, "Interview with Orson Welles"; and Bazin, "Return to Hollywood: 'Using Up
My Energy.'"

3. Comito's introductory essay appears to be based on an earlier publication, "*Touch of
Evil*," in *Focus on Orson Welles*, edited by Ronald Gottesman. Similar interpretations are
offered by Eric M. Krueger, "*Touch of Evil*: Style Expressing Content"; and Paul Arthur,
"Orson Welles, Beginning to End: Every Film an Epitaph."

4. Part of the perhaps unintended humor here is that Charlton Heston, fresh from playing Moses in *The Ten Commandments* and now done up in brownface, delivers his lines in an Anglo accent, even when he speaks Spanish. On the film's ambivalent representation of both racial and sexual difference, see Calvo.

5. Comito seems to have taken his cue from Bazin, who as noted earlier suggests that the film is "morally ambiguous," a reading that Welles "strongly resisted," as Peter Wollen recently asserted, "with absolute lack of ambiguity" ("Foreign Relations" 23).

6. See Heath, "Film and System: Terms of Analysis, Part II," 94; and "Part I," 76. See also Peter Cowie, who makes a similar observation in *The Cinema of Orson Welles*, 148.

7. As Nericcio points out, "We never hear his name, only the one Susan ascribes to him" (52). See also Heath, who first noted that "'Pancho,' indeed, is a slighting nickname given by Susan" ("Part I" 39).

8. In fairness, Naremore elsewhere notes how the appearance of such "cameo" players as Gabor and Dietrich provide self-conscious movie references that serve "to break the surface of the illusion," as with Dietrich's Tanya, who "keeps her German accent" (171).

9. For the definitive text on and of "code switching," see Anzaldúa.

10. "I judge a scene by how it sounds—a difficult acting scene. I almost *prefer* to turn away from the actors. I think the sound is the key to what makes it right. If it sounds right, it's gotta look right" (Welles and Bogdanovich 310).

11. Consider, for example, Michael Denning's argument that "Orson Welles was the American Brecht" (362) and fostered an antifascist aesthetic in his work, including *Touch of Evil*, which "ends with one of Welles's most remarkable allegories of the apparatus: Vargas's use of the 'listening' apparatus, as Welles called it, to trap Quinlan into admitting his guilt. The feedback of the listening apparatus turns it into an audio house of mirrors in which the final shoot-out takes place," when the "antifascist Vargas uses the machine against the fascist Quinlan" in what for Denning is a political allegory (402). As Welles remarks in the *Cahiers* interviews:

> At this moment, Vargas loses his integrity. So he's thrust into a world to which he doesn't morally belong. He becomes the crude type who deliberately eavesdrops, and he doesn't know how to be such a person. So I tried to make it seem that the listening apparatus is guiding him, that he's the victim of that apparatus rather than simply of his curiosity. He doesn't know very well how to use his recording machine; he's able only to follow and obey it because this device doesn't really belong to him. He isn't a spy, he isn't even a cop. (Bazin, Bitsch, and Domarchi 212)

Welles also addressed how "the apparatus" compromises Vargas in the Bogdanovich interviews, where he also discusses the political theme of the film, stating that it seems to him "if anything *too clearly* antifascist" (Welles and Bogdanovich 298–301; emphasis in the original).

12. Dudley Andrew, in an auteurist reading of Welles's work—which I am trying to resist here—argues that "Welles was able to break decisively with Hollywood sound practice," especially through overlapping dialogue, that "is first perceived as a problem to be overcome" ("Echoes of Art" 165).

# 13

## "Sound Sacrifices"

### The Postmodern Melodramas of World War II

**DEBRA WHITE-STANLEY**

President Bush's recent comparisons between Operation Iraqi Freedom and World War II have demonstrated the power and ubiquity of World War II as a convenient historical model.[1] However, World War II remains available as a usable military past in part because media censorship shaped public understanding of the conflict as a Good War rather than "insensate savagery" (Fussell 284). Paul Fussell argues that "the real war will never get in the books" because a "traumatic amputation" has severed representation of the war from the complexities of wartime behavior and motivation (268–90). For instance, World War II enlisted American women as combatants, factory workers, and pinups, as it exposed women worldwide to the brutality of total warfare (De Pauw 248–49). However, despite this participation of women in the war, "the belief that women were not sent into harm's way, so essential to the myth of war as an exclusively male activity, persisted among Americans" (De Pauw 248). Soldiers risked their lives to protect a homeland imagined as female, domestic, and mythological, reifying a bourgeois division between war and the home front (Elshtain, *Women and War* 181). The World War II film represents these contradictory attitudes about the involvement of women in military conflict, rehearsing militaristic impulses to shield women from sacrifice, even against the way that women shouldered the work of empire building. The contemporary use of World War II as a representational model for Operation Iraqi Freedom suggests a resurgence of such backlash against the involvement of women in the war. Such contradictory attitudes suffuse, for instance, images of Private Jessica Lynch's melodramatic hospital rescue, as well as the scapegoating of Army Reservist Lynndie England after her appearance in torture photographs.

The contemporary World War II nostalgia film mobilizes the figure of woman, especially through the soundtrack, in such a way as to emphasize women's involvement in and support for the "Good War." The World War II nostalgia film re-creates the world associated with the period of World War II, in a "multitudinous photographic simulacrum" that approaches "the 'past' through stylistic connotation, conveying 'pastness' by the glossy qualities of the image, and '1930s-ness' or '1950s-ness' by the attributes of fashion" (Jameson, *Postmodernism* 19). Jameson famously argues that the nostalgia film appears in a postmodern "crisis in historicity" (22), providing a way that we can "seek history by way of our own pop images and simulacra of that history" (25). World War II nostalgia films re-create World War II by manipulating the figure of woman to emphasize women's involvement in and support for the "Good War." Three recent World War II nostalgia films—*Saving Private Ryan* (Steven Spielberg, USA, 1998), *The Thin Red Line* (Terrence Malick, USA, 1998) and *Pearl Harbor* (Michael Bay, USA, 2001)—utilize memories of wives and mothers and domesticity to transform home into "a fantastical space that is utterly bereft of the complexities, ambivalences, and incoherence of daily U.S. life both in those war-torn years and today" (Biesecker 397). This nostalgia for home is central to the World War II melodrama, perhaps the most popular yet understudied type of World War II film.[2] And, as my analysis of *Saving Private Ryan*, *The Thin Red Line*, and *Pearl Harbor* will demonstrate, the World War II nostalgia film fuels its retropatriotism using aural devices that represent the figure of woman in ways similar to the military enemy, in a convergence of cultural codes defining the way that audiences hear the imagined community of nation, through gender and race.

## Saving Private Ryan

As the elderly James Ryan (Harrison Young) walks into the Normandy American cemetery, near Saint-Laurent-sur-Mer, John Williams's "Hymn to the Fallen" swells, and numerous white crosses marking the graves of the fallen surround Ryan's lone figure. These white crosses quickly outnumber the notes of the musical motif, in an "external logic" suggesting that the sacrifice of the soldiers exceeds the capacities of visual and aural representation (Chion, *Audio-Vision* 46). In another melodramatic soundscape dramatizing the interment of a fallen soldier, soldiers of Captain Miller (Tom Hanks) are photographed in sepia-tone profiles as they bury one of their own. This use of melodramatic scoring in celebration of the sacrificed male body suggests film music's ability to materialize a sense of national, political, or cultural collective identity (Flinn 22; Kassabian, "Twilight" 258). "Hymn to the Fallen" dramatizes the sacrifices of the "greatest generation" as a lost national unity—one that fits in with Caryl Flinn's discussion of how film music can create a "generalized sense of nostalgia" for a utopic past (110). It may be no coincidence that "Hymn to the Fallen" is used in *Saving Private Ryan* to commemorate not only fallen soldiers but also their mothers and wives left behind on the home front. Psychoanalytic critics such as Kaja Silverman and Mary Anne Doane have viewed such nostalgic imaginings for a lost unitary and imaginary plenitude to be a longing for the lost maternal object. Such engendered memory is rendered, for instance, through an Edith Piaf record to which the soldiers listen before the battle for the bridge at Ramelle. Two Piaf songs, "Tu es partout" and "C'était une histoire d'amour," function as signatures of erotic memory—signatures translated from French into English, in much the same way that the dialogue of a German prisoner of war (POW) is translated into English. Piaf's voice remains the only female voice we hear in the film, singing lyrics representing a woman longing to hear her absent lover one more time ("You speak softly in my ear / And you say things that make my eyes close").

"Hymn to the Fallen" also dramatizes Mrs. Ryan's (Amanda Boxer) emotional reaction to the arrival of an army officer and a priest at her farm to inform her of her sons' deaths. Although the character of Mrs. Ryan has no dialogue in the film, her plight—the possible loss of all four of her sons in the war—sets the plot into motion. As Cynthia Weber notes, "She never speaks. She collapses. She is never really who Ryan seems to be saved for. She's sort of a plot function. . . . [S]he doesn't figure into the film" (Cohn 464). The musical motif of sacrifice represents her grief, as her body falls noiselessly onto the porch. The viewer neither sees Mrs. Ryan's facial expression nor hears her cries, in a scene invoking Kaja Silverman's argument that "whenever the female voice seems to speak most out of the 'reality' of her body, it is in fact most complexly contained within the diegesis, and most paradigmatically a part of Hollywood's sonic *vraisemblable*" (70–71). Mrs. Ryan's voiceless embodiment demonstrates the type of confinement that Silverman is pointing toward: "confinement to the body, to claustral spaces, and to inner narratives" (45). To further illustrate Silverman's argument that the listening experience is built around the voice of the castrated woman, a voice-over of General George C. Marshall (Harve Presnell) cussing "Goddamn it" functions as a sound bridge connecting this scene with the next.

As the musical theme engenders the memory of traumatic war loss, male voice-overs comment on the complex relationship among the history of war, elite military planners, and the sacrifices of the average citizen. The voice-overs subsume the unknown "foot soldiers" of the war—both male and female—underneath the voices of powerful commanders such as General Marshall. Michel Chion utilizes the term *acousmêtre* to describe the voice-over that has been uncoupled from the image of the body and hence develops powers of "ubiquity, panopticism, omniscience, and omnipotence"—the ability to be everywhere, to see all, to know all, and to have complete power (*Voice in Cinema* 24; *Audio-Vision* 129–30). In *Sav-*

*ing Private Ryan,* the *acousmêtre* glorifies and condemns the official hierarchy of elite male military voices preeminently positioned on the soundtrack. Accepted versions of military history—and artistic representations of war—are thrown into question, by foregrounding the way that ordinary men and women are sacrificed to the ruthless ambitions of military commanders. This purposeful usage of the *acousmêtre* is perfectly exemplified in one representative scene in which female typists in the war office silently type letters of condolence to American families whose sons have been killed in the war. Over close-up shots of the women typing, unseen military commanders read the letters in disembodied voice-overs, praising the heroism of the fallen. As the musical motif of sacrifice continues, one female typist closely examines several of the letters, eventually bringing them to her supervisor. What is most interesting here is that the woman's actual dialogue is dubbed over by the "Hymn to the Fallen." The viewer sees her "explaining" the content of the letters to her supervisor, and a montage of images presents her and her supervisor ascending the hierarchy of command.[3] This muting of the speaking woman in *Saving Private Ryan,* in which a musical motif drowns out her words, is tantamount to an "aural blackout" that, as Silverman explains, is a "defensive reaction against the migratory potential of the voice"—a defensive reaction that *Saving Private Ryan* mobilizes in each act of engendering of memory (*The Acoustic Mirror* 84). After this female war worker climbs the ladder of authority, a male commanding officer narrates the dilemma of the death of the Ryan brothers directly to General Marshall. Marshall then reads a famous letter written by Abraham Lincoln during the Civil War to Mrs. Bixby in Boston, praising her for the sacrifice of her sons.[4] Whether Marshall then orders the rescue of the fourth Ryan brother out of genuine concern for Mrs. Ryan or out of a misplaced desire to enter history on par with Lincoln ultimately remains a matter of interpretation. At the close of the film, Marshall's voice-over praises the heroic sacrifices made by Americans like Mrs. Ryan who have laid "so costly a sacrifice at the alter of freedom." However, such usage of Lincoln's words occludes the story of Miller, whose death is acknowledged in a series of glances between him and James Ryan that go unrecorded by Marshall's cultivated and measured tones. Marshall's military ambition is realized through the death of yet another anonymous soldier.

Such intricate work in muting the female characters and fixing them within the crosshairs of the visual prefigures the efforts of the American soldiers to "fix" the Germans in the crosshairs of their weapons. Enemy forces whose threatening guns are heard but not seen menace the American soldiers in each battle scene. Chion's term *acousmachine,* used to describe threatening machine noises that cannot be located in the visual field (*Voice in Cinema* 37), aptly describes this German threat that must be located and neutralized. In each battle scene, the encroaching enemy is audible but not yet visible to the American soldiers; the heard is at war against the seen in "active off-screen sound" that prompts fear and curiosity (Chion, *Audio-Vision* 85). Thus, the film represents an ongoing meditation about the primacy of the visual and the unexpected contribution of the aural to the narrative "information" being processed by the viewer. Anchoring the aural onto the visual—pinpointing the visual source for any given sound—becomes an important fighting skill. For instance, at Omaha Beach, German machine-gun fire emanates from an unseen emplacement to menace the American forces attempting to land. The Americans are caught in the visual crosshairs of the Germans, with no way of radioing for reinforcements and no escape from Omaha Beach. Captain Miller overcomes this radical isolation by ingeniously locating the German emplacements and bombarding them.

Conversely, although the Americans have ocular mastery at the battle for the bridge at Ramelle, they are surrounded and overwhelmed by the Germans, who are able to take that position of ocular mastery away from them. Jackson and Miller, who communicate through silent hand signals to provide visual information about the movements of the German soldiers, are visually located and neutralized by the Germans. The illusion that Captain Miller has blown up a Nazi tank by shooting his revolver at it is disrupted by the sounds of reinforcements arriving. By acting as sacrificial placeholders for a mission ordered by a General Marshall eager to join the ranks of the history books, Miller and his soldiers ironically save more "G.I. Joes" from deaths at the hands of the Germans. Although James Ryan cannot compensate the soldiers who have sacrificed their lives to rescue him, their sacrifices are justified by shifting to the larger picture. Although the film's tagline reads "This time the mission is a man," the effort to shield Mrs. Ryan from the loss of all four sons ends up satisfying the most important military objective: defending the bridge at Ramelle, which unexpectedly becomes a key point of siege in the war against Nazi Germany. As Carol Cohn writes, "In the film, then, saying 'the mission is a man' exacerbates the moral dilemma of men dying for a man, but simultaneously justifies it, because the mission always comes first" (470). The "mission that is a man" takes the soldiers to the heart of the military battle, rehearsing the analogy between home front and battle front, and demonstrating yet again that "the nation is home and home is mother" (Elshtain, "Sovereignty, Identity, and Sacrifice" 169).

## Thin Red Line

In a remarkable use of cinematic voice-over in Terrence Malick's *Thin Red Line*, a dead Japanese soldier, partially buried in a shallow grave, addresses his enemies, in English with a Japanese accent: "Are you righteous, kind? Does your confidence lie in this? Are you loved by all? Know that I was, too. Do you imagine your sufferings will be less because you loved goodness, truth?" Even as the soldier lies in his shallow grave, American soldiers collect wartime souvenirs, including gold teeth, from captured and dead enemy soldiers. As each side believes itself to be morally superior, it commits war atrocities that belie its good and just self-image. On the American side, anti-Japanese sentiment drew America into the war, and became so widespread that some American soldiers mailed polished Japanese skulls back to the United States (Fussell 116, 138). Terrence Malick's "Good War" demonstrates not only racial animosity but also strategic error, political ambition, sexual violence, and fear. Soldiers mobilize evasion or denial as reaction to these realities using the "I-Voice"—Michel Chion's term for a voice-over that is "both completely internal and invading the entire universe" (*Audio-Vision* 80). Such complex psych-ops demonstrate the "rhetorical lack" or silence of Allied soldiers, many of whom may not have had a clear sense of "why we fight" (Fussell 134).

The *Thin Red Line* portrays the Battle of Guadalcanal through criminal activity, rather than heroic sacrifice, in a vicious campaign of war atrocities and desperate bids for the moral high ground. Such self-deception raises a question posed by Shoshana Felman: "What does it mean to inhabit history as crime, as the space of the annihilation of the other?" (Felman and Laub 189). As trauma text, *The Thin Red Line* straddles "the impossibility of saying the inside to the outside," in an attempt to represent interior soundscapes of battle fatigue and cultural shock (Felman and Laub 249). *The Thin Red Line* dramatizes an aspect of World War II that serves as corrective to the narrative of victory and sacrifice offered by D-day and Victory in Europe. The film's concern with multiple perspectives and historical redress recalls Felman's argument that the very possibility of writing a unified historical narrative has become shattered by the "missed encounter" of historical forgetting (268). Voice-over narration, Hans Zimmer's electronic score music, and ambient noise work in synch to foreclose any unified or sanitized narrative of the Guadalcanal campaign. Even as aural techniques manufacture a sense of the enemy, Japanese soldiers and women are given the opportunity to inject their counternarratives into the story.

Malick's film funnels the novel's social commentary about the closeted gay culture of the military[5] into philosophical ruminations about peace, war, and the figure of woman. The character of Private Witt (James Caviezel)—in the novel, a patriotic and racist southern fighter and sharpshooter—morphs into a sensitive pacifist. At the opening of the film, Witt has gone AWOL on a Melanesian island; his I-Voice contemplates the island natives and wonders whether war is at the heart of human nature. These speculative voice-overs about the meaning of war are superimposed over images and memories of two maternal figures. He asks one female villager (Polyn Leona), who holds a small child, whether she fears him. Her maternal image triggers a montage chronicling his mother's

(Penelope Allen) death, stressing her self-sacrificing love. Witt's engendering of peace, through these maternal images, inflects his understanding of the war as a fall from the mother's grace. When the villagers begin to shun him because he is an American soldier, he ruminates, "We were a family. How did we break up and come apart so that now we're turned against each other, each standing in the other's light. How'd we lose the good that was given us, let it slip away, scattered careless?" Captain Bosche (George Clooney) echoes this conception of the army as a family of brothers at the end of the film when he identifies himself as the father at the head of that family. By the end of the film, Witt sacrifices himself to save this army of brothers. This sacrifice is a radical departure from Jones's novel, in which Witt refuses to sacrifice himself and cannot restrain himself from denouncing Captain Bosche and the mission, abruptly leaving the company.

Private Bell (Ben Chaplin), an anonymous foot soldier in *The Thin Red Line*, fantasizes about his wife, Marty (Miranda Otto), in scenes that mobilize her image in conjunction with music and odes to her sensuality and attractiveness. For instance, as Bell and other soldiers move to take out a Japanese emplacement, we hear his words: "We together, one being" with the musical motif over the image of Bell's wife. He continues, "flow together like water till I can't tell you from me" and "I drink you now," "now now"—emphasizing the links between his wife, water, music, and sensuality. Like Witt's island paradise, Bell's memories of his wife construct a utopia of belonging to which he can retreat in the midst of battle. Whereas Witt fantasizes about a native woman on a tropical island, Bell choreographs the image of the white woman with water imagery and music. As native women and children inspire Witt's philosophical ruminations on human nature, Bell imagines his wife on a beach and asks, "How do we get to those other shores, to those blue hills?" Juxtaposing the battle around him to the serenity of her voiceless image, he ponders, "Love, where does it come from? Who lit this flame in us? No war can put it out, conquer it. I was a prisoner, you set me free." The memory of her sets him free from war, in a classic "rhetorical amputation" of the many subject positions beyond the simple juxtaposition of home front and battlefield (Elshtain, "Sovereignty, Identity, and Sacrifice" 171). This rhetorical amputation, however, allows Bell relief from the incessant traumatic reality of the battlefield and a set of gendered memories into which he can escape.

Malick's version of *The Thin Red Line* deletes the novel's commentary on the closeted homosexual culture of the military, and depicts Bell and his fellow soldiers as heterosexuals. In a number of voice-over scenes, Marty serves as a conventional cinematic representation of the faithful wife waiting for her soldier to return. However, having constructed Marty through stereotype, the film allows her a presence on the soundtrack, in an acute rebellion of the female subject used to justify "why we fight." As a revisionist representation of the wartime wife, Marty refuses the wartime sacrifice, and instead sends her husband a "dear John" letter. As Marty reads the letter asking him for a divorce, Bell's image is treated rather like the image of Mrs. Ryan in *Saving Private Ryan*. As he reads the letter, film music and the sound of her voice drown out the sounds of his re-action to the letter. He cries and at one point laughs, in emotional reactions that are excluded from the soundtrack, positioning her voice as dominant, enfolding his image within its tones. Malick has successfully turned the tables here; the woman caught in the crosshairs of the visuals speaks. This aural device recalls Chion's discussion of the power of the voice-over as "a double, a conscience, an instrument, a reproach, a doubt" (*Voice in Cinema* 98)—for Marty no longer is willing to be the person for whom John Bell fights his war. In each version of the *Thin Red Line*, the *acousmêtre* expresses troubling links between the enemy, the figure of woman, and the voice.[6]

Whereas the musical motif in *Saving Private Ryan* marks the sacrifices made by Captain Miller and his men to save the young James Ryan—and all of West-ern civilization—the musical motif in *The Thin Red Line* marks the terror of soldiers trapped by the unseen enemy and betrayed by their own commanding officers. In these scenes of entrapment and mechanized horror, the men have most completely fallen from grace—a grace aurally rendered by maternal and sensual memories of women in the film. Melodramatic synthesizer music and the rustling of the tall grass dramatize the horrifying scenario of being killed or maimed by an unseen and vicious enemy. Just as often, the music dramatizes the vagaries of the military hierarchy, characterized at every level by pride and ambition. Ambient battle sounds and inept voices in command over the sound power phones—special portable telephones used during the war—work together with the music to express a standoff between Captain Staros (Elias Koteas) and Lieutenant Colonel Tall (Nick Nolte), his commanding officer. Tall's gravelly voice-overs, in particular, express his dilemma of official hypocrisy, as when, before even arriving at Guadalcanal, he contemplates the soldiers under his command and muses: "All they sacrificed for me, poured out like water on the ground." Tall, who communicates with the front lines through the sound power phone, becomes an unseen menace as potentially deadly as enemy soldiers. With the pro-gression of the campaign, his interior state of conflicted ambition made manifest by voice-overs, a "military horror motif" dramatizes his inward tumult even as he succeeds in bullying others. Against this military horror, the film score also constructs moments of utopic escape—moments that position women in the natural world in an idyllic way. For instance, a "native paradise" musical motif marks Witt's AWOL period on the Melanesian island. Witt has recurring aural flashbacks of his Melanesian AWOL paradise. One such aural flashback occurs after his death, in an underwater scene in which he is again swimming with the natives. The sounds of nature, including the lapping sounds of waves and the cries of island birds, testify to a peaceful world of love, existing only through the memory of women, forever destroyed by war.

Aural distortion renders the soldiers' traumatic dissociation from reality, in a manner that foregrounds the uselessness of the sacrifices they are making. Whereas in *Saving Private Ryan* aural distortion is used to demonstrate the personal cost of wartime heroism, in *The Thin Red Line* aural distortion poisons body and soul. Aural distortion is used in *Saving Private Ryan* to communicate Captain Miller's battle fatigue, and paradoxically to communicate his superhuman strength in overcoming battle fatigue to lead his men so effectively. Captain Miller experiences a physiological dissociation of time and place, but his dignity and wisdom are not damaged by war. Dissociation, used to depict the fragmentation and explosion of bodily integrity, filters the experience of battle, shielding the "self" of Captain Miller—a core of moral values and personal memories that war cannot pervert. Not so in *The Thin Red Line*, which uses aural distortion to communicate war's destruction of moral values, in which the soldier's very "self" is contaminated and erased. For instance, Malick uses aural distortion to represent the narrowing of Sergeant Keck's (Woody Harrelson) awareness, after an accidental grenade explosion pulverizes his sexual organs and buttocks. Keck's death is cast not in the heroic terms of *Saving Private Ryan* but rather in sexual terms, as when he laments that he has blown his ass off and will not be able to have sex anymore—"what a fucking recruit trick to pull." Aural distortion is also used to represent Sergeant McCron's (John Savage) descent into insanity; his altered perceptions of time and space begin when he must check his dog tag to learn his own name. And aural distortion filters the trauma of American war crimes inflicted on the Japanese, as when an American soldier uses a pair of pliers to extract gold fillings from the mouths of POWs at a captured Japanese camp.

These uses of aural devices such as voice-over narration, music, and distortion to trouble the sacrificial discourse of the war film are as subtle and powerful as the crocodile that, in the film's opening shot, slides noiselessly and powerfully into a jungle river coated with thick moss. Bored American soldiers later capture the reptile, and imprison it in a bamboo cage. This dynamic of entrapment and escape symbolizes the way that the war film seizes the image of woman to inscribe wartime heroics. As Susan White argues in her essay about *Full Metal Jacket*, even in war films that represent the "repression of the feminine," "a threatening (not simply a passive) femininity resurges to the forefront of the text" (229). White's observation that the "repression of the feminine" can provoke a resurgence perfectly describes Malick's use of the soundtrack to relegate figures of woman to— and allow them to "talk back" from—the margins of cinematic representations. As an extended meditation on the relationship between nostalgia as organized forgetting and the unwilled eruption of traumatic memory, *The Thin Red Line* energetically attacks our culture's fixation on the heroism of sacrifice.

## Pearl Harbor

A Hans Zimmer musical motif suggests the impending threat of Japanese pilots running to their planes and preparing to attack Pearl Harbor in the early morning of December 7, 1941. One Japanese soldier says a prayer in English: "Revered father. I go now to fulfill my mission and my destiny. I hope it is a destiny that will bring honor to our family, and if it requires my life I will sacrifice it gladly to be a good servant of our nation." Although we hear the sacrificial prayer in English, throughout the rest of the film the Japanese speak in Japanese with English subtitles, as in the film *Tora! Tora! Tora!* (Richard Fleischer and Kinji Fukasaku, USA, 1970). Before and after the prayer, the film manipulates the tones of the Japanese language to sound guttural and harsh. This juxtaposition of intelligible prayer and murderous tones fits with the American pattern of representing Japanese soldiers in World War II through "massacre violence" or "sacrifice violence" (or both) (Nornes 153–55). These elements of massacre and sacrifice were again put into play in the September 11, 2001, terrorist attacks, only four months after the release of *Pearl Harbor*. Using Pearl Harbor as a model through which to understand the September 11 attacks has created a false "sense of divine mission" and downplayed more cautionary military tales, such as the conflicts in Vietnam or the Persian Gulf (Landy 83–86). As a nostalgia film, *Pearl Harbor* is "condemned to seek the historical past through our own pop images and stereotypes about that past" (Jameson, "Postmodernism and Consumer Society" 118). The wartime nurse emerges in *Pearl Harbor* as the central pop image used to evoke a sense of the 1940s wartime mood. Melodramatic film music, aural distortion, and voice-over representation align the wartime nurse with the Japanese, in a pattern of sacrifice and threat.

Through a professional code of Nightingale-ism, the white-clad nurse suggests possibilities of grace, moral conversion, and redemption, bandaging wartime injury with insignias of salvation. Of the 350,000 American women who served in the U.S. armed forces during World War II, about 60,000 served as army nurses, and 14,000 as navy nurses (Tomblin 209). These women were subject to the "slander campaign" consisting of rumors that they engaged in sexual liaisons (even prostitution) with enlisted men (253). Since then, with few exceptions, nurses have been excluded from official accounts of the war, as well as newspapers, books, radio, film, and television. Elaine Scarry demonstrates that representations of war injuries generally fall into a similar pattern, tending to omit or redescribe the "fact of injury," or to use the "fiction-generating" function of the war wound for propagandistic purposes (81, 121). *Pearl Harbor* portrays the war wounded and the military nurse through silence, redescription, and propaganda. Musical motifs rather than dialogue convey the melodramatic dilemma of Nurse Lieutenant Evelyn Johnson (Kate Beckinsale), involved in a romantic triangle with Captain Rafe McCawley (Ben Affleck) and Captain Danny Walker (Josh Hartnett). Alternatively, aural distortion and out-of-focus shots filter the viewer's perception of Nurse Johnson's ability to take charge in the aftermath of the Pearl Harbor attack, when the carnage is at its worst.

Aural surveillance of the Japanese and Nurse Johnson establishes parallels between the Japanese Pearl Harbor attack and Nurse Johnson's romantic betrayal of Captain McCawley's affections. As the Japanese prepare to attack Pearl Harbor, Captain Thurman (Dan Ayckroyd) works in Naval Intelligence to decode Japanese radio transmissions. As in *Tora! Tora! Tora!* the code breakers know that "something is up" but cannot produce the necessary evidence to convince the authorities to act. Thurman predicts that the Japanese will attack Pearl Harbor, based on his "interpretation" of intercepted radio transmissions. As Naval Intelligence decodes Japanese radio transmissions, the viewer-auditor decodes aural signals of a transformation in Nurse Johnson. This changing portrayal can be understood through the lens of Klaus Theweleit's *Male Fantasies,* which explores how, for German officers during World War II, the nurse condensed complex political and psychological drives inspiring authoritarian military fixations. Hence, the image of the nurse figure is bifurcated as mother (white) and whore (red). The white nurse is the "mother, sister (-of mercy, nurse) and countess all in one person"—the good woman who protects (95). She is made to appear strong but at the same time must be protected from outside enemies (99). The red nurse is a proletarian woman and a prostitute—a castrating woman (85, 91). The character of Nurse Johnson swings between white- and red-nurse representational patterns, in a shift accomplished largely through the aural dimension of representation. As white nurse, Nurse Johnson has remarkably few lines of dialogue but exists in a "text of muteness"—a melodramatic work in which tensions and emotions are rarely spoken aloud but rather expressed by mise-en-scène and music (Brooks 48). She dances with Captain McCawley to swing music, as their relationship quickly unfolds in one evening before he leaves to join the Royal Air Force in England. A melodramatic musical motif dramatizes how, as white nurse, Nurse Johnson pines away for the pilots as they save the free world. Nurse Johnson silently parses a few letters from Captain McCawley—letters that Captain McCawley reads in voice-over, as she sits on the beach. His voice intrudes over hers, reinforcing her aural blackout.

As the Pearl Harbor attack begins, changing patterns of aural representation create a backlash against the presence of Nurse Johnson's voice in the soundscape of the film. In a departure from the muteness of her white-nurse persona,

Nurse Johnson assumes the castrating authority of the "red nurse." As if to mark this character transition, Nurse Johnson realizes that a fellow nurse—who was engaged to be married—has been killed. As red nurse, Nurse Johnson grabs her red lipstick to mark soldiers for admittance or nonadmittance to the hospital, after the debilitating attack on Pearl Harbor. Ordered to take command of the chaotic scene in front of the hospital, she directs the movement of bodies and resources. Vertical bars cross the visual field, screening out the trauma of the wounded patients, and she adopts a tone of professional command. She orders and directs, at one point even telling a doctor how to save a wounded patient. Aural dissociation funnels diegetic sound to the background, and brings music to the forefront, stressing her mastery of voice and action. As she sorts treatable patients from those who will die, she pretends to be trustworthy but then uses her red lipstick to write a big *F* on the forehead of soldiers who will be denied hospital treatment. As red nurse, Nurse Johnson rediscovers her voice. In a confrontation with Captain McCawley she reveals that she is pregnant with Captain Walker's child and will marry him. After this scene, Captain McCawley burns her letters; as flames destroy the letters, her voice-over is abruptly silenced and the film returns Nurse Johnson to the "white nurse" representational pattern.

During the Doolittle bombing raid on Tokyo, Nurse Johnson's voice is gradually drowned out, and she is again enveloped in silence. As Captain McCawley and Captain Walker fly over Tokyo, Nurse Johnson wants to listen to their radio transmissions. Although she is allowed to sit outside the door to the command post, she—like the female typists in *Saving Private Ryan*—is "seen but not heard." The command post is a small room in the center of a much larger space, and the commanding officers (all male) within the small interior room can hear the Doolittle-raid radio transmissions. The door is opened for a moment, leaking the sounds of the attack, but then shut again, leaving Nurse Johnson in suspense. This deliberate use of "point of audition"—the suturing representation of what a particular character can and cannot hear (Chion, *Audio-Vision* 90–91)—allows us to briefly share Nurse Johnson's sense of frustration at hitting the glass door of admittance into the film soundtrack. We leave Nurse Johnson's restricted point of audition, and are invited to fix Japanese targets in the crosshairs of the visual apparatus. As the American pilots bomb Japanese factories in Tokyo, Nurse Johnson is being "punished" for her betrayal of Captain McCawley. Her lover Captain Walker is also punished, bound to a mule harness and shot by Japanese soldiers. This sacrifice frees Captain McCawley to return from the mission whole and physically unscarred, a hero with his movie star smile and both legs still intact. It is worth noting here that Ted Lawson, the army pilot who participated in the Doolittle raid and wrote the original *Thirty Seconds over Tokyo*, suffered greatly, lost a leg, and was permanently disfigured and scarred.[7]

Now married to Captain McCawley, and mother of Captain Walker's child, Nurse Johnson begins to narrate the last scene of the film: "When the action is over and we look back, we understand . . ." However, since this is the only line of her voice-over narration, exactly what "we" understand is left to conjecture, as her voice trails off. With the war now over, she watches from below as Captain McCawley takes their child up in a crop duster. Not only does Pearl Harbor quote *Tora! Tora! Tora!* and *Thirty Seconds over Tokyo* (Mervyn LeRoy, USA, 1944) to mimic previous representations of the Pearl Harbor attack, but it also constructs a pseudohistorical narrative frame in order to quote its own cannibalized history. As a boy, Captain McCawley is abused by his father, a World War I veteran who clearly suffers from lasting psychological trauma and hits him for trying to fly the crop duster. However, as a man and a World War II combat veteran, Captain McCawley does not seem to have lasting battle scars. As the symbol of an incipient mapping of the new world order, the child flyer, soldier of the future, recalls Lauren Berlant's argument that "the fetal/infantile person is a stand-in for a complicated and contradictory set of anxieties and desires about national identity" (6). In Berlant's "theory of infantile citizenship," the naive citizenship of the child has surfaced as the ideal type of patriotic personhood in America, "on whose behalf national struggles are being waged" (21). *Pearl Harbor* provides a neat example of the innocent child soldier, the ultimate nostalgic representation of World War II, whose cognitive map is construed from the bomber's point of view while his mother waits far below on earth.

World War II nostalgia films contain and manipulate the female voice, linking gender politics to the military enemy. *Pearl Harbor, Saving Private Ryan,* and *The Thin Red Line* use aural devices such as voice-over narration and asynchronous representation of the image of women to represent gender through memory and absence. Mrs. Ryan in *Saving Private Ryan,* Marty Bell in *The Thin Red Line,* and Nurse Lieutenant Evelyn Johnson of *Pearl Harbor* engender the construction of memory, and the involuntary resurgence of traumatic memory. In each film, the dynamic soundscape of gender representation militarizes the figure of woman in ways comparable to that of the military enemy. World War II nostalgia films re-create historically conflicted attitudes toward the involvement of women in war, constructing gendered scenarios that simultaneously mobilize men to protect women from war while mobilizing women in war as a military enemy.

To "lower the boom" on the misleading use of World War II to justify nation building, we must mount a search for the human voice in the rubble of the war-film soundtrack, sifting through music, ambient noise, and voice-over narration. These aspects of the soundscape—but especially the human voice—represent and obfuscate the traumatic pain of the sacrifices that the nation-state demands of its citizens. When citizens are asked to sacrifice themselves in order to belong to the "imagined community" of the nation, the cinematic soundscape records the concomitant collective trauma. Under the rubble of militarism, the war-film soundtrack registers the damage inflicted by war on the human body and psyche.

## Notes

1. See, for instance, the following presidential remarks: "Winston Churchill and the War on Terror," delivered at the Library of Congress, Washington, D.C., on February 4, 2004, http://www.whitehouse.gov/news/releases/2004/02/20040204-4.html; Memorial Day radio address of May 29, 2004, http://www.whitehouse.gov/news/releases/2004/05/20040529 .html; and dedication of the World War II memorial, http://www.whitehouse.gov/news/ releases/2004/05/20040529-2.html.

2. I bridge conceptions of the war film as a masculine realm of representation and the appearance of the feminine through the melodramatic hybrid war film, which has proven to be more popular than the combat film (Schatz 102). Of the nearly four hundred war films produced between 1942 and 1944, maternal and romantic war melodramas such as *Mrs. Miniver* (Wyler, USA, 1942) and *Since You Went Away* (Cromwell, USA, 1944) enjoyed greatest success at the box office and dominated the war-film genre (Schatz 104, 111).

3. This scene quotes *The Fighting Sullivans*, in which a letter informing the parents of the death of all five sons is being typed by male office workers, as the film launches into a dramatic re-creation of the torpedoing of the Ryan brothers' ship. Another allusion to *The Fighting Sullivans* arrives with the silence of Mrs. Sullivan when told of her loss, which is expressed in part by the screaming of the other Sullivan women.

4. One critic notes that historians have debated whether the legendary letter written by Lincoln to Mrs. Bixby in Boston was ever actually written; historians have also discovered that Mrs. Bixby had two sons instead of five (Auster 101).

5. Readers of the novel *The Thin Red Line* see most of the events of the Guadalcanal campaign through the eyes of Fife and Doll, both of whom are involved in some type of gay sex over the course of the novel. Fife accepts his tent mate's argument that, deprived of the company of women, "guys could help each other out" through oral sex (Jones 121). Fife reasons that this "buggering" is common practice in the army, different from overt homosexuality and tolerated by the authorities (123). Doll, who is looking for a "relationship" (463), becomes attracted to a subordinate's "girlish ass" and makes overtures (415–16). This theme of gay sex is written out of the film.

6. In Jones's novel, Marty criticizes the American soldiers for committing war atrocities against Japanese prisoners. Her voice becomes her husband's conscience, and he resents this, comparing the eyes of a dead Japanese soldier to her "cunt" (187). Marty's "cunt" has eyes that do not see, reinforcing the power of her voice in letters, imagined conversations, and dreams. Bell then has a malarial dream that Marty has given birth to a black child—a child who becomes a miniature Japanese soldier, complete with uniform and a "tiny, baby iron star" (396–97). His dream demonstrates that the black man and the Japanese man are interchangeable sexual threats—sexual threats that are also aurally dramatized in the

1964 version of *The Thin Red Line*. In this earlier film, Bell dreams that his honeymoon is interrupted by a recurring phone call. Over the phone lines, he hears a Japanese voice saying, "Your number is up. We want your number"—as the image of him killing the Japanese man is superimposed over the image of Bell and his wife making love.

7. Captain Ted Lawson dedicates his memoir to fellow Doolittle-raid pilots who did not make it back, and writes not only a postscript but also an assessment of the raid after a year's time. His memoir is conspicuously absent of Pearl Harbor's "perfect smile" heroism. Although it argues for the value of the Doolittle raid, it is also frank about the lasting physical and psychological costs suffered by the participants. The film version of *Thirty Seconds over Tokyo* represents some of the length and difficulty of Ted Lawson's (Van Johnson) ordeal. Back on American soil, Johnson bears a representative facial scar and falls down in front of his wife while trying to stand on one leg to greet her.

# 14

# Real Fantasies

## Connie Stevens, Silencio, and Other Sonic Phenomena in Mulholland Drive

### ROBERT MIKLITSCH

> Listen again to Lynch's films.
> Listen with your eyes.
>
> —Michel Chion, *David Lynch*

David Lynch's *Mulholland Drive* (USA/France, 2001) is an unusual, and unusually complicated, audiovisual text. Originally shot as a two-hour pilot for ABC and Disney Touchstone, *Mulholland Drive* was conceived as an open-ended, continuing prime-time television series. The network hoped for an "event drama" (Hughes 239) on the order of the first sensational season of *Twin Peaks* (ABC, 1990–91), while Lynch—always the eternal optimist—was hoping against hope that it would not suffer the same untimely fate as *Twin Peaks,* which the network precipitously canceled, or *On the Air* (ABC, 1992), which was unceremoniously axed after only three episodes. Lynch's initial pessimistic response to the latter event, which he promptly painted on a piece of plywood, was: "I Will Never Work in Television Again." To invoke Sean Connery on reprising Bond: never say never.

The bad news is that, confirming Lynch's worse fears, *Mulholland Drive* never aired on ABC, which predictably enough found it to be too slow, too long—too, in a word, Lynchian. The good news is that Canal Plus, a French entertainment company, not only bought the rights to the pilot from ABC but also offered Lynch two million dollars to shoot a new "closed" ending. The result was what Mary Sweeney, the film's editor, called a "fantastic hybrid" (quoted in Hughes 240). The fantastic, hybridic character of *Mulholland Drive*—for example, the way the final part appears to provide a realist counterpoint to the first fantasmatic part—has, not surprisingly, been the focus of a rapidly growing body of critical literature. In this essay I propose to examine the sound design and, in particular, musical elements of *Mulholland Drive* in order to address the film's enigmatic or herme-neutic structure (dream-reality) from the perspective of Lynch's vexed relation to classical Hollywood cinema.

Given the complexity of the "sound environment" (Chion, *David Lynch* 44) in *Mulholland Drive,* my analysis will concentrate on two sequences. First, the sound stage where Carol (Elizabeth Lackey) lip-synchs Connie Stevens's "Six-teen Reasons" for the director, Adam Kesher (Justin Theroux), and where, at the end of which audition sequence, Betty Elms (Naomi Watts) suddenly appears, flush with her own postaudition euphoria, to momentarily meet the director's gaze. Second, the Club Silencio scene where, among other things, Betty and Rita (Laura Harring) sit in the balcony listening to Rebekah Del Rio sing an a cap-pella, Spanish-language version of Roy Orbison's "Crying" (Llorando). More to the analytic point, I will read these two sequences as two sides (recto-verso) of Lynch's audiovisual aesthetic, an aesthetic that, true to his self-designation not as a director but as a "sound man" (169),[1] manifests itself in *Mulholland Drive* as a set of linked binaries: image-sound and sound-silence.

## The Ear of the World

In his composite, alphabetic portrait of Lynch, Chion includes an entry titled, simply, "Oreille" (Ear). Referencing *Blue Velvet* (David Lynch, USA, 1986), Chion remarks that what Sandy Williams (Laura Dern) says at one point to Jeffrey Beau-mont (Kyle MacLachlan) "could be said by a great many characters in Lynch's films: 'I hear things'" (*David Lynch* 169). The same could, of course, be said of Lynch himself, whose ear is finely tuned, like so many antennae, to sonic phe-nomena that are beyond the ken of most directors. This auditory sense and sen-sibility is abundantly on display in his very first major filmic project, *Eraserhead* (USA, 1977), where Lynch worked closely with sound editor Alan Splet to fashion an acoustic environment that possesses its own peculiar logic and spell. It is as if Lynch, not unlike a child fascinated by the sheer fact of sound, always has his ear to the ground. The result is that his films—from *Eraserhead* to *Mulholland Drive*—frequently emit an "abstract cosmic murmur" (52). As Chion articulates this phenomenon with his characteristic mix of clarity and poetry, "This murmur is always in a precise register, which is Lynch's own, evoking intimacy, the world's voice speaking in our ear" (170).

Lynch's supersonic aesthetic receives perhaps its most literal expression at the beginning of *Blue Velvet* when the neo-noir detective-protagonist, Jeffrey, after having visited his ailing father in the hospital, spies a severed ear in an abandoned field. Whereas neither the coroner (Peter Carew) nor Sandy's father, Detective John Williams (George Dickerson), are plussed by its existence, a close-up of an ear before Jeffrey enters Sandy's house—accompanied by a low, roaring sound—suggests that the severed ear is not merely a missing part of some body (in this case, Dorothy Vallens's [Isabella Rossellini] husband, who has been abducted along with her son), but one piece of an exceedingly enigmatic puzzle the full extent of which Jeffrey is only just beginning to grasp. As such, the severed ear is a figure of the life-world in all its mystery and contingency, mutability and unpredictability.

If the severed ear in *Blue Velvet* may be said to be a figure, on a more formal level, of Lynch's profound investment in the acoustic dimension of cinema, this faculty is given free play in *Mulholland Drive*, which periodically invokes the medium of silence to comment on the relation between dream and reality, artifice and authenticity. Consider, for example, the first extended sequence in the film when the ominous quiet of Rita's chauffeured limousine ride on Mulholland Drive high in the Hollywood Hills is shattered by the screaming sound of teenagers drag racing, their hands thrown up in the air in wild abandon. This brief cutaway to the wild-at-heart teens, which acts as an analogue of sorts to the young people jitterbugging to big band music in the pretitle sequence ("Jitterbug"), signals the onset of that absolutely unique thing: the Lynchian universe—a topsy-turvy world where the unexpected becomes the expected. And where, for example, a crime syndicate–ordered "hit" is interrupted at the very last moment by an ear-shattering car crash (that eventuates, in turn, in two of Lynch's most resonant sound images: fire and smoke).

The absurd, aleatory nature of the world according to Lynch is not, however, without its regulatory principles, which tend to be that of repetition. Thus, the drag racers' scream in the first sequence foreshadows not only Rita's muffled scream of horror when she and Betty discover the decomposing body in bungalow 17 but also Diane's own hysterical screaming when, at the end of the film, the comic-demonic old couple invades her apartment and she flees to her bedroom to kill herself. Accordingly, exorbitant sonic phenomena—in this case screaming[2]—might be said to bookend the narrative of *Mulholland Drive,* introducing an internal rhyme into its asymmetrical, sonnetlike structure.

Chion observes that "certain changes in the voice . . . convey the force of an impression received by the ear" (*David Lynch* 170), and just as screaming can—depending on the context—signify fright or euphoria (or both), so crying can be a sign of sorrow or happiness. In *Mulholland Drive* the sequence set at the Club Silencio (whose name is itself an oxymoron of sorts) represents an especially complex example of the fraught relations that obtain among image, sound, and silence in Lynch's work. Suffice it to say that, unlike the scene where Rita is simultaneously crying and hacking at her hair with a pair of scissors because of the dead body she has just seen, the "cause" of Rita's and Betty's crying at the

Club Silencio appears to be a direct result of Del Rio's moving performance of "Llorando". In fact, since Betty's and Rita's reaction can be said to exceed its source, the sequence at the Club Silencio melodramatically foregrounds the intimate relation in *Mulholland Drive* between sound and fantasy. For Lynch, it is only through this relationship—what John Alexander calls the "artifice of banality" (29)—that one can access that fantasmatic "truth" at the very heart of the real.

The various forms of sonic phenomena that populate *Mulholland Drive* are not, needless to say, limited to the human voice. For example, the white noise that accompanies the aerial shots of downtown Los Angeles (and appears to invest the metropolis with a palpable animus all its own) is epitomized by a whooshing sound that is also associated with the homeless person behind Winkie's on Sunset Boulevard.[3] So when, near the beginning of the film, Dan (Patrick Fischler)—at the apparent, therapeutic instigation of his friend Herb (Michael Cooke)—ventures outside of the diner to reenact the dream he has been recounting, he encounters not mundane everyday reality but a horrific phantasm in the form of the "fungus-faced" tramp (Bonnie Aarons). This dream-within-a-dream-within-a-dream (where Dan's dream is a function of Rita's dream, which is ultimately a function of Diane Selwyn's [Naomi Watts] fantasy) closes on a striking note: the sound of Herb's voice as if it were underwater, or—from another, even more disturbing, disorienting perspective—as if we, the audience, were inside Dan's head as he is dying from shock.

The cut to Rita fast asleep under the art deco table in Betty's aunt's kitchen—a scenario that is itself set up by the amnesia Rita suffers after she escapes from the demolished limousine—intimates that Dan's sudden physical collapse constitutes a steep fall into the abyss of the unconscious. More important perhaps, Dan's collapse (and, later, Del Rio's on the stage of the Club Silencio) indicates that, ontologically speaking, the register of its action is fantasy. Here the fantasmatic, like silence, is inseparable from that persistent audition of the world that produces the real. Hence, when Dan collapses, the "fantasy collapses too," but not, however, in the conventional sense that we are now finally, safely, "back in reality," but rather in the sense—heightened by the vertiginous sound, like a siren in reverse, that accompanies his fall—that what we are witnessing "is the fantasmatic Real at its purest" (Zizek, *Organs without Bodies* 169).

## *La mode rétro:* Connie Stevens Redux

In the introduction to an interview with Lynch about his film *Blue Velvet,* Chris Rodley has remarked, in the context of Ben's (Dean Stockwell) "estranged" rendition of Roy Orbison's "In Dreams," that the director's images are not only "transformed by the sounds and sentiments of the music but these images . . . re-invent the music itself—twisting its meaning or complicating its often simple, emotive intent until the two become inseparable" (125). Lynch would no doubt be the first to point out that, pace Rodley, the pop and rock music of the 1950s is neither simple nor naive. (Characteristically, Lynch credits his passion for music as having originated from the moment he *missed* seeing Elvis's first appearance on *The Ed Sullivan Show* [Rodley 127].) The debt that Lynch's audiovisual aesthetic owes to that fantasmatic "structure of feeling" that is the '50s is audible everywhere. Listen, for example, to Lynch on the "source" (music) of *Blue Velvet:* "[Bobby Vinton's version of 'Blue Velvet'] sparked the movie! . . . It made me think about things. And the first things I thought about were lawns . . . and the neighborhood. . . . And in the foreground is part of a car door, or just a suggestion of a car, because it's too dark to see clearly. But in the car is a girl with red lips. And it was these red lips, blue velvet and those green-black lawns . . . that started it" (quoted in Rodley 134). This Proustian recollection establishes the terms of Lynch's cinematic modus operandi: *first sound, then image.* It also vividly conveys the almost originary status that the 1950s as a period has assumed for him.

The "Sixteen Reasons" sequence is illustrative in this regard. Betty has just completed—courtesy of her aunt Ruth's studio connection, Wally Brown (James Karen)—an electrifying audition for Bob Brooker (Wayne Grace), who, hack director that he is, is a master of circumlocution: "Don't play it for real until it gets real." (Brooker's postaudition remarks are equally, hilariously inane: "Very good . . . I mean, it was forced maybe, but still humanistic.") Although Betty is pleased as punch with her performance, to her surprise and dismay a casting agent, Linney James (Rita Taggart)—the "best" in the business, according to Wally—casually dismisses the aging Brown's ability to get a film financed. Linney does, however, promise to get Betty another audition with a hot, up-and-coming director named Adam Kesher who's "got a project," Linney tells Betty, that she "will kill [for]." Of course, the last claim acquires not a little resonance since by the end of the film, we realize that Betty/Diane has been perfectly willing to "kill" Rita to get what she wants.

As Betty, her agent, and assistant Nicki (Michele Hicks) wait for an elevator—and before the camera cuts away—we hear the first few "harmonized" bars of the Connie Stevens song "Sixteen Reasons" (written by Doree and Bill Post). Cut to a medium shot of a white female vocalist, Carol, attired in a sparkly pink dress and sporting rhinestone earrings and a '50s bouffant hairdo, singing into a stationary microphone:

(Sixteen reasons) Why I love you
(One) The way you hold my hand
(Two) Your laughing eyes

(Three) The way you understand
(Four) Your secret sighs

Nothing in the film up to this point—including the silhouetted "jitterbug" credit sequence that opens the film—quite prepares the spectator for this spectacle. It is as if we were suddenly transported back to a musical performance from some long-forgotten '50s teen musical starring Connie Stevens.[4]

As Carol continues to sing what appears to be a live version of "Sixteen Reasons," the camera slowly pulls back to a wide shot of a pair of "mixed" singers standing on either side of her, the women dressed in pink skirts with gold tops, the men in matching midnight-blue jackets and jet-black pants. Still, it is not until the crane shot rises to reveal the "window" frame of the recording booth, the recording console, and, eventually, the camera (with operator) and director of photography that it becomes clear to the audience that we are, in fact, on a film set soundstage. Having established this "master" point of view, the camera then cuts from a high crane shot of the set—populated "off-screen" with sundry personnel performing various production tasks—to a rapid track-in that concludes with a "star" close-up of Betty arriving at the set.

What happens next—or, more properly, does *not* happen—is decisive. The ensuing shot of Adam turning, cigarette and speaker in one hand, to meet Betty's excited gaze seems to be a prelude to that classic moment in Hollywood studio films when the ingenue is finally noticed by the very director who will transform her into a star. And in fact when Betty, having looked away for a moment, turns back and returns Adam's gaze a second time—the lush romantic sounds of "Sixteen Reasons" flooding the set—it would appear that Betty's dreams of stardom are about to come true.

But do not be misled, Lynch seems to be saying, by the age of innocence conjured up by the nursery colors and nostalgic music: in Lynchland, all paradises are lost. When "Sixteen Reasons" stops, so does the fantasy, for both Betty and the audience. Adam briskly returns to business, asking an assistant who the next actor is, at which point Camilla Rhodes (played in her first appearance in the film by Melissa George) expectantly takes her place on the set before the microphone without the backup singers. Adam says, "Playback," and Camilla promptly, mechanically, begins lip-syncing "I've Told Every Little Star":

O, baby, I've told every little star
Just how sweet I think you are
Why haven't I told you?
I've told ripples in a brook
Made my heart an open book
Why haven't I told you?

The difference between this and the "Sixteen Reasons" sequence is subtle but unmistakable. Unlike the Connie Stevens song, which begins in medias res (and therefore has, at least initially, little illusionist force), the word *Playback* immediately establishes a jarring disjunction in the sequence between sound and image, fantasy and reality. The performance is, in a word, "canned." This alienation effect is acoustically italicized by the volume of "Every Little Star" that—unlike "Sixteen Reasons"—changes perceptively to reflect the diegetic distance between the performer, Camilla Rhodes, and the director.

Moreover, the point of view or, more precisely, point of audition is not objective—as in the "Sixteen Reasons" sequence—but subjective: that is to say, it is Adam's point of audition that dominates. From this perspective, Adam is the apex of a triangle whose other points are Betty and Camilla, where Betty is associated with the visual and Camilla with the audio register. The irony, however, is that although Adam is wearing headphones—à la Lynch—and looking at Camilla (his black, thick-framed glasses emphasizing his specular proclivities), he is not really listening to her. In fact, right in the middle of her perky, too perfect performance of "Every Little Star," he calls for the studio manager, Jason Goldwyn (Michael Fairman), to inform him that Camilla, *not* Betty, is "the girl." This, then, is the fateful moment in *Mulholland Drive* when Adam—who had vowed earlier that he would never allow Camilla to appear in his film—finally capitulates to the studio's demands, at the behest of the crime syndicate, to cast her.

The concluding action of the sequence unspools like a dream. As if she can read his lips, Betty suddenly realizes she is late for her appointment with Rita. A final track-in to Betty, followed by a tight close-up of Adam—their eyes locked in on each other—suggests what might have been. Or, from another, reverse angle, a real creative and romantic "connection"—one significantly different from the autocratic sort that drives the film industry—"is pointedly *not* made" (Nochimson, "Mulholland Drive" 40). The sublime silence of Betty and Adam's reciprocal, albeit fleeting, look gives way to the panoptical gaze and "wrong silence" associated with the syndicate.

## (Club) Silencio: Superdiegetic Sound

I have already discussed the moment when, after having seen a decomposing body, Rita, sobbing, frantically tries to cut her hair before she is stopped by Betty: "I know what you're doing. Let me do it." The immediate consequence of this action—where Rita adopts a blond wig à la Betty and then gazes at her new self in a mirror ("You look like somebody else")—erases the difference between the two, as if Rita were now a simulacrum of Betty. (Note the dissolve from a medium two-shot of Betty and Rita standing in the mirror—now simulacral doubles of each other—to an oval close-up of Betty's face in bed.) Still later, when Rita and Betty make love and Betty repeatedly tells Rita that she is in love with her, the circuit is complete. Since we later discover that the entire first half of the film is narrated from the perspective of Diane's fantasy, it represents the apotheosis of her lethally wishful vision of how things should be at this exact moment. She has completely realized her romantic fantasies, fantasies that conveniently relegate Rita to the status of a bit or "B" player as opposed to the director-preferred "star" that, in retrospect, we know she is. Indeed, a substantial part of Betty-Diane's psyche is intimately bound up with her self-interested conception of the amnesiac Rita as someone who is almost wholly dependent on her. (By comparison, when Diane later tries to "top" Camilla—played in the second half of the film by Laura Harring—the dirgelike music "Go Get Some," a stripped, guitar-and-drum blues number written and performed by Lynch and John Neff, evinces a sex drive devoid of its romantic patina.) As in Hitchcock's *Vertigo* (USA, 1958),[5] if "envisioning oneself as the rescuer of one's love object is the ultimate fantasy scenario" (McGowan 76), then Diane, after Camilla has already left her for Adam, is hopelessly lost in the Oz-like dream that now possesses her.

Diane's fantasy does not—and cannot—last forever. While she, as Betty, lies sound asleep, another dream is speaking through Rita. Lynch's characteristic attention to the acoustic register is evident here, as in the "Every Little Star" sequence, in the way that Rita's repetition of the word *silencio* rises in volume from a whisper to a cry. The fact that Rita's eyes are already open as she utters the phrase "No hay banda" when Betty tells her to wake up is also portentous. Is Rita still dreaming, or is it, rather, that Diane does not want to face her own "reality"?

If Rita has been the passive object of Betty's desires up until this point in the narrative, both dream lover and idealized double, now Rita's fantasy takes precedence and will become a veritable nightmare for Betty-Diane, who will not be able to escape its inexorable logic. This turn or reversal is indicated by Rita's insistence on going somewhere "right now" in the early hours of the morning. As the two wait in the street for a cab, the image goes out of focus—which anticipates the blurred shots that accompany Diane's later unsuccessful attempt to masturbate—and is mirrored on the soundtrack by bursts of sonic deformation. Similarly, during the cab ride from the apartment complex to the Club Silencio, the white noise associated earlier with the metropolis of Los Angeles and its ubiquitous, malignant force field is not only invoked but reinforced on the visual plane by the rapid tracking shot past wind-swirled debris to the door of the club.

Unlike the pastel-colored studio set in the "Sixteen Reasons" and "Every Little Star" sequences—which are shot in high-key lighting and whose perfect playback acoustics evoke the pristine pop-rock musicals of the 1950s—the Club Silencio is a theatrical, crepuscular space reminiscent of the Red Room in *Twin Peaks*. It is not, however, a *film* theater. This distinction is critical, since this passé space—complete with proscenium arch, box seats, and red-velvet stage curtain—becomes the privileged place in *Mulholland Drive* for Lynch's exploration, via the trope of silence, of the relation between sound and image, as well as the limits and possibilities of classical sound cinema itself.

The sequence begins with master of ceremonies Bondar (Richard Green), a flamboyant, cane-twirling, Wellesian figure of (dis)enchantment who declaims, reiterating Rita's dream speech, "No hay banda" (There is no band) and "Il n'y a pas d'orchestre" (There is no orchestra). It is all, he declares, a "tape recording." Bondar's highly dramatic monologue literalizes the thematics of the earlier "Sixteen Reasons" sequence where the disjunction between sound and image was insinuated by the slow reveal of all the accoutrements of a live film set.

Still, the Club Silencio is not only or simply about disenchantment or disillusionment. Like the earlier musical passages, which revel in their music even as they deconstruct the production of synchronized sound, the "Llorando" sequence investigates the paradoxes of what Chion calls "audio-vision": the unspoken compact between sound and image that virtually defines classical sound cinema (*Audio-Vision* xxvi). "No hay banda" / "And yet, we have a band." So, as Bondar indexically puts it, if "we want to hear a clarinet" or "trombone," the sound of the instrument can be summoned up, like so much magic, out of thin air: the art of film sound as an art of prestidigitation.

At the same time, the appearance of a trumpet player (Conti Condoli)—*after* Bondar has defiantly thrown away his cane—confirms that there is no essential, "magical" link between sound and image. The trumpet player throws up his hands by way of demonstration while muted trumpet notes continue to fill the air. In this thoroughly demystified context, the turning point of the Club Silencio sequence ruptures the shot-countershot logic (stage-balcony) that dominates the sequence as a whole. A low-angle shot from off-stage left shows Bondar with a stationary microphone in the foreground and a woman with blue hair in a box seat high in the background. Subsequently, when Bondar—hands raised aloft like the trumpet player—says, "Listen," what follows is not the sound of an instrument but random sonic phenomena including and especially an acoustic "storm" signified by thunder and blue light(ning). The ensuing reaction shot of Betty shaking uncontrollably in the balcony, paired with the mysterious presence of the woman with the blue hair, initiates a fundamental "key" change where Betty's world will be shaken and literally inverted. The fact that Bondar's face, lit dramatically from below, assumes the contours of a devilish mask before he disappears in a puff of smoke like the Wicked Witch of the West[6] also hints that the forces at play here are not without their demonic or infernal aspect.

Though the lighting immediately shifts back to a more naturalistic, desaturated mode, the final part of the Club Silencio sequence is clearly under the auspices of the woman with blue hair rather than the magician and, triggered by the change in lighting, revolves around Rebekah Del Rio's performance of "Llorando":

> Yo que pensé que te olvidé
> pero es verdad, es la verdad
> que te quiero aún más
> mucho más que ayer
> Dime tú qué puedo hacer
> ¿No me quieres ya?
> Y siempre estaré
> llorando por tu amor
> llorando por tu amor.

The first thing to be said about Del Rio's version is that it is sung in Spanish, therefore representing a linguistic estrangement of the original English lyrics (composed by Orbison and Joe Melson). Since Rita is Latina (like Rita Hayworth, from whom she takes her name),[7] it also references the "Mexican folklore story of *La Llorens,* The Weeping Woman," who—to attract her husband's attention— killed her two children by throwing them into a river and then killed herself, only to haunt the river with her inconsolable cries for her dead children (Staiger 11).

The second thing to be said about the "Llorando" sequence is that Del Rio, like the trumpet player, is picked out by a spotlight. This lighting cue indicates that despite how "real" her singing seems (compared to, say, the Camilla Rhodes version of "Every Little Star"), it too is subject to the machinations of film production. In other words, Del Rio appears to be singing live (hence the microphone, which "trembles" in the swirling blue light before her entrance); however, true to Bondar's earlier demonstration, her performance is just as much an illusion as Carol's or Camilla Rhodes's.

The most important thing to be said about Del Rio's performance is also the most obvious: Betty and Rita, sitting hand in hand, are moved to tears by it ("It's hard to understand / But the touch of your hand / Can start me crying"). How-ever, even this apparently "real," genuine aesthetic effect is positioned as reflexive by Lynch since, as the painted tear on Del Rio's cheek indicates, Betty and Rita are crying while listening to a song about crying ("I'll always be crying for your love"). Thus, the manifestly illusionistic nature of Del Rio's "act" is laid bare when she collapses in the midst of singing and, while stagehands remove her body, her voice continues to be heard.

Finally, the "Llorando" sequence is key to the film's narrative structure because it leads to the discovery of the "blue box" that Betty finds in her purse while look-ing for a handkerchief. Now, given the fantasmatic logic of causality that governs

*Mulholland Drive,* it is safe to say that the box has appeared, like the trumpet solo, out of thin air. Put a more emphatic way, the blue box has materialized *because* of Del Rio's moving performance. The significance of this particular fantasmatic action is that it subtly reframes the preceding play between sound and image, illusion and reality, even as it reiterates Lynch's surreal conception of the aesthetic.[8] In other words, the point is not so much that what appears to be real turns out to be a matter of contrivance but that artifice does not necessarily represent the opposite of the real. In Lynch's films, in fact, the two are intrinsically coupled.

One consequence of this proposition for Lynch's audiovisual aesthetic is that a lip-synched song may well be *more* powerful than one performed "live." According to this logic—the same logic that, not coincidentally, regulates the classic American musical (Altman, *American Film Musical* 67–71)—sound not only trumps image, thereby reversing the traditional hierarchy, but what I call superdiegetic sound trumps the diegetic track itself.[9] The primary figure of the superdiegetic in *Mulholland Drive* is the word *silencio,* which is first spoken by Rita before she and Betty go to the Club Silencio and is later uttered by the woman with the blue hair at the conclusion of the film. This word is not only part of the diegetic soundtrack but a material "acoustic signifier" (Miklitsch) for that which subtends or transcends sound. In fact, from this standpoint, silence can be said to be an abstract figure for the ultimate discontinuity between sound and image, reality and fantasy, where continuity—whether with respect to image or sound—is the sine qua non of classic realist cinema. In other words, the word *silencio* underwrites the constitutive, irreducible "fact" of fantasy. Just as there is no essential relation between sound and image (as Chion says, "Il n'y a pas de bande-son" [There is no soundtrack] [*Audio-Vision* 39]),[10] so the superdiegetic figure of silence bespeaks art's potential, via artifice, to fantasmatically capture the real.

## Coda: The Box, the Tramp, the Woman with Blue Hair, and the "Acoustic Gaze"

In discussing his method for assembling the score of his films from music composed by Angelo Badalamenti along with bits and pieces of sound, Lynch has commented that "there are sound effects, there are abstract sound effects," and then, somewhere along the way, "music turns into sounds, and sounds turn into music" (quoted in Kenny, "Cruising with David Lynch" 133). One net effect of Lynch's transformational process of "action and reaction" is that there is no contradiction in postproduction between music and effects. In fact, the final mix tends to blur the line between the two so completely that the attendant sound-music design functions not so much as "background" but as "atmosphere" or soundscape.

There is, of course, one other element in the classical trinity of film sound—in addition to music and sound effects—and that is dialogue.[11] Though it would be easy to underestimate the importance of this component in Lynch's films, especially given his celebrated ability to produce startling image as well as sound-music collages, he has been intensely concerned with the materiality of language since the beginning of his career (see, for example, his first short film, *The Alphabet* [USA, 1968]). Lynch himself is on record saying that he had to "learn to talk" (Rodley 32), and this "resistance"—his inability or unwillingness to articulate the passions and obsessions that animate his art—has, over the years, conferred a special magical status on verbal language.

It is only within just such a discursive context that one can begin to appreciate the "silent" dénouement of *Mulholland Drive*. The "added value" of this perspective is that it simultaneously underscores the significance of the music and effects in the film as well as—evidenced in the "Sixteen Reasons" and "Every Little Star" sequences—the technology of sound production itself. Reflecting on the impact of Dolby sound on contemporary cinema, Chion has declared that it "introduces a new expressive element" into the medium, what he calls the "silence of the loudspeakers." For Chion, this silence exposes the audience (and here one might note that the word *audience* derives from the Latin *audire*, "to hear"), "laying bare" our very listening "as if we were in the presence of a giant ear" ("Silence" 151).

One figure of this superambient silence in *Mulholland Drive* is the tramp, not least as he is the only major character in the film who does not speak and is therefore, literally, silent. The tramp, of course, is not only the "awful" Wizard-like person who is pulling the strings in the dream that Dan recounts at Winkie's ("he's doing it") but the person who is last seen in possession of the mysterious blue box, an object that can itself be seen as a signifier of silence. Thus, when near the end of the film the camera cuts straight from Diane, contracting a hit man to kill Camilla, to the tramp, the attendant smoke and flickering red light recollect—in a different but equally phantasmagorical color key—the superdiegetic space associated with the Club Silencio and the woman with blue hair.

Unable to open the blue box, the tramp puts it in a plain brown paper bag and discards it like so much trash. (Such, Lynch seems to be saying, is one possible fate of the imagination in that indeterminate dream space or place called "Hollywood.") The unexplained materialization—in miniaturized form from the blue box—of the elderly couple, who befriended Betty at the beginning of the film, recollects a critical earlier passage. After Betty and Rita have returned from the Club Silencio and Rita retrieves the matching blue key from her purse, Betty suddenly vanishes. After calling out for Betty—saying both "Where are you?" and "¿Dónde estás?"—Rita opens the box. As the camera tracks into the abyssal emptiness of its interior, the whooshing sound previously associated with the tramp is heard and the screen goes black. This too, I submit, is a kind of silence as it forms the lacuna signaling the end of Diane's fantasy.

Despite the film's narrative return to the "real," this does not signal the end of the fantasmatic in *Mulholland Drive*. Cut to the scene immediately after the emergence of the miniature couple, both laughing hysterically, from the discarded blue box. A plain blue key rests on the coffee table in Diane's drab apartment, a predesignated sign from the hit man that indicates, like a reproach, Camilla is dead. There is loud knocking at the door (an aural motif that recurs throughout the film), and the couple's "hysterical" laughter on the soundtrack—combined with the shot of Diane immobile on the couch—signals the onset of psychosis and the extreme auditory hallucinations that are one of its telltale symptoms. Screaming, Diane tries to escape to her bedroom—the place coded from the beginning of the film as a space of fatal assignation—and, as the laughing-screaming reaches a fevered pitch, she shoots herself in the mouth with a pistol. Smoke pours upward from the bed, blue lightning flashes, and the muffled echo of thunder resounds in the distance.

Although this scenario is unrelievedly grim, it is imperative to add (especially with respect to the film's lesbian thematics)[12] that *Mulholland Drive* does not end on this dystopian note. The music changes to a serene, pacific register, and for a final phantasmal time the tramp's visage reappears like a mask in the wavering darkness. As the "Mulholland Drive" theme segues into Betty and Rita's "Love Theme," their faces—ghostly, overexposed mirages—float superimposed over the city of Los Angeles. They are silent, but the "phantom sound" (Chion, "Audio" 218) of their visualized laughter reverberates with the bright optimism that first brought Betty to the City of Angels. The final moment of *Mulholland Drive,* a long shot of the woman with blue hair sitting in her box seat in the Club Silencio, crowns the film's shift from the "terrible" to the "marvelous" (Nochimson, "All" 177).

If, as Chion says, we no longer listen to the film but *"it listens to us,"* then the woman in the balcony box—whose deep blue hair links her with the tramp and the mysterious box—is one visible figure of the superdiegetic sphere: a utopian place (located, literally, "nowhere") that should be distinguished from the debased places where the necessary business of making movies takes place. Yet even as the blue-haired woman's last utterance indexes the superdiegetic, those "ideal" moments (whether magical or terrible) when the world of the dream or fantasy invades mundane reality, her position in the box seat bespeaks her position as a surrogate not only for the spectator but also for that which transcends our limited point of view—the fantasmatic link between the superdiegetic and the point of audition. The Lady in Blue thereby embodies what one might call the "acoustic gaze" (Miklitsch): that other place from which the film listens to us as we watch— eyes and ears wide open—listening as if our very lives depended on it.

## Notes

1. Lynch has acted as a sound designer and rerecording mixer on his films since *Twin Peaks: Fire Walk with Me* (USA/France, 1992).

2. In a different context, the car crash also looks forward to the sound of broken dishes that "sutures"—via a sound match—the crucial cut late in *Mulholland Drive* from the party at Adam's house in the Hollywood Hills, where he announces his engagement to Camilla, to Winkie's, where Diane is contracting a hit man to kill her.

3. Billy Wilder's *Sunset Blvd.* (USA, 1950) is—as the Havenhurst courtyard with its echoes of Norma Desmond's (Gloria Swanson) property attests—one of the primary classical Hollywood sources for Lynch's film (Nochimson, "Mulholland Drive" 39–40).

4. In fact, "Sixteen Reasons" is associated with the television spy drama *Hawaiian Eye* (Robert Altman et al., ABC, 1959–63), which featured Connie Stevens as photographer and nightclub singer Cricket Blake.

5. In addition to the gray form-fitting suit that Betty wears to the Paramount audition (shades of Madeleine), the theme of feminine duality in *Mulholland Drive*—in particular, the class difference between Rita and Diane's former lover who lives in apartment 12—evokes *Vertigo*. The track titled "Diane and Camilla"—composed, as is the majority of *Mulholland Drive*'s score, by Angelo Badalamenti and performed by him with the City of Prague Philharmonic—also echoes Bernard Herrmann's score for Hitchcock's film.

6. As in *Wild at Heart* (Lynch, USA, 1990), there are numerous references in *Mulholland Drive* to *The Wizard of Oz* (Victor Fleming, USA, 1939): Winkie's (a reference to the Wicked Witch of the West's flying monkeys), Coco's Miss Gulch–like spiel about Wilkins's dog, and so on.

7. Rita Hayworth—née Cansino—was not only the "classical-era movie star most famous for arriving in Hollywood via the *frontera*" but, as Williams notes, her femme fatale role in *Gilda* (Charles Vidor, USA, 1946)—which is set in Buenos Aires—"foregrounded her Latin heritage" (8).

8. Lynch's films are marked by their affinity for what Buckland calls paradigmatic as opposed to—as in classical Hollywood cinema—syntagmatic structure (135–36). This compositional trait is one aspect of Lynch's "surrealism." Lynch, true to form, has both denied and acknowledged its influence; see, respectively, Kaleta 12–13; and Hughes 24.

9. I mean to distinguish my understanding of the superdiegetic from both Altman's conception of "supradiegetic sound" as well as Claudia Gorbman's equally provocative notion of "metadiegetic sound" (for example, voice-over narration [*Unheard Melodies* 22–23]).

10. Although this pronouncement has rightly been criticized by Altman et al. ("Inventing the Cinema Sound Track"), Buhler has also remarked that implicit in such a proposition is the notion that film analysis "necessarily takes as its object music *for* cinema rather than music *in* cinema" (58; emphasis added).

11. For an insightful reading of, inter alia, sound-music "segues" and "constellations" in *Mulholland Drive,* see Davison.

12. Staiger, for example, contends that *Mulholland Drive*'s conclusion (that is, Diane's suicide) signifies "heterosexual condemnation of [her] desires" and that the film would

be more sympathetic were "it not so ironic in its critique of Hollywood as a site of failed dreams" (8). A properly dialectical reading of these issues is beyond the scope of this paper, but suffice it to say that Diane's suicide reinstitutes the heterosexual romance narrative between Adam and Camilla that is associated in the narrative with the "mob kingpin," Mr. Roque (Michael Anderson). To wit: Diane and her alter ego, Betty, remain—as the "silent" conclusion to *Mulholland Drive* suggests—figures of utopian desire. For a reading of Lynch—in terms of *femm(e)rotics* and the *faux-féminin*—that does justice to the "real" play of his films, see Braziel.

# PART V

# Case Studies of Film Sound

# 15

# Selling Spectacular Sound

## *Dolby and the Unheard History of Technical Trademarks*

### PAUL GRAINGE

In 1982, Dolby Laboratories produced a demonstration of sound technology for public exhibition. Eight minutes long and given the promissory title *listen . . .* , the film was intended for screening as a short prior to theatrical features. Incorporating live action and animated sequences, the film was designed to accentuate the presentational possibilities of stereo technology; it interweaved a number of scenes that could exemplify the range and heterogeneity of multichannel sound. Cast by Dolby as a "sight and sound journey," the demonstration included scenes of a trickling brook, a saluting cannon, a string quartet, a church organ, the sounds of children at play, and the thundering liftoff of an Apollo rocket. Together with a Jiffy Test film, made available to theater owners and projectionists to evaluate the performance of theatrical sound in particular venues, these promotional initiatives responded to ongoing industrial concerns about the quality of film exhibition in the 1980s. With the emergence of rival technologies of home video, and a sense among exhibitors and distributors that theatrical venues in the United States were inadequate to the needs of the market, companies like Dolby were instrumental to the discourse of "quality presentation" that would transform the history of the multiplex in the 1980s and 1990s and, with it, the status of cinema as event.

*listen . . .* was produced with two promotional functions in mind. Primarily, it was designed to embed and naturalize expectations around sound quality. Rather than deliver an explicit sales message for Dolby Stereo (although Dolby was virtually synonymous with multichannel sound technology at this point), the demonstration film sought to attune audiences to the intense sensory experiences that new sound technologies could deliver. At the same time, it enabled theater owners to market potential investments in exhibition technology, the

widespread adoption of Dolby Stereo in the 1980s bringing with it new imperatives for marketing and promotion, both for Dolby as a licensed trademark and
for distributors and exhibitors seeking to retool the place of cinema in relation
to alternative media forms.

The expansion of home video created a particular impetus to reconstruct the
spaces and spectatorial conditions of the motion picture theater. listen . . . was in
many ways a forerunner of the promotional place that sound technology would
develop in what Charles Acland calls the "permanent marketing campaign" for
cinema-going undertaken by the film industry during the 1980s and 1990s. Specifically, he suggests that 1986 was a key year, heralding an agreement between
the National Association of Theater Owners and the Motion Picture Association of America to invest twenty-two million dollars in a joint effort to sell not
individual films but moviegoing as a whole. This developed on the cusp of the
majors' return to theatrical exhibition. At a basic level, the permanent marketing
campaign involved the likes of Cineplex Odeon showing all trailers attached to
prints and displaying posters for films even if they were not booked at particular theaters. More generally, however, it was symptomatic of broad industrial
reconfigurations extending the principle of entertainment across cultural and
consumer markets. Acland suggests that a concept of "total entertainment" defined cinema and cinema-going from the mid-1980s, linked to the aesthetics of
synergy and manifest in the multiplex boom and the ensuing development of the
megaplex. If the widening life cycle of film texts across territories and mediums
has led to a reformulation of film as commodity—as well as what it means to go
to the cinema—the permanent marketing campaign for movies and moviegoing
represents, in Acland's terminology, "a mode of experiencing and understanding
a wider environment of entertainment, and a world of new images, sounds, and
specially fabricated sites" (79).

Dolby Laboratories has played an integral part in the substrata of marketing
strategies that promote, and set expectations for, contemporary Hollywood film
in this context. Specifically, sound has become a mark of the "sensory surge" associated with technological revolutions deployed, and discursively construed, in
relation to the form and understanding of contemporary entertainment experience (for technological, industrial, and economic perspectives on the encompassing nature of this experience, see Chion, Audio-Vision; Acland; and Wolf). This
has been especially apparent with the emergence of digital technologies in the
1990s that have placed (surround) sound at the center of theatrical claims of "big
screen" experience while, at the same time, amplifying the centrality of television for film viewing as linked to the emergence of home theater. As Gianluca
Sergi notes, "Sound has never been such an important marketing force as it is
today" ("A Cry in the Dark" 163). Not only has it become central to the visceral

aesthetic of the contemporary blockbuster, sound has become a key factor in the reconstruction of cinematic space. In significant ways, the cinema complex, and its acoustic architecture, has been redesigned to maximize the "lure of sound." In the post-Dolby era, this represents for Sergi "an evident shift in the weight given to the figure of the spectator as listener" ("Sonic Playground" 124), the expectations and enticements of sound technology becoming a certifiable mark (literally in the case of theaters carrying the THX logo) of quality presentation.

This essay examines the place of Dolby's multichannel sound technology in the formulation of cinematic "experience," or what Rick Altman otherwise calls cinema's "event-oriented aesthetic" ("Material Heterogeneity" 15). Rather than explore specific instances of sound technology in particular filmic examples, however, it concentrates on the means by which Dolby has become an instructive logo for the film industry and its audiences. This approaches Dolby as a promotional and discursive venture as well as something measurable in terms of technological and aesthetic innovation. First, the essay considers the capacity of sound technology to create "added value" in the synchronism between sound and image, situating the brand capital of Dolby Laboratories through early industrial and promotional initiatives. Second, it explores the function of Dolby as a licensed trademark, including its value as an intermedia sign. These provide a basis for examining, finally, the brand battles fought over digital sound in the 1990s. Analyzing a number of Dolby trailers that update the sensory principle of *listen . . .* I consider the way that Dolby Digital has sought to enliven its market presence in a competitive audio field. Critically, this essay considers different stages in the trademark life of a key sound technology, concentrating on the way that Dolby formats have been naturalized, standardized, and commercially imagined. In so doing, the essay asks questions of how the sensory affect of sound has been made increasingly visible in brand terms.

## "Making Films Sound Better": The Added Value of Sound

In theorizing dimensions of sound analysis, Michel Chion adopts a term beloved by account planners and brand managers: *added value.* In the mercurial world of corporate branding, this describes "assembling together and maintaining a mix of values, both tangible and intangible, which are relevant to consumers and which meaningfully and appropriately distinguish one supplier's brand from that of another" (Murphy 2). For Chion, it means something more specific, based on the formal history and reciprocal development of sound-image synchronism. Put simply, it describes the "expressive and informative value with which a sound enriches a given image," a principle based on the impression of "an immediate and necessary relationship between something one sees and something one hears"

(*Audio-Vision* 5). In each case, qualities of affect are set forth, measured in relation to the experiential value of particular products, services, and technologies. In the history of contemporary film, Dolby sound technology has an especially significant place at the intersection of these commercial and technological definitions of "added value"; the tangible and intangible value of Dolby as a technology and corporate name has become central to the patterning of cinematic expectation and pleasure.

Before considering the circulation and refinement of Dolby as a technical brand, it is helpful to locate the "added value" of stereophonic sound in ways that draw attention to the seeding of "Dolby" as a recognized trademark. The foundational impact of Dolby Laboratories on the development of multichannel sound has, of course, been far-reaching, impacting the layering and directionality of film sound and the very hierarchies (industrial and aesthetic) governing relations *between* sound and image. According to Michael Cimino, "What Dolby does is give you the ability to create a density of detail of sound—a richness so you can demolish the wall separating the viewer from the film. You can come close to demolishing the screen" (quoted in Schreger 351). The idea of "demolishing" or "breaking through" the screen was, and remains, an important metaphor in the aesthetic and marketing potential of postclassical sound. In seeking to shape industrial wisdom about audio technology in the late 1970s, however, Dolby Laboratories took initial steps to assuage misconceptions that "Dolby" had a characteristic sound of its own, often conceived as a particular brand of loudness. Rather, information campaigns cast Dolby Stereo as a process rather than a specific effect, playing down the impression (absorbed within the work of Charles Schreger) that multichannel sound was principally a means "of making the moviegoer think he has a typhoon between his ears" (351). Emphasizing the creative and commercial implications of heightened "sound realism," attempts were made to draw out the *use value* of stereophonic technology for directors, sound mixers, projectionists, and other relevant personnel. Indicative here is an advertorial "Progress Report" written by Dolby Laboratories and published in *Variety* in August 1978. In being associated primarily with the recording business, the Dolby name and logo were at the time largely associated with stereo components and as a built-in extra for home cassette recordings. The report in *Variety* put forward a series of facts, assurances, guarantees, and promises, organized around a vision of Dolby Stereo as a threshold technology for the motion picture industry.

The rhetorical moves of this advertorial report are suggestive in the formation of Dolby's early technological and trademark status, constructing a set of appeals based on artistic control, technical efficiency, and appeal to Hollywood's core youth audience. By creating value in "the life-like reproduction of sound" ("Dolby Stereo" 7), Dolby Laboratories sought to move beyond the science fiction and rock music genres to which its sound technology had become coupled in the late 1970s, instead investing audio quality with envisioned meanings of cinematic possibility and prestige (Terrence Malick's *Days of Heaven* [USA, 1978] was one of the first major "quiet" films to see the value of Dolby Stereo). As a privately owned company located in San Francisco rather than Los Angeles, Dolby Laboratories nurtured an ethic of independent consultancy, projecting its singular commitment to sound technology in meeting a range of industrial needs. Within the "Progress Report," Dolby establishes both an informative portrait of multichannel-sound technology—what it can do, who it can benefit, how much it costs—and of Dolby's status as a technical laboratory unaffiliated with any major studio. The brand appeal of Dolby Stereo only really emerges, however, when the "added value" of sound is mapped onto the expectations and economics of the contemporary movie audience. It is perhaps unsurprising that the "Progress Report" ends on this point:

> [Dolby] is becoming so widely accepted primarily because good sound is good business—and Dolby Stereo makes good sound possible in a technologically and economically practical format. Today's relatively young and affluent moviegoers have grown up with high-fidelity stereo sound as part of their lives. They appreciate it when the sound in a movie theater matches the picture. According to a recent article in *Daily Variety*, "Of those responding to a random telephone survey last July in cities where *Star Wars* played in Dolby Stereo, 90% said it added depth to the picture . . . [and it] was rated equally as important as the visuals." ("Dolby Stereo" 7)

*Star Wars* (George Lucas, USA, 1977) is generally credited with initiating a large-scale conversion to four-channel Dolby Stereo. However, this did not happen as a matter of course; Dolby Laboratories continued to invoke wider transitions in audio sensibility to construct an industrial *argument* for multichannel sound as the untold future of film. Far from an established or inevitable force within industrial lore in 1978, the "Progress Report" indicates the cultivation of common sense regarding the capacity of multichannel sound for producers, distributors, and exhibitors. Signed with the symmetrical double-D logo (representing the perfect symmetry of the encode-decode noise-reduction process), the report casts Dolby Stereo as a tool that "will take a while for everyone to use to its most exciting advantage" (7) but has the capacity to realize itself as an industry standard.

Informational campaigns of this sort were designed to shape prevailing wisdoms about trade practice. They set the ground for marketing strategies, however, addressed more specifically to the cinematic spectator. If, as Gianluca Sergi suggests, "the combination of heightened expectations and increased aural sophistication has produced a highly demanding, active and discerning listener of

Hollywood films" ("Sonic Playground" 126), Dolby Laboratories has shaped these expectations in quite deliberate ways. As the company's first theatrical trailer, *listen . . .* was a strategic means of investing multichannel sound with *sign value*, a marketing initiative responsive to the particular means by which multichannel sound was being developed in relation to the visual field, and in relation to new ideas about the active listener. This brings into focus the reciprocal relationship between (cinematic) sound and (screen) image at the level of promotion.

Optical stereo technology was a pivotal development for Hollywood cinema in the 1970s and 1980s, but this does not mean to say that choices were not made about its use and development. Michel Chion contends that Dolby Laboratories favored the development of "passive offscreen sound," extending the balance and activations of sound in audiovisual space but avoiding aberrant sounds that might exist disconcertingly "in-the-wings" (*Audio-Vision* 85). This effect is not, in fact, inherent within Dolby Stereo but describes the relational compact between sound and image as it has been drawn and codified within particular styles of editing, scene construction, and sound design. Essentially, sound in contemporary Hollywood has been given far greater density and precision but has been anchored to the screen image in providing the "gathering place and magnet for auditory impressions." It is for this reason that Dolby Laboratories has employed cinematic trailers such as *listen . . .* to announce the launch or refinement of particular audio formats. If, as Chion puts it, the "sound-camel continues to pass through the eye of the visual needle" (143), Dolby has used screen visuality to magnify the company's audio achievements, nurturing expectations of sound quality among audiences as a means of standardizing the use and appeal of its technologies for exhibitors. Though relatively inexpensive to install by industrial standards (costing approximately five thousand dollars in the late 1970s), exhibitors were initially reluctant to commit themselves to multichannel-sound conversion. Robert Altman was quick to note the paucity of investment in reproduction technology in the late 1970s, commenting, "Most of the problems with sound in film today are in the reproduction. Sound in theaters—the overwhelming majority of them—is just terrible" (quoted in Schreger 354). In this respect, however good a recording might be, sound reproduction would often be compromised without requisite equipment and forethought to acoustic design.

Initiatives such as *listen . . .* catered to the specific needs of theater owners in encouraging full multichannel-sound conversion. Specifically, it enabled theater owners to *market* sound as a form of experience and pleasure, providing a basis for audience choice about the location of film consumption. Dolby vice president Ioan Allen wrote in 1983, "Maintaining and promoting the very high quality attainable in theaters right now, with today's practical soundtrack and formats and playback equipment, can contribute significantly to the growth and stability of our industry" (19). This was part of a larger argument—current within industrial talk of crisis and reconfiguration—based on the premise that "to remain competitive, the movie-going experience must be unique." According to Charles Acland, attempts to stabilize cinema-going in the 1980s gave rise to a specific understanding of theatrical exhibition; it led to a particular organization of audience space where "upscaling, comfort, courteousness, cleanliness, total entertainment, and prestige emerged as qualities to be offered through the services provided and through the design of auditoriums" (106). Synonymous with prestige sound, Dolby Stereo developed a significant trademark life in this context. As sound technology was taken up in the reformation of the reproductive environment, the Dolby name became a means through which exhibitors could help audiences meaningfully differentiate one theater from another and appropriately distinguish the immersive potential of cinematic sound from the flatter audio effects of domestic television.

listen . . . helped figure sound quality as something that could be openly advertised and exploited. It set the precedent for a number of subsequent trailers and theatrical marketing initiatives by Dolby Laboratories and the likes of THX that established the cinema complex as a space for new kinds of sensory experience. This has been captured in specific logos that pointedly address the active listener, from the "deep note" crescendo of THX with its tag line, "The audience is listening," to the various trailers by Dolby gathered under the rubric, "We've got the whole world listening." Before analyzing these in relation to digital sound, it is necessary to widen the discussion of Dolby as a licensed trademark. At one level, the dissemination of the Dolby name across the terrain of entertainment media and electronic consumer goods demonstrates the interconnections of hardware and software industries, and the means by which different industries have drawn upon Dolby sound technology (and its logos) to sell products through a notion of audio quality. At a more immediate level, however, the case of Dolby invites questions about the circulation and regulation of technological trademarks, especially as it relates to the accretion of brand value across cultural and consumer markets.

## "Breaking Sound Barriers": The Circulation of Technological Trademarks

With the widespread adoption of multichannel sound technology by audiovisual industries in the 1980s, Dolby Stereo began to proliferate as a licensed technology, traversing industrial borders as a marketing device for the added value of high-fidelity sound. In many ways, Dolby has come to function as an intermedia sign, appearing in and between different sectors of the entertainment industry.

From music and gaming to film and television, Dolby is a ubiquitous name attached to various forms of entertainment software and hardware. If Dolby is unusually present as a logo in the cultural terrain, it is the result of systemized licensing agreements that include specific guidelines about when and how the Dolby name can and should be used. These agreements serve a mutually beneficial promotional function for licensor and licensee: they allow Dolby Laboratories to extend itself as a recognized trademark, naturalizing its position in the field of audio technology, while enabling electronics and consumer industries to exploit Dolby's accumulated brand renown.

It is through the process of trademark standardization that Dolby Laboratories has worked to assert and regulate its brand authority. Indeed, the protection and monitoring of corporate trademarks is linked inextricably to the maintenance of brand value. If trademarks serve to indicate consistency in the particular qualities or associations attached to goods and services, trademark standardization is a process of building and protecting the cultural and capital value attached to corporate signs. Significantly, the control of trademark identity prevents goods and services from becoming generic names and therefore vulnerable to cancellation as intellectual property on these grounds. Dolby has been careful to protect itself from generic drift by maintaining its status as a brand sign, establishing a number of provisions within licensing agreements that protect the value-added status of its name, logos, slogans, and soundmarks. These include rules about the use of "Dolby" as a word, about the size and placement of logos when used to sell encoded material (DVD discs) and consumer products (DVD players), and about the correct use of specific logo variations.[1] Together, these foreground Dolby as a brand name related to specific technological innovations ("Stereo," "Surround," "Digital") and prevent it from being used as a *generic* term for enhanced sound quality, or confused as a manufactured consumer product in its own right.

If the history of Dolby can be read through its use and introduction of trademarked technologies, these have been organized around promotional initiatives that link Dolby sound technology to new sensory and affective possibilities for experiential listening. Dolby Laboratories has fashioned itself as an instantly recognizable brand name that makes "cutting-edge, in-demand audio technologies and renowned trademarks available to all sections of the electronics industry" ("Licensing"). Whenever Dolby has introduced a new technological innovation, the basic "double-D" trademark has been refined, announcing developments at the "cutting-edge" with new visual and audio signatures. This has been particularly marked with the development of digital and surround technologies, where the proliferation of logos—Dolby Digital, Dolby Digital EX, Dolby Surround Pro Logic, Virtual Dolby Surround—has served to create an array of distinctions in the type and degree of sound experiences available in public and domestic space.

While setting expectations of sound quality in theatrical space, these logos have also become significant in the particularized economies of taste associated with developments such as home theater.

In particular, Dolby Surround (the consumer designation for the home-theater version of Dolby Stereo) has become central to the sensory reinvention of film culture in domestic space. According to Barbara Klinger, this is joined to a discourse of elite television viewing. Klinger suggests that, in attempting to transform domestic space into that of a movie auditorium, home-theater technology "depends on the aestheticization of digital technologies for domestic use, a development which parallels the contemporary rebirth of the motion picture theater through the promotion of digital advances in visual and audio representation and consequent upgrading of theater facilities" (16). As a key digital trademark, Dolby Surround has become a hallmark of viewing distinction in the domestic fusion of sound and image, instrumental to what Klinger calls a "semiotics of class superiority, refinement and good taste," or what she otherwise labels home theater's "aristocratic techno-aesthetic." Using Dolby Pro Logic or Dolby Digital AC-3 processors to separate the audio signal into five channels, passed through five speakers, the Dolby Surround trademark offers up a promise of active spectatorship for particular social groups. Technological capacity, in this sense, is inextricably linked to the cultural capital associated with bringing cinematic audio technologies into the home.

Much could be written of home theater's re-creation of big screen sound. In discussing the brand value of Dolby, home theater demonstrates that trademark status is never "standardized" in discursive terms, but is given meaning through socially determined frameworks of consumption and spectatorship. Moreover, home theater provides a potent example of the increasingly blurred boundaries between forms of audiovisual culture and the contributory role that sound plays in the "total entertainment environments" created by and for particular kinds of audiences. In a period when hardware and software industries have moved closer together, and cultural industries have based operations on the expansion of media platforms and exhibition windows, Dolby sound technology resonates in the relations *between* film, television, music, video, gaming, and electronic media, drawing on the sensory promise of sound for broad audiences and specialized taste segments.

As a licensed technology associated with processes of recording (software) and reproduction (hardware), the Dolby name has acquired brand capital across a spectrum of entertainment machines and media. Though often parsed as a "third-party trademark"—occupying a residual position in logo terms "to ensure that consumers do not mistake another company's advertised product or service as one provided by Dolby Laboratories" ("Trademark")—sound has come to prominence in the delineation and marketing of audiovisual "experience." This has encouraged proactive strategies by Dolby Laboratories to calibrate and brand this experience, becoming especially significant in the 1990s with the competitive developments of digital sound. Sound has not only become more visible in discursive terms but has also been visualized quite literally by Dolby Laboratories in brand initiatives designed to accentuate the industrial and affective significance of its technologies.

I want now to analyze this in the form of the cinematic audio trailer. Although Dolby is an intermedia trademark, cinema has become the privileged form through which new audio technologies are showcased, reconstituting the active listener—or at least the expectations of cutting-edge sound—in ways that reverberate through the wider channels of moving-image culture. Specifically, I will consider a number of digital trailers that were produced by Dolby Laboratories in the 1990s for use in movie theaters but also "played" on applications such as DVD, broadcast television, and video games. Announcing the audio effects of spectacular sound—with titles such as "Train," "Aurora," "City," "Canyon," "Egypt," "Temple," and "Rain"—these trailers convey a particular scenic inhabiting of the Dolby universe.

## "We've Got the Whole World Listening™": Advertising the Sound Event

In June 1992, Dolby Laboratories launched a striking trailer for its new Dolby Digital format. Entitled "Train," the trailer was designed and mixed by Randy Thom and Skywalker Sound, the visuals undertaken by an independent digital workshop called Xaos. Thirty-two seconds in length, the trailer focused audiovisual effects through the scene of a sepia steam train, its wheels slowly emerging and then gathering momentum through billowing clouds of gray smoke. The choice of a train was not incidental. In practical terms, it offered up a range of illustrative audio effects. Beginning with a hollow clanking (set to a blank screen), the heavy movement of carriages (gray smoke) gives way to the compressed movement of steam and metal (train wheels), the billowing smoke of the funnel (pistons) matching the rhythmic acceleration of a heavy locomotive across a track (train moving into the near distance). Fading to black, the scene is followed by a silver Dolby Digital logo, turning gradually to gold.

In discussing the heterogeneity of sound, Rick Altman invokes the railroad to explain how molecular sound is essentially a vibration carried by a transmitting medium (for example, air, water, rail) in the form of changes of pressure ("Material Heterogeneity" 17). "Train" was designed to amplify this heterogeneous quality through the power of digital and surround technology. For all its promise of the digital audio future, however, the trailer went representationally backward in time. In significant ways, the monochromatic "Train" invoked the legacy of Lumière and the first public actualité presentation around which legends of audience wonder have grown. Whereas the famous *L'arrivée d'un train en gare de le Ciotat* (1895) created sensory thrall, and with it the apprehension that a locomotive would break through the screen, so "Train" rehearsed this moment in sonic terms, producing dense layers of sound effect and pressure in ways that gave the audiovisual scene a sublime physical presence. According to Vivian Sobchack, the Dolby trailers (and especially "Train") were "purposefully oneiric—'dream devices' that constitute both an intimate and immense poetic space in which one can wonder at, as [Gaston] Bachelard puts it, 'hearing oneself seeing . . . hear[ing] ourselves listen'" ("When the Ear Dreams" 3).[2]

Unlike *listen . . .*, which in 1982 sought to normalize expectations around sound quality, "Train" was a more determined brand initiative for audiences accustomed to high-definition sound. "Train" was, in short, a calculated advertisement for Dolby multichannel technology, set squarely in the "sound war" being waged in the move from analogue to digital. Indeed, digital sound systems developed by DTS (in which Universal swiftly bought a share after the technology was fully developed) and Sony (SDDS) quickly rivaled the Dolby Stereo Digital System, introduced with *Batman Returns* (Tim Burton, USA, 1992) and promptly rolled out in prestige theaters. These systems were not interchangeable. Although each system could perfectly reproduce the range and variety of sound captured by sophisticated multitrack film recording, the digital formats produced by Dolby and Sony were on-film systems, encoded on the movie soundtrack itself, compared with DTS, which synchronized a movie's visual track with sound from a CD-ROM. As a result, fierce corporate battles emerged to control the industrial and technological promise of digital sound, creating new competition for what the *Los Angeles Times* called "the hearts and ears of U.S. moviegoers" (Natale D1).

These struggles were propagandist in nature, incorporating information campaigns and marketing initiatives designed to persuade exhibitors to convert to their own system. In doing so, the motion picture theater became an enlivened space for the constitution of the sound event. Marketing the technical apparatus of sound played an especially significant part in defining what cinema, and cinema-going, might represent in experiential terms during the 1990s. In its pitch to exhibitors, Dolby Laboratories described how digital sound would give theaters a "competitive edge" in "keeping 'going out to the movies' a unique entertainment experience" (Jasper 1). Reformulating claims made for multichannel sound in the 1970s, Dolby argued that audience expectations were being shaped by familiarity with improved sound quality (notably through use of CD players) and by the desire for a heightened sense of "show business" relegated by domestic video.³ "When it comes to film sound," the company proclaimed in a promotional leaflet in 1992, "no name is more familiar to audiences than 'Dolby.' When it comes to the latest technology, nothing means more to audiences than 'digital.' Put them together—in your theater, on your marquee, in your ads—and you have a winning combination." In the industrial endeavor to formulate expectations of cinema-going pleasure, Dolby Laboratories has continued to sell the lure of sound both to producers and to exhibitors. As the above statement suggests, this is linked to marketing initiatives designed to accentuate the value of (Dolby) sound as a renewable form of consumer enticement. It is for this reason that Dolby Laboratories, in its own words, supplies exhibitors "with a *choice* of exciting digital trailers that audiences regularly applaud, and issues new ones on a regular basis."

Dolby Laboratories does not oblige filmmakers or exhibitors to use its tailored brand materials. Marquee signs, lobby posters, and trailers are made freely available to cinemas that have installed Dolby equipment. The promotion of sound in theatrical space must be understood within contexts of industrial competition and alliance, including the inscribed rivalries and collaborative relations of extant technologies. Whereas rivalry can be witnessed in the struggle between Dolby and DTS for dominance in the market for worldwide theater installations, collaboration can be seen between Dolby Laboratories and LucasFilm THX in the development of Dolby Digital Surround EX.⁴ If this represents a tension between studio-sponsored initiatives in sound development and those developed by specialist audio and effects-based companies, leverage has been sought through sound-reliant event movies. Whereas Universal's *Jurassic Park* (Steven Spielberg, USA, 1993) launched DTS and obliged theaters to upgrade to the studio's own sound system, Dolby Digital Surround EX premiered with *Star Wars: Episode 1, The Phantom Menace* (George Lucas, USA, 1999), along with strong recommendations from George Lucas about the necessary sound technology required by theaters to exhibit the film. Trailers, such as those made available by Dolby, are a means of asserting market presence and of distinguishing one name in sound technology from another. In the process, however, the affectivity of sound has become manifest within certain kinds of brand aesthetic, significant in demonstrating both the contours of sound marketing in the 1990s and the patterning of audio experience.

According to Michel Chion, contemporary sound quality is based on the "realization of the modern ideal of a great 'dry' strength." This is organized less around fidelity than on the "technical capacity to isolate and purify the sound ingredients" with very little reverberation. This is captured for Chion in the short THX sound trailer that he summarily describes as "a bunch of glissandi falling towards the low bass register, spiralling spatially around the room from speaker to speaker, ending triumphantly on an enormous chord. And it's all at an overwhelming volume that leads the audience to instinctively react by applauding in a sort of physical release" (*Audio-Vision* 100). The soundmark of THX is made up of a "deep note" crescendo that highlights the capacity of THX to reproduce sound with the quality and power intended at the point of production. This is matched, in aesthetic terms, by the solid physical presence of the THX initials that encapsulate a visual feeling of clarity and bulk. For Ann Brighouse, director of marketing at THX, "THX cinema trailers have become tremendously popular among moviegoers. Audiences anxiously await our signature Deep Note crescendo; it creates a sense of anticipation and eagerness for the feature presentation" ("Press Release"). In a familiar gesture, particular claims are made here for trailers in shaping the active (and applauding) listener; trailers are seen to act physically upon the audience in establishing the sound event. Vivian Sobchack develops the point, suggesting that "the trailers are exciting to watch and often applauded by audiences less for their computer-graphic bravura than for their primary function of visibly articulating sound—and, more importantly, of visibly imagining and articulating *sound as such*" ("When the Ear Dreams" 8; emphasis in the original).

Whereas THX has come to rely in its trailers on the dynamic strength of theatrical sound, Dolby has used a number of atmospheric scenes to project auditory impressions. These can be seen in defined stages during the 1990s that correspond with the launch of Dolby Digital (1992) and Dolby Digital Surround EX (1999). In considering how the affectivity of sound has been made visible in brand terms, it is worth considering the supportive visual elements that have been given to the sound mix in Dolby trailers. Broadly speaking, these have come to present sound either in terms of natural elements (light, water, fire) or through forms of embodied space (canyons, temples, cities). In each case, digital sound is figured as a world to be explored. Gianluca Sergi suggests that the technologically advanced space of the film theater has become "a sonic playground in which the spectator actively participates, making sense of what is around him or her, and discovering new pleasures" ("Sonic Playground" 121). Although the idea of sound as discovery is nothing new in promotional or discursive terms, it was parsed in striking audiovisual forms as Dolby sought to assert its claim on the digital market.

"Train" was in many ways anomalous in this respect. Having established the significance of Dolby Digital in terms that echoed the very birth of cinema, subsequent trailers played with themes of exploration, enabling audiences to behold Dolby through the exotic vistas of "Canyon," "Egypt," and "Temple," all made in 1996. In each case, the audience moves through monumental space; the trailers present imposing ancient structures that, filled with contrasts of sunlight and shade, reveal the Dolby Digital logo as an imposing tablet or hieroglyph. Created by the same company in each example (Digital Artworks), these trailers provide a supporting visual environment for a sound mix designed "to demonstrate that Dolby Digital does not make only loud sounds more impressive, but also makes subtle, quieter sounds crisper and cleaner" ("Explore Our World"). This must be set in the context of complaints (by exhibitors) about the loudness of theatrical trailers. Precipitating the formation of the Trailer Audio Standards Agreements in 1997, forged between sound engineers, exhibitors, and major studios, Dolby Laboratories created scenic trailers as a direct response to industrial concerns about the imposing physicality of sound. Where "Train" relied on bold contrasts in the level and directionality of sound, linked to the movement in narrative space of a heavy mobile object, Dolby's scenic trailers spatialized sound around rhythmic and kinetic sensations. The use of syncopated drums, cymbals, bells, and other percussive instruments—all marked as exotically non-Western—add to the discreet sound design in this respect, reinforcing the impression of Dolby as a world to inhabit rather than a simple technology to employ.

"Canyon," "Egypt," and "Temple" rely on the figurative representation of lost worlds, invoking ancient myths and anthropological fascinations to make sound itself an object of discovery. If the moniker "Explore our world" has become a banner for Dolby Laboratories, it has become further manifest in trailers that deepen Dolby's status as a brand. In 1998, Dolby launched a trailer called "Rain" with *Star Trek: Insurrection* (Jonathan Frakes, USA). This saw colliding water droplets passing through a strobe light. Designed to illustrate how "clear," "pure," and "quiet" a movie trailer could be, it also introduced a symphonic soundmark that identified Dolby Digital with an auditory signature. Unlike the deep-note crescendo of THX, this was made up of a delicate five-note theme, a continuation of Dolby's primary association with sound definition rather than dry strength.

The introduction of the soundmark was conceived in a moment when Dolby Digital was beginning to attain screen dominance over DTS. Indeed, 1998 was the first year since the release of DTS that Dolby surpassed its rival in terms of worldwide theatrical installations. Whereas Dolby had a total of 13,073 auditoriums, DTS had 12,800, leaving SDDS with a mere 5,201. With the release of Dolby Digital Surround Ex, Dolby would increase this number to 20,000 installations in 1999, compared with 15,881 for DTS (largely based on the launch of its DTS-ES upgrade) and 6,675 for SDDS (see Hindes "Universal," "Theaters").[5] The adoption of a soundmark was significant in pressing home Dolby's growing advantage over DTS in a period when the majority of major film studios (including Universal from 1998) were distributing films in all three formats. According to *Variety*, this left "exhibitors able to make their equipment purchasing decisions based on a system's price and performance, rather than on distributor's corporate allegiances" (Hindes "Dolby"). Cost and functionality became a more determined basis for theatrical adoption. In selling sound investment to audiences, however, use value was rarely divorced from a technology's sign value, and the means by which installation could be advertised by theaters as a recognized symbol of commitment to high-quality presentation.

Accordingly, the soundmark would accompany the Dolby Digital logo in all future trailers and on other Dolby applications. This included "Aurora," designed specifically for the launch of Dolby Digital Surround Ex and the much hyped release of *The Phantom Menace*. Swiftly adopted in the scramble to secure bookings for the most eagerly anticipated film of the decade, Dolby Digital Surround Ex represented a new surround system employing a 6.1 format using an additional rear center channel. For Gary Rydstrom, the aim was "to open up new possibilities and place sound exactly where you would hear them in the real world" (quoted in Olsen n.p.). Consequently, "Aurora" was developed (by a digital workshop called yU + Co) to emphasize the rear-channel effect, taking the viewer through the development of an aurora borealis in space. In this, sound came from behind the camera, sweeping down and around the audience so that, beginning in the rear channel, it filled out from back to front. Rather than the "organic" sound of water droplets, the trailer was distinguished by a choral rendition of Dolby's soundmark, audiences encircled by the celestial fanfare of a trademark that first came to prominence with *Star Wars* in 1977 and marked its association with the franchise twenty years later as a stellar brand of its own.

Dolby has continued to refine and reinvent the visual tropes of materialized sound. In 2003, it launched "Perspectives," a live-action trailer made in collaboration with theatrical percussion group Stomp. This was based on a vibrant rhythmic demonstration of its Dolby Digital Surround Ex format. Through different forms of an audiovisual trailer, digital sound has been framed as a ranging and exploratory dimension of the cinema-going experience. Trailers draw attention to the sound event, enabling, in the words of Michel Chion, "sensations to be perceived for themselves, not merely as coded elements in a language, a discourse, a narration." For Chion, this can be seen as a defining aspect of sensory cinema. He continues: "Cinema is not solely a show of sounds and images; it also generates rhythmic, dynamic, temporal, tactile, and kinetic sensations that make use of both the auditory and visual channels. And as each technical revolution brings a sensory surge to cinema it revitalizes the sensations of matter, speed, movement, and space" (*Audio-Vision* 152). Capitalizing on the possibilities of digital technology, Dolby Laboratories has been quick to brand the affective potential of the contemporary sound event. In the effort to maintain Dolby at the forefront of cinematic audio technology, brand strategies have moved toward a more concerted vision of Dolby as a source of experiential wonder. If branding is based on emotive, rather than informational, appeals in the commercial patterning of value, Dolby has sought to distinguish itself not only through demonstrations of the depth and directionality of digital sound but also through a concept of the audiovisual sublime, inscribed variously in cinematic ("Train"), monumental ("Temple"), and even cosmic ("Aurora") terms.

To speak of cinema as an "experience" or "event" brings with it a number of questions about the very space of cinema as a site of social interaction and cultural practice. James Hay suggests that film is "practiced among different social sites, always in relation to other sites, and engaged by social subjects who move among sites" (212). Sound is constituted in and between these sites. It has been used to promote the "unique" experience of the motion picture theater, just as it has been figured as the ultimate commodity of home cinema. These are linked by an industrial concept of "total entertainment," specifically the creation of environments that, according to Barbara Klinger, see "multiple technological and aesthetic economies [creating] a kind of *Gesamtkunstwerk* of the possibilities of fusing sound and image with the utmost veracity and impact" (7). Within this framework, sound has become something to brand, a form of "added value" that has developed new (or renewed) significance in the constitution of film's "event-oriented aesthetic."

By concentrating on the sign value of Dolby technology, one can situate the developing place of film sound within the industrial and affective economies of contemporary Hollywood, moving beyond the question of film aesthetics to the wider commercial environments in which sound has come to function. Critically, this essay expands the horizon of entertainment branding to include the technologies that lie at the core of contemporary media spectacle, instilling specific kinds of sensory promise. My interest has been in the way that Dolby has developed life as a technical trademark. If the challenge of sound criticism has been to wrest the analysis of film from its dependence on visual and text-based analysis, I have sought to investigate the way that sound technology has a promotional and discursive history, linked to conceptions of the active listener. Although Dolby is an intermedia sign, forged through a range of hardware and software licensing deals, its status as a brand has been especially refined in the battleground of digital sound. Having naturalized multichannel sound for the film industry and its audiences, Dolby Laboratories had to imagine its relation to sound in more determined ways in the 1990s as a means of asserting itself against its rivals. This meant developing new digital systems but also a provision for selling (and standardizing) its trademark technologies within new theatrical and consumer markets. With the widespread adoption of digital sound technology across film and electronics industries, Dolby has remained a pivotal trademark but one that has also tried to stabilize its hold in the audio market through promotional initiatives that territorialize the affectivity of sound.

Selling sound is nothing new; it can be seen as part of the film industry's enduring use of technology to heighten sellable notions of "realism" and "spectacle." Yet the significance of sound has been increasingly realized through a regime of branding in which audiences are solicited not simply through a company's name recognition but through an invitation to inhabit and figuratively enter the very world of the trademark. This goes to the heart of contemporary brand practice, what Elizabeth Moor characterizes as a set of strategies designed to expand the potential spaces of marketing and where "consumer experience itself is increasingly both the object and subject of brand activity" (42). Rather than function as a sign of quality, branding relies on linking products, services, and technologies to "sensual and memorable" experiences. This provides the principal means of engaging with consumers, forging durable affective connections in the interplay between branded objects and consumer bodies, or what might equally be seen as the relationship between sounds and spectators. If "total entertainment" describes a mode of experiencing a "world of new images, sounds, and specially fabricated sites," Dolby has sought to visualize itself *as* a world for audiences that have come to recognize sound quality as a component of choice in their encounter with film and its ancillary entertainments. If, indeed, Dolby Laboratories has got the "whole world listening," this has been achieved in no small part through strategic efforts to maximize and enliven Dolby's brand identity as a way of maintaining industrial and audience appeal.

## Notes

I would like to thank Ioan Allen for reading a draft of this chapter and for providing a number of helpful comments and suggestions. Thanks also to Gianluca Sergi for pointing me in the direction of *Dolby News,* and other useful material.

1. In short, Dolby must always be used as an adjective; its logo must always be in black, white, gold, or silver and be recognizable as a third-party trademark; and it should always match the correct "Stereo," "Surround," "Digital" format being used.

2. Similar to my own interests, Sobchack is concerned with the means by which Dolby trailers visibly articulate and imagine sound as an expressive tendency of contemporary cinema. Her argument is phenomenological in approach; she provides a detailed analysis of the "sonority of being" that each trailer works to produce. In attending to the poetic, rather than promotional, function of the Dolby trailers, Sobchack provides an incisive textual reading of how "sound *hears* and then *imagines* itself temporally sounding" ("When the Ear Dreams" 11; emphasis in the original). My approach is to open out these concerns in industrial terms, to consider how the Dolby trailers relate to particular trademark strategies and to wider corporate attempts to structure the contemporary entertainment environment.

3. In industrial terms, Dolby Digital was sold in ways that bore a marked similarity to

the positioning of Dolby Stereo in the late 1970s. Emphasis was placed on the fact that Dolby Digital was a creative means, not an end, that it could be used for more than special or dramatic effects, that it did not render obsolete existing cinema installations, and that it could help filmmakers and exhibitors "enhance that very special experience of going to the movies."

4. Dolby Digital Surround EX added a third surround channel to digital film sound, reproduced by the speaker array at the back of the theater. This enabled greater use of front-to-back and back-to-front transitions, especially useful for creating the effect of spaceships flying over the audience (DTS would also introduce a 6.1 format using a center surround channel, launched as "extended surround" or DTS-ES). Although LucasFilm THX collaborated on the development of Dolby Digital Surround EX, the brand name of THX describes not a format but instead a theater standardization system. As a trademark licensed to theaters and manufacturers, the THX certificate identifies compliance with performance parameters for the playback environment established by LucasFilm Ltd.

5. As a measure of cost, the price of upgrading to Dolby Digital Surround EX stood at $2,500 in 1998, compared with DTS-ES at $1,875.

# 16

## (S)lip-Sync

### Punk Rock Narrative Film and
### Postmodern Musical Performance

**DAVID LADERMAN**

## From Live Lip-Sync to Slip-Sync

According to a recent article in the *New York Times* (February 1, 2004), a majority of popular music fans have come to accept lip-syncing at live performances. Contemporary audiences bred on MTV are now well accustomed to seeing singers lip-sync; the mainstreaming of rap and hip-hop during the 1990s has contributed, too, since it features prerecorded music and sampling at live performances. But the advent of live lip-syncing also seems yet another dramatic symptom of contemporary postmodern culture blurring the boundary between reality and its reproduction. Focusing on Super Bowl half-time shows, televised awards ceremonies, and concert "extravaganzas," the *Times* author observes that "these concerts are about spectacle and sheer star proximity, not the miracle of live music production" (Nelson 30). As one might expect, the article situates the issue of live lip-sync in terms of debates around authenticity. Recorded music for some time now has embraced all manner of electronic simulation. Yet, for many, a performer lip-syncing live represents a betrayal of both the integrity of artistic voice and the presence of the audience. It can seem like colluding with something out of *Brave New World* to applaud a live singer mouthing to his or her own recorded voice.

To be sure, authenticity has always been a contentious site within rock and roll culture, and popular music in general.[1] On the one hand, rock music invokes an earnest spirit of rebellion, aiming to achieve the aesthetic status and social impact of art. At the same time, rock music has always been shaped and driven by the corporate interests of the entertainment industry, interests that often lead to the substitution of simulated for human-produced music. Lawrence Grossberg has conceptualized this tension between authenticity and commercialism in terms of sound and image. In a nuanced and detailed discussion of rock music's relationships with film, television, and visuality itself, he asserts that "the authenticity of rock has always been measured by its sound and, most commonly, by its voice" (204). The visual realm—including live stage shows, television, film, and all manner of marketing and packaging—ultimately corrupts, or at least compromises, such aural authenticity. "The eye has always been suspect in rock culture; after all, visually, rock often borders on the inauthentic" (204). Conceding that "the ideology of authenticity is increasingly irrelevant to contemporary taste" (203), Grossberg thus proposes the term *authentic inauthenticity* to characterize postmodern rock culture, where "authenticity is itself a construction, an image, which is no better and no worse than any other" (206). As live rock performance answers more and more to the demands of visual spectacle, the sound of the music itself becomes increasingly susceptible to simulation. The infiltration of lip-syncing into live rock performance seems another step in the direction of such authentic inauthenticity.

Yet it would be difficult to maintain the argument that rock and pop music *as sound* enjoy a privileged and pure status, and that visualization represents some fall from grace, some secondary articulation. Surely, rock music has always been something exciting *to watch* as much as to listen to. From album art to the concert spectacle, distinctive and often outrageous fashions, and countless films and television appearances, rock and pop music seem essentially visual on some level (which is not to say the music matters any less). Two formidable examples that immediately come to mind are Elvis Presley and the Beatles who both arguably relied on films and television performances as much as music for their cultural impact.

Would granting authenticity to the visual dimension of rock be the same as approving of live lip-sync? Philip Auslander might say yes, in saying no to the very notion of authenticity. Building on Steven Connor's conception of "postmodern performance," where, for example, new technologies of reproduction became brazenly integrated into the megaconcerts of the mid-1980s (172–77), Auslander persuasively argues that "liveness" itself possesses meaning only relative to nonliveness, or recorded sound. Put differently, live performance has always been "mediatized" on some level, therefore constituting yet another category of simulation. "Live performance cannot be said to have ontological or historical priority over mediatization, since liveness was made visible only by the possibility of technical reproduction. . . . [T]he live can exist only *within* an economy of reproduction" (54). Auslander insists, for example, that the Milli Vanilli lip-sync scandal (where the group was caught lip-syncing to someone else's vocals) was no scandal at all, but rather symptomatic of both traditional pop commercialism and new approaches to music production. The music industry's subsequent effort to punish Milli Vanilli testifies to the ideological need for the trope of authenticity, which glosses over the fact that there is none (94–96). The apparent advent of live lip-sync bears out Auslander's perspective that rock authenticity is "a matter of culturally determined convention, not an expression of essence" (70).

With all manner of sampling, remixing, digitalization, and simulation rampant throughout pop music production today, the hip performer and audience member comprehends that new performance and recording technologies render obsolete such distinctions between the real and the artificial, the authentic and the inauthentic. Auslander recounts a Laurie Anderson concert, where she "wandered on- and off-stage, as if to suggest that the computerized, audiovisual machine she had set into motion could run itself, that *it* was the show, with her or without her" (59). The *New York Times* article on live lip-sync concludes with a similar anecdote, implying that pop music audiences have caught up with postmodern simulation theorists and avant-garde performance artists: "During a concert at Madison Square Garden in August, the R & B singer R. Kelly did not even bring a backing band with him, working strictly with prerecorded tracks. At one point, he put down his microphone in the middle of a song and let his recorded vocals keep singing. By all accounts, the audience loved it" (Nelson 30).

Such a euphoric embrace of live simulation can be interpreted optimistically, as heralding new creative performance possibilities, or negatively, as evidence of technology superseding the human voice (and, implicitly, human identity and presence). This negative attitude toward live lip-syncing, in fact, achieved a frenzied pitch in the entertainment media about six months after the *Times* article appeared, when Ashley Simpson was "caught" lip-syncing on *Saturday Night Live*. In an embarrassing example of a lip-sync slip-up, the pop singer started mouthing lyrics to a different song than the one audiences heard. The beating she took from fans and the press might derive partially from her perceived persona as an already essentially inauthentic pop singer. But it also demonstrates the staying power, in certain contexts, of the audience's desire for a truthful and honest voice—especially live.

In this essay, I aim to provide some historical context to the recent phenomenon of live lip-sync by discussing a handful of punk and new wave narrative films from the early 1980s. Their portrayals of disembodied lip-sync within "live" performance sequences prove to be remarkably visionary. As narrative films from a particular moment in rock music history (punk, new wave, the onset of MTV), they also furnish a fertile site for considering new tensions between sound and image, and between the performing body and performance technology. Films such as *Jubilee* (Derek Jarman, UK, 1978), *Breaking Glass* (Brian Gibson, UK, 1980), *The Great Rock 'n' Roll Swindle* (Julien Temple, UK, 1980), *Starstruck* (Gillian Armstrong, Australia, 1982), *True Stories* (David Byrne, USA, 1986), and

*Sid and Nancy* (Alex Cox, UK/USA, 1986) signal a shift in cinematic sensibility regarding the representation of musical performance—a shift that continues to play itself out. They articulate rock music performance through a peculiar sound-image fissure, a playful anxiety around the performing body both fusing with and refusing mass media technology and celebrity-culture commercialism.

I will refer to this performance tension as "(s)lip-sync" (hereinafter slip-sync), since the performer stops lip-syncing, or at least slips out of sync, alienated from yet caught up by the performance spectacle. Undermining both the commodity fetishism and the underlying trope of authenticity intrinsic to conventional lip-sync, the slip-sync moment in these punk films engenders a split in the show, a rupture involving a double excess: the performer exceeds the performance spectacle by disengaging; the performance spectacle exceeds the performer by continuing—consuming, in a sense—the authenticity signified by the performer's dissociation. To better convey this notion of double excess, as well as the contentious political ambivalence around authenticity in the context of punk, I respectfully substitute "in/authenticity" for Grossberg's "authentic inauthenticity." The latter, to my mind, applies more accurately to mainstream acts that are "honest" about their "false" commercialism. Punk, in contrast, partially defines itself by achieving authentic subversion of such "authentic inauthenticity." Indeed, punk slip-sync seems a postmodern riff on Jane Feuer's notion of the "myth of spontaneity" in classical musicals, which usually privilege authentic "bricolage" performance over those that are "canned" and "prepackaged" ("Self-Reflexive" 444). The films mentioned above offer distinct variations on the slip-sync performance, sometimes as a moment of rupture, other times as an overall unhinged sound-image design.

Before turning to these films, I would like to suggest a more general cultural framework for situating punk slip-sync performance: the Reagan-Thatcher social transformation. For different, perhaps antithetical, reasons, punk and Reagan-Thatcher share some oddly collusive seams. For example, both ideologies denigrated the 1960s counterculture; likewise, both celebrated a certain anarchy of consumption in the marketplace. There is also a way in which both Reagan-Thatcher and punk articulate an unnerving ambivalence around authenticity and inauthenticity, fiction and truth, spectacle and spontaneity, dissimulation and simulation. Reagan's performance of authentic "commonsense" populist values within the context of an inauthentic conservative propaganda machine (one driven by entirely unpopulist economic policies) coincides more than chronologically with punk's interlocking of vehement rebellion and shameless media sham. By focusing on a specific and consistent representational strategy that exaggerates in/authenticity, this essay aims to at least raise the possibility of thinking through the relationship between punk's nihilistic pastiche of popular culture and

Reagan-Thatcher's deregulation of market forces, not to mention their shrewd manipulation of misleading images.

## The Historical Context of Slip-Sync

Due no doubt to elaborate high-tech developments in sound technology, as well as intensified marketing "synergy" around the soundtrack, sound since the 1980s has become increasingly central to film art.[2] Yet synchronized sound continues to be the dominant film sound design principle. Michel Chion's notion of "synchresis"—the appearance in film of a "spontaneous coupling of sound and image"—proves to be the persistent, pervasive underlying concept: most film sound seems to come from, or be linked to, something visual (even if off-screen or subjective) (*Audio-Vision* 63). Even today, synchresis is especially well evidenced by the musical genre, where actors seem to break into gloriously synchronized, effortlessly choreographed song and dance.[3] But Chion advises that "there are degrees of synchronism," and that "loose synch gives a less naturalistic, more readily poetic effect" (65). His reference to the conspicuously "off" postsynching of Italian films begins to suggest punk slip-sync's more overt dismantling of synchresis: not nonsynch or asynch sound but a rent or rupture of synchresis from within, that is, a narrative moment that renders palpable the precarious materiality of film sound technology itself.

In fact, one notable if surprising historical precursor to punk slip-sync is *Singin' in the Rain* (Gene Kelly/Stanley Donen, USA, 1952), which, let us recall, dates from just before the dawn of rock 'n' roll. It also coincides with the more widespread use of lip-sync in film and television. In the scenes where Don Lockwood (Gene Kelly) and Lina Lamont (Jean Hagen) attend the preview screening of their first sound film, *The Dueling Cavaliers,* disaster strikes when the soundtrack goes "out of synchronization," as the assistant to the producer exclaims in terror. Hilarious as the sound already is for other reasons, the unsynched soundtrack provokes the most comic anxiety, culminating in the bit when Lina and the villain exchange "no no no" (head nodding yes) and "yes yes yes" (head shaking no). Commiserating over the preview debacle, Lockwood, Kathy (Debbie Reynolds), and Cosmo (Donald O'Connor) mock this unsynched sound gag, then come upon the "brilliant idea" of dubbing Kathy's voice for Lina's (another surprising precursor to the Milli Vanilli scandal).[4] These scenes from *Singin' in the Rain* play with slipped synchresis in the narrative context of an unsettling performance, here reflecting the nervous transition from silent to sound film technology. More self-conscious and sardonic, the punk films similarly reflect an anxious transition from live to simulated performance.

Another interesting precursor to punk slip-sync occurs in the Beatles' *Hard Day's Night* (Richard Lester, UK, 1964). Early in the film, on their way by train to a television performance rehearsal, the Fab Four escape to the luggage car, where they break into "I Should Have Known Better." After they sit down to play cards, the song begins playing on the soundtrack, presumably nonsynch. However, we see both Ringo and George rocking their heads in time with the music; this playfully blurs the boundary between synch and nonsynch sound. Suddenly, as the first vocal begins, an abrupt, handheld jump cut shows John "singing" the song, with the other members pretending to play their instruments. I say "pretending," not merely because their instruments have magically appeared: George strums an electric guitar, though we hear an acoustic on the soundtrack; likewise, Ringo pounds his drumsticks on something, but we never actually see the drum kit. Most conspicuously pretending is Paul, who also lip-syncs with John, but this is peculiar, as the song only has one vocal track (John's). Whereas John's lip-syncing is fairly convincing, Paul's becomes a kind of illogical if charming excess: smiling, he is not too concerned about correctly mimicking the lyrics. Such a reflexive departure from traditional synchresis no doubt derives from director Lester's overall hyper–new wave film style. Although the utopian tone owes something to classical Hollywood (highly formalistic and idealized), the scene sets the stage for both the celebratory, promotional aesthetic of MTV (fast cuts, playfully self-conscious lip-sync) and the more unnerving punk slip-sync.

A significant early example of punk slip-sync characterizes *The Blank Generation* (Amos Poe/Ivan Kral, USA, 1976), a no-budget documentary on the mid-1970s New York punk music scene. Eschewing all but the thinnest veneer of organization, the film comprises simply raw footage of various punk bands performing at small clubs like Max's Kansas City and CBGBs. Its punk approach to sound foresees slip-sync, but as more extreme, a kind of antisynch. Here, all the concert footage is presented with the music slightly out of sync, creating a perpetual disconnect between sound and image. Thus, there is really no slip (or, there is a constant slip), as sound and image from the start are never synchronized: the visuals are at a different point in the song than what we hear on the soundtrack. *The Blank Generation* establishes a fundamental punk sound design that rejects everything about the commercial music industry, including synchresis and lip-syncing. Noting in passing the film's "non-synchronous songs," Gina Marchetti links the film's "technically crude" formal approach to punk's preference for the intimate performance venue, where "the punk audience is as important and active as the musicians" (273).[5] The moment of punk slip-sync I will be discussing in early '80s narrative films owes its tone of resistance to this DIY (do-it-yourself) approach.

However, punk's DIY reinvigorates rock music just prior to a cultural and commercial development with much more pervasive, and in many ways contrary, aftershocks: the advent of MTV, perhaps the most crucial catalyst in the wide-spread dissemination of lip-sync. Andrew Goodwin's discussion of the emergence of MTV, notable for its empirical, historicizing perspective, furnishes an illuminating framework for appreciating the slip-sync gesture in punk films. Insisting that lip-syncing itself is not what distinguishes music videos from past pop music image-texts, he argues that MTV's "break" with the past revolves around the coincidence in the early '80s of "the perceived 'failure' of punk rock" and the transition to new music production technologies (*Dancing in the Distraction Factory* 31). As another concurring writer puts it, "The emergence of punk rock and its symbolic affront to the commercial 'insincerities' of the music industry coincided with the growing displacement of the skills of the musician, as new computer-based technologies of musical production made it possible to produce and replicate sounds both in the studio and on stage" (Mundy 233).

The "new pop" that grew out of (and against) punk transposed emphasis from live amateurism to electronic dance music. For Goodwin, this shift in musical sensibility is inextricable from the success of MTV as a promotional (and artistic) vehicle that celebrates simulated music. Goodwin goes on to argue that the new pop more comfortably embraces image and marketing (which punk viciously parodied), rendering the romantic ideology that links creativity and authenticity with performance no longer relevant (*Dancing* 33–35). Beginning in the early '80s (and culminating in 1990 with the Milli Vanilli controversy mentioned earlier), this "displacement of the musician" (32) seems to me a key factor of the slip-sync gesture in these punk films. Apparently with a finger on the pulse of the promotional, electronic, and visual turn pop music is taking, punk slip-sync reflects negatively, but also ambiguously, on these cultural trends as a kind of betrayal, or at least a contradiction.[6]

Let us look a bit more closely at the punk-MTV coincidence regarding sound and image. Despite being so playfully artificial, MTV lip-sync operates mainly through synchresis, or the pretense of synchronous sound: the voice we hear (recorded separately, earlier) does emit from the lips pretending to sing. For all the talk of the music video's visual "incoherence," there is in fact usually rigorous coherence between music and image, where rhythm and cutting are synchronized, and especially where for the most part singing and lip movements are synched to the vocal soundtrack (Goodwin, "Fatal Distractions" 45–48). David E. James distinguishes this sound-image aesthetic from punk "hardcore" videos of the early '80s. In the former, "the image track is controlled by the soundtrack; the visuals are assimilated to the song's rhythms. . . . Whatever autonomy is temporarily allowed to the visuals is reigned-in as the clip returns for closure to the performers' bodies and especially to lip-synch shots of the singer's face and lips" (236–37). In contrast, the hard-core punk videos reinvent the "scorched-earth production values" Marchetti noted in late '70s punk documentaries. One aesthetic strategy James observes in these hard-core punk videos is the synchronous superimposition of band-performance shots with audience shots, reproducing "the rejection of the distinction between audience and band that is central to punk's politics and overall alterity to corporate consumer culture" (225). What I aim to show below is that such "alterity" (along with punk's obverse complicity) gets played out through slip-sync in various overdetermined musical performance moments of punk narrative films.

## Jubilee

A low-budget, mostly experimental film, Derek Jarman's *Jubilee* came out before MTV, but seemed to see it coming. Working squarely within an avant-garde and modernist narrative-film practice (one that would include Warhol, Godard, and fellow Brits Russell and Roeg), *Jubilee* envisions the corporate, image-oriented rock that prevailed throughout the 1980s; it also offers an incisive critique of the emerging punk subculture. Noting its punk "home movie" aesthetic and mode of production, as well as its "paranoid and hysterical" visions that came true, Michael O'Pray situates the film at the center of the burgeoning London punk scene (94–97). The film's "English Renaissance structuring device," which irked some punk pundits (95), is as follows: Queen Elizabeth, seeking knowledge of the future, is escorted through time by alchemist John Dee, with the aid of Shakespeare's angel Ariel, to present-day England of 1977.[7] This time-traveling trio observes the exploits of a group of punks surviving in and around postapocalyptic London, where buildings appear bombed out, baby carriages are left burning, crime runs rampant, and the police are fascist thugs. Most significantly, a demonic media czar has apparently inherited the top position of government.

An early punk music performance in *Jubilee* furnishes a vivid example of the slip-sync moment, in the context of an audition for this media czar. Borgia Ginz (Jack Birkett), who oversees a rock music television station the punks listlessly watch, has agreed to hear the film's leading punk character and narrator, Amyl Nitrate (Jordan). Before the audition, in tight close-up, Ginz articulates a bemused, pernicious discourse. Apparently speaking to no one in particular (thus directed at the film viewing audience), he pontificates on "the generation who forgot to lead their lives . . . too busy watching my endless movie. . . . [T]he media is their only reality; without me, they don't exist. . . . I sucked and I sucked and I sucked; I don't create power, I own it. . . . If the music is loud enough, they won't hear the world falling apart." Probably inspired by Malcolm McLaren, the notorious punk impresario who made similar pronouncements a few years later in *The Great Rock 'n' Roll Swindle,* this hyperbolic monologue frames the performance in terms of punk authenticity being "short-circuited by a culture of Devourers" (James, *Power Misses* 201). But it also intimates punk's complicity in its own exploitation.

Once the audition begins, aspects of the mise-en-scène emphasize in/authenticity, likewise framing the slip-sync moment to come. Amyl appears on an oversized theater stage, with tacky fog machine and bombastic flashing lights, a huge, warped mirror behind her. No musicians are visible; Amyl "sings" (lip-syncs) to recorded music—a punk-metal version of "Rule Britannia" (actually sung by Suzie Pinns). The only sense of this being "live" derives from the disturbing presence of her body—a wonderfully excessive, pretentious masquerade, to put it mildly. Decked out in a plastic Union Jack flag around her waist, black underwear, pink dish-washing gloves, a gold gladiator helmet, lime-green tights, and garish Ziggy Stardust makeup, she struts about dispassionately, striking absurd, contorted poses. Her dance gestures combine striptease and military marching, mocking both. Such costume and gestures forcefully convey how punk pits its own excess of artifice against that of mainstream-nationalist commodity culture.

Close-ups reveal that she makes little effort to lip-sync in a convincing manner. She looks almost into the camera during the enigmatic, ominous "air raid" roars entering the hall, engulfing the soundtrack. At one point, visibly irritated, she completely loses track of the vocal track, and stops moving her lips, yet her voice in the song continues without her. Although the camera style is relatively straightforward, the set design, acting and audio mix invoke the double excess underlying slip-sync: the one, her exaggerated, ironic costume and body gyrations; the other, the technological artifice of the spectacle proceeding without her, rendering her a kind of extraneous, decorative accessory. Amyl ambivalently embraces this status: she snarls and glowers, annoyed at the performance exceeding her. But it is hard to distinguish such frustration from every other pose she strikes. She walks offstage before the song is done, her voice lingering without a body.

The ensuing scene further frames this slip-sync performance moment through in/authenticity, that is, as a collision between a subversive avant-garde attitude and the highly accommodating and insidious media culture of the encroaching 1980s. This time in a studio soundstage, Ginz observes Amyl's friend Mad (Toyah Wilcox) and her band auditioning behind a glass booth (yielding the peculiar but appropriate effect of watching a monitor). During their song, another friend, Crabs (Little Nell), arrives with her "discovery," Kyd (Adam Ant), an aspiring punk singer she picked up in a diner. It is ironic and telling that, whereas Mad's performance is rife with typical punk rage, pounding music, and screaming into the camera (at Ginz), Ginz himself has his hands coolly on the controls of the sound mixing board, sliding the knobs up and down, his eye on the prize Kyd might yield. As in the previous slip-sync performance, "live" music, here at its most raw and violent, is coded as highly mediated and manipulated. Although

Grossberg might view this and other punk films as co-opting authentic punk music by inauthentically visualizing it, one could argue that musical co-optation is precisely what the "Rule Britannia" performance critiques, through slip-sync.

*Jubilee*'s slip-sync scene encapsulates some of the political contradictions within and around punk, but it also sketches a sound-image relation that dramatizes a disturbing erasure of any original sound presence.[8] Appearing more than ten years prior to the authenticity "scandals" of the 1990s (and the postmodern theoretical essays they inspired), *Jubilee*'s slip-sync mobilizes a deconstruction of film musical performance by rupturing lip-sync's implicit original-copy hierarchy. In conspicuously slipping out of sync, Amyl exposes "her" voice as a copy, one that precedes and produces her "live" performance. The fact that it is not Jordan's but another singer's voice further deconstructs any originality or authenticity—again, Jarman presciently intimating within punk a complicitous version of the media czar. At the same time, her momentarily unsynched performing body becomes a *remainder,* a punk excess that stains the spectacle of simulation and the simulation of spectacle.[9]

With this theoretical context and *Jubilee*'s example in mind, I would like to now skim through some subsequent instances of slip-sync.

## Breaking Glass and The Great Rock 'n' Roll Swindle

Two films from 1980 further extend the stylistic, thematic, and political stakes suggested by *Jubilee*'s slip-sync. Brian Gibson's *Breaking Glass* follows the "rise and fall" rock star–film formula, focusing on aspiring punk–new wave singer Kate (Hazel O'Connor, who also sings all the songs). Savvy about the need to promote her image, Kate also clearly aspires to effect social change through her music, specifically by singing about the impact of new computer technologies. Kate's persona and performance style is technorobotic, yet her songs critique such a mode of identity. Slip-sync becomes a theme throughout the film, as Kate struggles to retain authority over her (literal and figurative) voice while the music and celebrity industry gradually consumes her.

The film's opening credit sequence employs a variation on the slip-sync performance moment. As the camera dollies forward down the aisle of a tube train, to a close-up of Kate, the song "The Shape of Things to Come" begins playing on the soundtrack. Yet, as with "I Should Have Known Better" (also on a train), there is a notable ambivalence regarding sound. She seems to hear the song (in her head?), nodding and moving down the aisle in time with its rhythm; the other passengers show no sign of hearing it. With no fictional music source, she is nevertheless somehow "with" the music. Then, when she moves into the next train car, she is "singing" the song, obviously lip-syncing, looking straight into the camera, performing for us, the audience of the film. Here the camera dollies back, "pushed back" by her forward movement, which exaggerates the vocal track she now possesses (or is possessed by). In the next few train cars, the camera is outside, tracking alongside her as she moves down the aisle, brazenly putting up stickers and spraying graffiti; she has stopped lip-syncing, though her voice continues in the song on the soundtrack.

This tracking camera, now exterior and more distant, enhances the slip-sync double-excess effect. Exuding considerably more agency than Amyl in *Jubilee*, Kate's "slippage" in and out of sync is more choreographed, alternating in syncopation with each train car. Although this moving camera underlines the modernist challenge to convention Kate enacts, the sound mix is more postmodern, looking forward to music videos through its overdetermined relation to fictional space: the music (including her voice) exceeds her acting-performing body. Further illustrating this point is the last shot, from behind the last car as it pulls away into a dark tunnel. Barely visible in the extreme low-key light, she lip-syncs the final lyrics, receding from the camera into blackness, but also slipping away from the soundtrack. Treated with a slight reverb to convey the echo effect within the tube tunnel, her voice exceeds her as her lips disappear, foreshadowing her traumatized (and mediatized) fate.

To further appreciate the slip-sync double excess illustrated by *Jubilee* and *Breaking Glass*, let us consider Michel Chion's notion of the *acousmêtre*. Designating a nonsynch voice not yet seen, a voice that does not yet possess a body but possesses a magical power of narration, Chion describes it as follows: "It's as if the voice were wandering along the surface, *both inside and outside*, seeking a place to settle" (*Voice in Cinema* 23; emphasis in the original). Typically either a "voice of god" narrator or a privileged voice-over from within the fiction, this "ubiquitous" bodyless voice can also be sourced in "media," which "send acousmatic voices traveling" (24). Punk slip-sync is related to but not fully accounted for by Chion's *acousmêtre* (though his emphasis on "both inside and outside" resonates). In the above slip-sync examples, a voice that belongs to a body becomes acousmatic by virtue of the performance and recording technology within the narrative. In a more disturbing riff on the acousmatic, the singer's voice becomes disengaged from the body, reversing Chion's trajectory of "a voice that seeks a body," positing instead a voice that leaves a body, or a body that seeks the voice it has lost, now an acousmatic excess.

A different kind of slip-sync excess appears in *The Great Rock 'n' Roll Swindle*. Irreverently mixing documentary, fiction, and animation, hosted by Malcolm McLaren, the film is a macabre walking tour through the phenomenon of punk, offering a rather harsh and acerbic view, painting it as intrinsically a sellout. Early on, the film's title song (presumably written for the film) is performed by three of the original Sex Pistols (sans Johnny Rotten), plus a couple of backup singers, as a "1978 Audition," with the subheading "Anyone Can Be a Sex Pistol." Filmed after the Sex Pistols' dissolution in 1978 as a gag spectacle, the performance is intercut with the credits: a punk dwarf scrambles to put together the letters of the film title (last correcting "swine" to "swindle"). During the song, one groupie or fan after another comes onstage, takes the mic, and sings a verse or two.

This sequence relates to slip-sync as a kind of inverse or flip side: instead of one singer losing her vocal track, here multiple singers create a rotating vocal track, "slipping" in and out of sync. In another relevant twist, it appears that the band is "play-syncing," or miming, to a prerecorded instrumental track, whereas the singers sing live. At various points we clearly see drummer Paul Cook and bassist Sid Vicious blatantly faking it (one quick shot has Vicious on the drums, Cook on bass). The wild, spontaneous energy of the amateur lead vocalists lends authenticity, dramatizing the DIY theme so prevalent in punk, as noted earlier. Yet the sporadic rotation of singers, combined with the miming band, evokes a more cynical flaunting of artifice through sound-image disjuncture. The droning refrain "rock 'n' roll swindle" reinforces the ironic tone underlying (and undermining) such "audience participation": the last singer states directly into the camera as the song ends, "It's a swindle." We might also note the way the singers parody Johnny Rotten and Elvis Presley, further contextualizing punk in/authenticity as a gag, a gimmick—a swindle.

Other performances later in the film featuring Sid Vicious likewise articulate a relevant variation on *Jubilee*'s slip-sync. Sid lip-syncs to his cover versions of different Eddie Cochran songs: the first time in a bedroom in his underwear, mainly before a mirror; later on a motorcycle riding through the English countryside. Aiming apparently to both mock and emulate classic 1950s rockers, Sid's lip-sync is so reckless it constantly slips. The blatantly nonsynchronous soundtrack he not only hears but "sings" further exaggerates a playful disconnect between sound and image. Like the Johnny Rotten mask the first fan rips off his face in the film's credit sequence described above (only to be "doing" Johnny again, beneath the mask), here Sid's performing "singing" body becomes a mask that willfully slips across the surface of the soundtrack.

## Starstruck and True Stories

An odd Australian film from 1982 offers yet another angle on the punk slip-sync approach, though one would be hard-pressed to justify describing the film as punk: Gillian Armstrong's first feature, *Starstruck*. Even "new wave" seems too hard-core for a film with such a heartwarming, lighthearted tone. But the musical styles and fashion of *Starstruck* clearly reflect the influence of punk and new wave. More intriguing is the film's uncanny and truly quirky sense of self-parody, which might owe much to the spirit of punk, while at the same time satirizing it. *Starstruck* focuses on Jackie Mullins (Jo Kennedy), an aspiring new wave singer who like Kate in *Breaking Glass* employs a technorobotic performance style. With her fourteen-year-old cousin for a "manager" and numerous cheesy new wave musical dance numbers, the film recalls the early rock musical *The Girl Can't Help It* (Frank Tashlin, USA, 1958) for its interweaving of camp, parody, and excessive glee.

Yet there is a slip-sync moment in *Starstruck*, admittedly short-lived, that conveys some resistance to mass culture, being set in the context of television. Through an outrageous publicity stunt orchestrated by cousin Angus (Ross O'Donovan), Jackie has finally landed a chance to appear on a pop music television showcase. But the TV studio will not let her do the song she wants to, and will not let her band perform with her, substituting their own song and a studio band. In contrast to Amyl putting on her own phony persona, here we observe Jackie being made over by makeup artists backstage, into a more classical pop diva (undoing her own garish postpunk look). Yet she does not seem to mind, cognizant of the showbiz opportunity knocking. Further underlining the in/authentic setting, the television audience is set to go wild with admiration on cue.

Near the middle of the song (which has a pop-disco sound uncharacteristic for her), she gradually loses track of the vocal track. Probably not familiar enough with the program's song choice, Jackie becomes visibly confused and distracted by the arm signals directed at her by the camera technicians. Distressed and a little angry, she points back at the cameras, slipping out of sync. The song's refrain "I believe in you" becomes especially ironic, since no one believes in her, and she obviously does not believe in what she is doing. Like Amyl in *Jubilee*, she reacts negatively to the performance context exceeding her, and manifests her own excess of resistance. As the song comes to a close, pubescent screaming girls rush the stage, goaded by television host Terry Lambert (John O'May). All join her for the final chorus, but she flees the stage, dramatically removing herself from the performance, even as her voice faintly continues on the soundtrack, now buried beneath all the other voices.

Additionally, the mise-en-scène emphasizes such in/authenticity with various fragmented images of her on television monitors. One of these monitors is perched in the pub where she works, owned and operated by her family. Here, her rejected, "authentic" band pokes fun at her performance: they know right away what she comes to realize during the song—that it's a sham and a sellout. However, backstage we observe that she is distraught, not over the invasive performance technology and entertainment industry that has transformed her but over how "awful" she was. This moment of slip-sync drama is unique, as she blames herself;

unique too is how Terry comforts her by inviting her to a party in his penthouse pool, where another goofy and bizarre musical number simply washes away the anxiety of the slip-sync scene. In the tradition of the classical musical, *Starstruck* goes on to resolve its slip-sync in/authenticity conflict through a harmonic integration that ultimately celebrates showbiz (Feuer, "Self-Reflexive" 447–49).

David Byrne's 1986 film *True Stories* offers a very different yet complementary slip-sync approach to that of *Starstruck*. In the spirit of *This Is Spinal Tap* (Rob Reiner, USA, 1984), *True Stories* is a self-consciously modest, tongue-in-cheek mock documentary, focusing on a "specialness" celebration in the small (fictional) town of Virgil, Texas. But the film also functions as a promotional and artistic cinematic vehicle for the Talking Heads, one of the most important and innovative new wave bands to emerge from the late 1970s New York punk scene.[10] In *True Stories* director Byrne serves as the first-person tour guide who ferries us into the lives of the quirky inhabitants of the town. The film's particular version of slip-sync performance corresponds with its understated satire of the town's naive expectation of economic benefits coming from a new local computer company.

After introducing us to the computer-chip plant where most of Virgil's new global villagers work, Byrne drives us out that evening to the local nightclub. Here, various audience members, mostly characters we have already met, run onstage in rotation, each lip-syncing a different verse of the Talking Heads' song "Wild Life" (the single from the *True Stories* album released simultaneously with the film). Perhaps an accidental remake of the Sex Pistols' "audition" scene in *The Great Rock 'n' Roll Swindle,* this slip-sync scene mixes elements of karaoke culture, MTV's postmodern reflexivity, Warhol's fifteen minutes of fame (here fifteen seconds), but also punk's DIY philosophy. In contrast to the "Anyone Can Be a Sex Pistol" scene, where each singer actually sings over the recorded musical tracks, here Byrne's original recorded voice remains, over which each "singer" lip-syncs a few lines. The performance is coded as a live event, yet the rotating lip-sync to Byrne's voice undermines this.

Once again, television figures prominently, with a wall of TV monitors behind the band. During the performance, the film cuts periodically to various close-ups of unrelated TV images, creating an appropriate effect of high-art video meets MTV. The performers' close-ups in turn exaggerate how the lip-syncing slips, as each "singer" mouths the words, sometimes a little off. But there is also a sense of slip-sync in the conspicuous shift from one face to the next, each face inhabiting (or inhabited by) the same voice. This, combined with the intercutting of the TV monitor images, invokes the slip-sync theme where simulation technology rearticulates liveness and authenticity. Moreover, the rapid-fire star-turn onstage for each "singing" audience member itself becomes doubled, through their various impersonations of other stars: Elvis, Prince, Neil Young—even David Byrne himself. Most notable is the first impersonation, by Talking Heads' keyboardist Jerry Harrison, of Billy Idol—who of course represents an early '80s pop mainstream version of punk. Having a Talking Heads member "do" Billy Idol, decked out in caricatured punk regalia, is the scene's only direct, but perhaps overinvested, reference to punk.

Bearing even less visible traces of punk than *Starstruck, True Stories* is nevertheless postpunk, not so much by virtue of the Talking Heads but in its cool, parodic embrace of pop culture. The film's slip-sync performance translates punk's DIY into a goofy moment of "audience participation"—one that harbors a faintly condescending tone toward the quirks of middle America. In this way *True Stories* links with *Starstruck* for its optimistic, perhaps ironic, mainstreaming of punk and new wave.

## Sid and Nancy

Quite a different form of audience participation occurs in the slip-sync scene in *Sid and Nancy,* Alex Cox's swan song rock biopic of Sid Vicious (Gary Oldman), Nancy Spungeon (Chloe Webb), and the Sex Pistols. After achieving substantial critical acclaim with his first feature, the Los Angeles cult punk film *Repo Man* (USA, 1983), Cox was hired by a major studio to direct this more grandiose and polished punk film. Told in flashback just after Nancy's body is discovered, the film follows the mythical arc of Sid joining the Sex Pistols; his encounter, romance, and drug addiction with Nancy; the disastrous American tour of 1978; Sid's short-lived fame as a solo artist; and the final days leading to first Nancy's, then Sid's, death.

Like that of *True Stories,* the slip-sync scene in *Sid and Nancy* also derives from *The Great Rock 'n' Roll Swindle,* but more explicitly. At the center of the film, at the height of the most notorious punk rocker's popularity, Sid performs Paul Anka's "My Way" for a television studio audience. More a grotesque parody of Sinatra's celebrated rendition, this performance re-creates a scene from Temple's film. Typical of *Swindle's* staged-documentary approach, the original sequence features the real Sid Vicious pulling out a gun and "shooting" members of the upper-crust, sophisticated audience—thus undermining any sense of this being authentically "live." Cox and Oldman shift the scene from a French nightclub to a television studio, embellishing their fictionalized version with numerous expressionistic details, creating a postmodern hallucination. Once again, we encounter the spiral of simulation and in/authenticity, as the one film version re-presents a previous artificially "live" production. Possibly with *Jubilee* in mind, Cox further emphasizes this theme of in/authenticity by structuring the revised performance around a rather exaggerated instance of slip-sync.

Donning his classic spiked hair, Doc Marten boots, black leather, and dog collar, Sid struts in shadow down a pristine, semifuturistic stairway toward the camera. As the lights click on, he begins the first verse of "My Way," dementia style, accompanied presumably by an off-screen smooth jazz orchestra (though it is not clear whether live or recorded). Suddenly, for verse two a driving punk rock sound kicks in; he pogos, flails, and snarls the rest of the song. We should note that the orchestra continues to accompany the punk rock instruments, symbolic of in/authenticity, the co-optation of and within punk. As with most previous examples, there are no musicians visible on stage, invoking a sound design that equivocates regarding sync-nonsync. Up to this point Sid appears to be actually singing; it is a rather convincing lip-sync.

Then, as the last verse begins, the slip-sync gesture occurs, as the singer disengages from the soundtrack. Recalling Amyl in *Jubilee*, he strikes an angry pose at the key moment of dissociation; but unlike her, or Jackie in *Starstruck*, he does not seem even remotely perturbed by the slip-sync. In fact, more like Kate in *Breaking Glass*, he appears to have deliberately caused it, rendering "slip" perhaps not as accurate a prefix as "quit" (-sync). Kate slip-syncs as a kind of formalized, somewhat indirect affront to her audience of train passengers. Sid's slip-sync is more directly confrontational. He seems to suddenly recognize the audience for what they are: contrived for and by television, composed mostly of middle-aged, upper-class, theatergoing types. He eventually pulls out a gun and starts firing on members of the crowd, which includes Nancy and McLaren (David Hayman). McLaren looks on, impassive behind dark sunglasses; Sid does not shoot him. But he does exchange an angry glare with Nancy, who seems to dare him with her look—so he shoots her as the culmination of spectacular violence and rage. After "dying" in bloody slow motion, she wakes up, shakes it off, gets up, and joins him onstage for a kiss, just as the song ends. They march back up the steps together as the lights switch off and the orchestral denouement fades out.

Within the context of television production, here signifying mainstream assimilation and artifice, Sid's slip-sync as gangster becomes a sign of refusal, a destruction. But this negation of television in turn (and at the same moment) becomes recuperated through the artifice of the film medium's textual effects. That is, the film's visual and audio tracks collude with the television performance to rearticulate rebellion as commodity spectacle. The exaggerated blood splatter, the horrified gestures of audience members, the slow motion all underline how this "spontaneous" outburst of anger and disengagement from performance is staged and mediatized. Although Sid's performance itself is hardly subtle (and of course viewers in the know will recognize its revision of the scene in Temple's film), there is an ambiguous hallucinatory quality to the sequence, in terms of its place and function in the overall narrative. Did this really happen? Was part of it in Sid's mind? How did we get to this scene?

Such quality derives partially from how the "My Way" sequence occurs "inside" a sound-music design that suspends narrative logic and linearity. First, the scene is framed on either end by conspicuous nonsynchronous sound bridges, an eerie atonal sliding scale of notes (which Cox faithfully borrows from *Swindle*). During the first bridge, we cut elliptically to the television stage from a somewhat unrelated restaurant scene; then, as the "My Way" sequence closes with the same sound bridge, we straight cut from a black frame to a new scene: a bird's-eye view of Sid and Nancy at home asleep. Like a music video but also a classical musical number, "My Way" is set off, self-contained by the predominance of the music track, yet it is not entirely explained or justified by the narrative structure. The music overdetermines the narrative ambiguity of the scene (there is no dialogue), pushing the visuals into the Other, aural, register.

*Sid and Nancy*'s slip-sync articulates punk's co-optation by mainstream media; the contrived, artificial (whether dream or simulation) staging of the television performance; and a disturbing, indiscernible sound practice that undoes synchresis. Sid enacts excessive punk rebellion, appearing to break out of the contrived performance. On the other hand, given the scene's hallucinatory context, blurring the boundaries between real-unreal and live-recorded, such spontaneous rage becomes "spontaneous rage," a contrived effect of the performance spectacle. In contrast to both *Jubilee* and *Breaking Glass*, the double excess of slip-sync becomes truly absorbed into the overall performance mode: his dissociation, his excess as angry punk, seems itself part of the staged excess of the performance spectacle. Look carefully at the way he pulls the gun out from its holster neatly folded into the lapel of his tux coat, or at the mystical reunion of the romantic couple onstage, "beyond death." These are Hollywood clichés of gangsters and lovers who never die as screen personas. Its violent rupture all a sham, a show for and of television technology, the slip-sync does not challenge the performance but becomes a "challenge" that ultimately serves the performance.

## Slip-Sync, Postmodern Doubling, and Reagan-Thatcher

By way of a conclusion, let us tentatively situate such in/authenticity, and the double excess of slip-sync performance, within (or alongside) the larger conservative culture and postmodern politics embodied by Reagan and Thatcher. After all, the realpolitik of slip-sync involves the operation during performance whereby a manufactured, commodified voice displaces the actual voice and body of the performer. The double excess of slip-sync encapsulates a more general doubling that Linda Hutcheon identifies as key to postmodern art, culture, and politics. In a chapter titled "Political Double-Talk," she could be describing punk when she characterizes "postmodern challenges" as "paradoxically both inside and outside, compromised and critical," and "open to appropriation" (205).

Moreover, her pithy maxim, "There is contradiction, but no dialectic in post-modernism" (209), helps bring Reagan and punk under the same umbrella. Reagan, too, affected a kind of slip-sync within his presidential political identity, by seamlessly incorporating his previous acting roles, thus embodying contradiction as doubleness: "he presents political events of his own making as if he were somehow not responsible for them." Michael Rogin situates such "easy slippage between movies and reality" in terms of the postmodern shift from production to consumption, where the opposition between "the authentic and the inauthentic" no longer makes sense (8–9). Certainly, the phenomenal media blitz surrounding Reagan's death reached a nearly hysterical pinnacle of such antihistorical, hyper-mediated confusion around in/authenticity. Goaded on by the Bush administration, mainstream media accounts eulogizing "the great communicator" consistently deployed the trope of authenticity to characterize (and mythologize) the collective memory of perhaps the most shrewdly inauthentic president ever.

It is not my intention to conflate such "bad" postmodern politics with the post-modern negation that punk's self-conscious and reflexive in/authenticity promises. However, perhaps Sid Vicious doing "My Way" is not such a "grotesque parody" as I suggested earlier. Perhaps it contains an insidiously affectionate allusion to Frank Sinatra—who, of course, paralleled Reagan's own political swing from left to right. Put differently, perhaps punk harbors an elusive nostalgia that looks back not to the (parental) counterculture of the previous era (which it consistently derides) but to the (grandparental) Eisenhower fifties: the era of James Dean, the Beats, and the birth of rock 'n' roll, but also of Sinatra's prime, the birth of television, and the cold war ideology and military-industrial power nexus that would crucially shape Reagan's political persona (and that thrives today, resurrected for the inauthentic interventions of a new "globalized" American empire).

Yes, Sid's performance of "My Way" is full of irony and mockery, but it also conjoins a Reaganesque individualist bravado with punk's anarchic DIY ethos. Although Derek Jarman's subversive punk mining of Elizabethan literary England is radically opposed to Margaret Thatcher's strong-arming of Elizabethan imperial England, there is at least a sliver of contact. "Anarchy in the UK," in some ways, is just what Thatcher said and did to the British welfare state; "my way" returns through punk as a code phrase for privatization, deregulation, and rogue militarism. The double excess of punk slip-sync thus offers a critical mirror, a double, for the conservative double talk of postmodern political culture.

## Notes

1. For an early essay that raises many of the pertinent questions around authenticity, technology, and rock music, see Simon Frith, "Art versus Technology."

2. See two seminal works on this point, Jeff Smith's *Sounds of Commerce* and Anahid

Kassabian's *Hearing Film,* as well as the recent anthology edited by Ian Inglis, *Popular Music and Film.*

3. Neither *Chicago* (Rob Marshall, USA, 2002) nor *Moulin Rouge* (Baz Luhrmann, Australia/USA, 2001), vanguard postmodern musicals in many ways, depart much from synchresis.

4. *Singin' in the Rain* further envisions the slip-sync gesture: apparently, Debbie Reynolds's singing voice was dubbed in by other singers throughout the film (Wollen, *Singin' in the Rain* 56). Her "authentic" voice, both on the audio track and in the narrative, turns out to be in/authentic. Thanks to Jay Beck for bringing this point to my attention.

5. Stacy Thompson has recently imported Barthes's term *writerly* to characterize this punk film aesthetic, for its "open formal structure" that encourages the audience (of the music or the film) to produce meaning (50–51). Whereas he champions Don Letts's *Punk Rock Movie* (UK, 1978) in this regard, *The Blank Generation* to my mind offers a better example of a "writerly" text, due to its uniquely slip-synced soundtrack.

6. The fact that slip-sync occurs more prominently, and perhaps more dramatically, in British punk films suggests that the stakes regarding live performance on television are higher in Britain than in the United States. The history of pop music performance on the BBC differs somewhat from the American tradition, where historically the authenticity of live performance may not be as much of an issue. Simon Frith has alluded to the formidable influence of the British Musician's Union in establishing a certain quota of live performance on the BBC (*Sound Effects* 67–68). Elsewhere, Frith shows how the BBC, going back to its radio days, consistently resisted electronic enhancements and innovations (microphones, electric guitars, drum machines), partially due to the sway of the union's political preference for the live and human. Interestingly, Frith identifies punk as a moment in rock history coded as "more truthful" and authentic because it reacted against technology ("Art versus Technology" 266). However, he goes on to argue that technology often makes rock authenticity possible (269–77)—a notion explored by punk slip-sync. Thanks to Tony Grajeda for bringing this article to my attention.

7. The film's title refers to the Queen's Silver Jubilee celebration of that year—scandalized periodically by various punk outbursts.

8. Anticipating and complementing Auslander's notion of "liveness" above, James Lastra has engaged a compelling deconstruction of "the originality effect" in film sound. From his angle, original, unrecorded sound is itself "constructed" through "signifying marks," and can be recognized as "original" only through its repetition (as recorded sound). This reverses the conventional original-copy hierarchy by positing original sound as an "effect" of recording ("Reading, Writing, and Representing Sound" 83–84). Steve Wurtzler has likewise theorized contemporary popular music in terms of "copies without originals," focusing on new representational technologies that pose "a series of challenges to notions of the centered subject, and the binary opposition, live/recorded" ("She Sang Live, but the Microphone Was Turned Off" 92–93).

9. An interesting film that articulates some of these issues in a context more connected to *Singin' in the Rain* would be Julie Dash's *Illusions* (USA, 1983). Set during World War II, it focuses on a black female singer employed by a movie studio to "repair" a synchroniza-

tion error in a musical under production, by dubbing in the singing voice of the musical's white actress. A powerful political critique of Hollywood's racism, the film possesses its own slip-sync double excess: the black woman's voice itself has been dubbed, as she obviously lip-syncs during her own dubbing performance scene. Thus paralleling punk's slip-sync, the sequence intimates a provocative undoing of the notion of original voice.

10. A couple of years earlier, the Talking Heads made a splash in film, as the subject of Jonathan Demme's much acclaimed concert documentary *Stop Making Sense* (USA, 1984). Yet their music always possessed a keen cinematic sensibility, perhaps dating from the invocation of Hitchcock's Norman Bates in their first single of 1977, "Psycho Killer."

# 17

# Critical Hearing and the Lessons of
# Abbas Kiarostami's *Close-Up*

### DAVID T. JOHNSON

Abbas Kiarostami's *Nama-ye Nazdik* (Close-Up) (Iran, 1990) follows the trial of Hossein Sabzian, a man accused of impersonating a famous Iranian film director, Mohsen Makhmalbaf, in order to take advantage of an unwittingly trusting family. The whole story begins with a simple moment. Sabzian, riding a bus, reads Makhmalbaf's screenplay for *Bicycle-ran* (The Cyclist) (Iran, 1987). A woman sits next to him on the bus and asks him if he is Makhmalbaf; without hesitation, Sabzian replies in the affirmative. And so begins a long series of incidents, including Sabzian's moving in with the woman's family (in order to "scout" the location for his next film), rehearsals with the family's sons, and even removing a tree from the back garden for a better set design. Sabzian's "film," of course, is a ruse, and the family is understandably hurt by Sabzian's deception. He is arrested and brought to trial; the film moves back and forth between reenactments of the actual events (played by the actual participants) and a presentation of the trial itself.

In the final scene of the film—not, we are led to believe, a reenactment—Sabzian meets, at long last, the object of his affection: Makhmalbaf, who, unbeknownst to Sabzian, is wearing a lapel microphone. As the two approach one another, Sabzian embraces the real director, weeping as he does so. From the beginning of the scene, however, there has been a problem: Makhmalbaf's microphone has been cutting out. Narrating this problem are the film crew inside the van shooting the sequence. As the microphone starts going in and out, one of the filmmakers, perhaps Kiarostami himself, says very simply, "We can't retake this." Too late to reshoot the documentary scene, Kiarostami's crew watches this first meeting between the men, but with their conversation coming in and out in snatches. We hear snippets of dialogue as we watch them through the windshield, but we are never close enough, visually or aurally, to see and hear the entire exchange between the two. They ride together on a motorcycle to the family's house; when the family answers the intercom, asking who is there, Sabzian answers, "Sabzian," and then, hesitating, adds another name: "Makhmalbaf."

When I first saw this film, I found this scene, where the sound cuts out, to be the most moving of the entire film. My response is hardly unusual, since basic dramatic structure dictates that I should respond strongly (the scene is the climax of the film). But what I found most affecting was that the sound had gone out by accident. Somehow, the failing lapel microphone was providing a grace note to a question raised throughout the film: why did Sabzian do it? The meeting between the men was set up to provide this answer. Once the microphone failed, however, the burden of answering this question shifted from film to viewer, and instead of leaving the film feeling satisfied from a clear explanation, I was forced to raise further questions—in this case, an even more satisfying experience, since any explanation at this point in the film ultimately would have been reductive and incomplete. I read the scene as a happy accident, because I felt as though Kiarostami's film benefited enormously from the sound cutting out when it did.

Imagine my surprise when I discovered, quite a while after watching the film, that the accident had been faked (Margulies 238–39). A scene we are supposed to read entirely as documentary realism is, instead, part documentary and part fiction, the dialogue between the filmmakers—"We can't retake this"—clearly a lie. Upon thinking about this problem, I began to wonder if this moment has any ethical implications, and, if so, what are they? Kiarostami has said that he cut the sound because Makhmalbaf, wearing the lapel microphone, knew he was being filmed, whereas Sabzian did not. Had he kept the exchange unbroken, the scene would ultimately play to Makhmalbaf's favor, since he knew he was being recorded (Margulies 238–39). We get a hint of this when Makhmalbaf says to Sabzian, in one of the unmediated clips of dialogue, "I am tired of being me too." The potential lyricism of the statement is undermined by the fact that Makhmalbaf knows he is being recorded; it seems a little too prepared next to Sabzian, who can hardly speak at all, sobbing in the presence of his idol. One can understand why Kiarostami would choose to mediate this scene for us and remove the potential problems of a clean recording.

Nevertheless, the scene is a lie in the sense that its fictional aspects never announce themselves as such. Perhaps I am motivated as much by my disappointment (and clearly, my naïveté) that the accident could be faked in the first place. After all, the scene is meant to be *what actually happened*. But viewers of documentary have long ago abandoned the idea of direct cinema, since cinema at its least-tampered with has yet to capture the phenomenological experience of reality (whether that experience is one of heightened sensitivity or out-and-out boredom). So it is not so much that Kiarostami fakes certain aspects of his film, even those we assume are *real*. We know, for instance, that even beyond the fictional reenactments of earlier events, the documentary trial sequences bear Kiarostami's influence, one that mitigates the possibilities, ephemeral as they are, of a totally observational cinema. But what we might take issue with in this final sequence is that the faked aspect involves sound. Why this might present a problem has to do with a viewer's experience of sound and the kinds of choices about sound that filmmakers must necessarily make.

If the scene I am discussing seems hardly worthy of ethical considerations, we might consider an example where the stakes are higher. In his book *For Documentary*, documentary editor and theorist Dai Vaughan recounts working on a film that included a female circumcision, and the scene was shot outside the hut where it was taking place. As the filmmakers worked on the scene in post-production, a three-way argument developed about how to handle sound. One person suggested that a scream should be included to signal the woman's pain. Another replied by remembering that a scream had actually occurred during the ceremony and that it would be fine to use it within the scene. But a third person then countered, saying that victims of this ceremony rarely screamed at all and that to include it, though perhaps correct in its intention, would be a misleading illustration of the culture. Vaughan recounts this story to show how all three views represent, in his words, "the claim documentary stakes upon the world: in the first case, symbolic (a scream stands for pain); in the second, referential (this is what our equipment actually recorded); in the third, generalisatory (to include the atypical is misleading)" (xiv).

Although Vaughan's designations are useful, they also do not exist outside of a specific sequence. In other words, for the documentary viewer (if not the viewer of fiction film), sound exists first referentially, with the other two functions following that primary function. We do not read the symbolic value of the sound unless we assume the sound actually occurred; we do not think that the sound is part of the ceremony normally unless we think that, again, someone has actually made this sound. Yet, here, clearly the ethical implications of such an extreme case—representing a practice that would necessarily elicit a strong opinion in both filmmaker and viewer—require the ethical questions that Vaughan's colleagues attempted to work through.

Placed next to this example, Kiarostami's transgressions are innocuous and perhaps even playful. We may be fooled, but no one is suffering here. Still, there is no getting around the fact that most people will read the faltering sound referentially (the microphone is failing), and the symbolism follows only after that referentiality. Referentiality, after all, signals a kind of recorded "liveness," and the faltering microphone is the grit, the dirty tones that fool us into thinking we are hearing an unmediated recording.

I would venture to say that I know almost no one, aside from people who work in film production or a similar function, who does not read sound referentially before moving to what Vaughan calls "symbolic" or the "generalisatory." Often, our tendency to read sound referentially is because we are too busy focused on our visual stimuli. This privileging of the visual over other senses has been traced out by Martin Jay, Jonathan Crary, and other scholars. Regardless of our reasons for doing so, we are all too willing to accept the fidelity of auditory stimuli, particularly when combined with reception of the visual. Sound theorist Michel Chion calls this process "synchresis," whereby the brain fuses the seen and the heard into one experience. Walter Murch gives the example of recording an ax hitting a tree and then using that sound with a baseball bat. The result will not produce confusion for the audience but will rather signal "a particularly forceful hit rather than a mistake by the filmmakers" ("Foreword" xviii). Imagining, however, the reverse, whereby the visual of the scene is replaced, is almost impossible, and is, indeed, absurd. A baseball bat hitting a ball looks different from an ax hitting a tree. They sound different, too, but no one, with the exception of sound experts or people with especially attuned hearing, could tell the difference in the example Murch gives. Whether this phenomenon has to do with our cultural history (as Jay and others suggest), our physical makeup as a species (as Chion notes, "the absence of anything like eyelids for the ears" [*Audio-Vision* 33]), or something else, synchresis prevails with the ax and the baseball bat, especially when the narrative is hurrying us toward the bottom of the ninth inning, bases loaded.

This tendency to let sound pass by unnoticed has even affected the criticism we write. Despite many correctives in recent years, we most often see images over sounds in ideological critiques of a given cinema. Undoubtedly, this has to do with the philosophical tradition that criticism draws on, a tradition merely outlined here; as far as I know, the Lacanian gaze has no equivalent in the realm of hearing. Most film theory, in fact, since its inception, has focused on the visual. As Rick Altman has observed, "There is one claim on which we can all agree: the image has been theorized earlier, longer, and more fully than sound" ("Sound's Dark Corners" 171). In considering some of the high points in film theory, by way of example, one might mention Sergei Eisenstein's "Dramaturgy of Film Form," devoted almost exclusively to visual montage; André Bazin's "Ontology of the Photographic Image," clearly placing emphasis on the visual; and Laura Mulvey's "Visual Pleasure and Narrative Cinema." Think of the cinema itself, as well, as it has depicted the aspect of film most likely to reach the unconscious; one of the most memorable moments from Stanley Kubrick's *Clockwork Orange* (UK, 1971), after all, is the reaction shot of Malcolm McDowell as he is exposed to images of violence, his eyelids grotesquely pried apart. In fact, given this brief outline of image versus sound theorization, we should probably be *more* wary of the relationship of sound to ideology and perception. Michel Chion makes this exact case: "Sound more than image has the ability to saturate and short-circuit our perception. The consequence for film is that sound, much more than image, can become an insidious means of affective and semantic manipulation" (*Audio-Vision* 33–34).

Read outside the context of the entire film, Kiarostami's final *Close-Up* sequence seems to capitalize on a perceptual deficiency in most cinemagoers for its effect. All cinema, it could be argued, capitalizes on perceptual deficiencies, but we are often well enough aware of such deficiencies not to believe, as we exit the theater, that we will be greeted by a horde of zombies (though we may look like one). With documentary, however, the stakes are much higher, especially when the filmmaker presents the sounds of that scene as unmediated. Is Kiarostami, then, manipulating us, and if so, has he crossed some kind of ethical boundary? If not, what is the alternative? To explore these questions, we must turn to a term that comes up frequently in relation to his cinema: *pedagogy*.

Given Kiarostami's background, it is hardly surprising that many critics have commented on his pedagogical tendency. He spent much of his early filmmaking career with the Kanun-e Parvaresh-e Fekri Kudakan va Nojavanan (Institute for the Intellectual Development of Children and Young Adults), or Kanun. Kanun was, as Hamid Dabashi reports, "part of a general pattern of cultural development that the Pahlavi regime had initiated to engage the Iranian youth in politically harmless entertainment" (44). These short films were essentially conceived as small didactic exercises, and Kiarostami excelled at making them. Interestingly, they often instruct not only on the ostensible subject but on the process of filmmaking itself. Consider this description of *Orderly or Disorderly* (1981), written for the filmography in Mehrnaz Saeed-Vafa and Jonathan Rosenbaum's study on the director: "In separate takes, Kiarostami films the orderly and disorderly behavior of children at school—leaving class, going to a water fountain, boarding a bus—while he talks with his crew about some of the aesthetic qualities of each camera setup. Then he films orderly and disorderly traffic at a busy intersection" (132). Many other films operate on similar, simple formal ideas that Kiarostami plays with for as little as three or four minutes (though many are longer), and they seem as concerned with aspects of the medium as with teaching children the value of, for instance, being orderly.

Interestingly, these early films rarely fall under scrutiny as being complicit with a repressive regime. In fact, Dabashi has argued that Kanun was actually a "Trojan horse for the government" that allowed for "symbolically subversive counterculture" under the guise of education (44–45). Nonetheless, it may be that this early direct association with the government has lent credence to more recent attacks on Kiarostami's politics. It may also be that the combination of Kiarostami's explicit formal experimentation and political concerns that are less so have made critics wary. Since this essay is about the relationship between a formal aspect of the medium (sound) and ethics, I want to address briefly some of these critiques. Of those to appear in English, perhaps no other makes as compelling an indictment as Azadeh Farahmand's "Perspectives on Recent (International Acclaim for) Iranian Cinema." In the essay, she addresses the international film-festival circuit and the way it encourages unproblematic imagery of Iran to circulate on the world stage. This circulation is actively sought by the current regime as an indirect means of gaining access to economic possibilities without addressing internal social problems. As she writes, "Iranian cinema has been rediscovered as a promising means through which to renegotiate the imagery of the nation, and gradually to reclaim a place for the country within the global economy in the name of art" (87). Farahmand argues that Kiarostami's international popularity has facilitated romanticized images of rural culture, wise elders, innocent children, and exotic women. In this way, his films hide rather than make plain the actual political realities of Iran; in so doing, they allow the current regime's political mandates to go unchecked on the world stage, even if only in a film-festival context. This effect extends to plot structure as well, as the films often use a "mediating character" with whom viewers cannot identify but only sympathize (100). As a result, "the viewer is thus protected from any shock, unpleasant encounter or guilty conscience. He can maintain his distance and remain uninvolved, be fascinated, securely appreciative of the exotic locales, as though viewing an oriental rug, whose history he does not need to untangle" (101).

Farahmand's essay makes Kiarostami complicit with a repressive political regime. Extending her argument, in fact, makes us all complicit: not only do Kiarostami's films hide actual social realities, but we choose not to seek them out. Against this charge, Laura Mulvey offers an excellent counterargument. As she sees it, Kiarostami's films always raise many more questions than answers—and perhaps even call for the very untangling Farahmand's rug metaphor evokes. "Unlike any other medium," Mulvey writes, "the cinema is able to construct and inflect the way a spectator relates to its images" ("Afterword" 257). Farahmand most likely would not disagree. From here, however, Mulvey chooses instead to defend Kiarostami's films, which she sees as especially important in

> explor[ing] the narrow line between illusion and reality that is the defining characteristic of the cinema. Avoiding an either/or approach, his interest lies in boundaries and in the tension between the cinema's ability to register and print the actual image in front of the lens and its ability to transform and transcend it. This "what is cinema?" approach to filmmaking affects the spectator's relation to the screen. Here, issues to do with the gaze and ways of seeing are extended beyond ideological content into a wider demand to question the nature of the image itself. (260)

In other words, no image is ever so simple in Kiarostami's film as to be easily read, since the formal complexities always prevent any overly easy answers to questions the films might raise. And this form of "interrogative spectatorship," as Mulvey puts it, is at the heart of Kiarostami's aims: "To ask the spectator to think—and to think about the limits and possibilities of cinematic representation—is to create a form of questioning and interrogative spectatorship that must be at odds with the certainties of any dominant ideological conviction—in the case of Iran, of religion" (260). The hope, of course, is that "any dominant ideological conviction" will extend into our own ideology as well.

The reader will note, however, that almost all of the language in both Farahmand's and Mulvey's essays revolve around imagery. Whether it is "viewing an oriental rug, whose history [the viewer] does not need to untangle" or "the cinema's ability to register and print the actual image in front of the lens," both arguments invoke our sense of sight over our sense of sound. What might these arguments be if they discussed the possibilities of sound? In exploring this question briefly, I will take up Mulvey's sense that Kiarostami's films offer a kind of interrogative spectatorship, though Farahmand's argument would probably find support as well in the films' aural aspects. Nonetheless, this interrogative spectatorship might account for my initially naive readings (hearings?) of the final scene of the film—and my subsequent desire to critique them. If we are to invent a rough aural equivalent for interrogative spectatorship (since "spectator" is clearly rooted in visuality, as so much of our critical language is), what might it be? Interrogative aurality? Critical hearing? Whatever it is, Kiarostami's film perhaps intends to teach us about sound in calling our attention to it. In fact, in this film, these impulses are rarely separate.

Take the case of an earlier scene, probably the other most discussed one in the film, when a cab driver waits outside of the family's house. Nothing happens in this scene. Imagine if a theater director decided to stage *Hamlet* by showing what every character is doing when they are supposed to be offstage—Ophelia eating breakfast; Claudius tying his shoe. Here, a minor character, the cab driver, kills time by kicking an aerosol can off a leaf pile. We watch the can roll down the street, but the sound is so loud that we are made aware of that aural materiality. The loud sound breaks the relative silence that the scene has enjoyed. Of course, as if to underscore his awareness of sound, Kiarostami has his reporter go from house to house, asking to borrow a tape recorder to record Sabzian's testimony as the police take him away. This moment is clearly indicative of an aural pedagogy at work. Sound is so often at the behest of the narrative drive (even in a documentary), and it is often used, in continuity editing, to hide the cuts, where a sound will extend over a transition (or even signal one—a kind of eyeline match for the ears). Sound is, simply put, meant to be heard but unnoticed for its own aural qualities, in the sense that a sound that draws too much attention to itself will pull us out of the narrative. But here, in a scene emptied of narrative drive, we are asked merely to perceive; we are asked to listen to this sound and consider it in its own right. What does an aerosol can really sound like? *Like this,* the scene seems to say, much like an old children's toy ("A cow says, 'Moo'").

At a more thematic level, sound also interacts with the film, especially in regard to the idea of testimony. We are constantly trying to figure out why Sabzian did what he did, largely through actual court testimony but also through confessional testimonial moments (and even the testimonial lies he tells the family). Is Sabzian a con artist, a cinephilic dreamer, a poor man in need of financial assistance, or a bit of all three? Whatever he is, we move between trial scenes and reenactments, each moment promising a final testimony—a true, complete one—that the film never ultimately delivers. And throughout these moments, sound often overtakes the visual in its simultaneous desire and inability to verify, to give us a final truth. In the reenactments, like the scene with the aerosol can, it is often simply the crisp sound of the silent, empty room that we are asked to contemplate. In these cases, the ambient silence—never, of course, the absence of sound so much as a low microphone hiss, along with a noise or two—tells us as much as the dialogue among the speakers. For instance, late in the film, we have the same scene we had watched earlier from outside (the aerosol can scene), but now we are inside the house. Now, we think, is the payoff for not being allowed inside the house earlier. What we find instead, however, is no more of an answer to the question *Why?* than previously. To be sure, we do get a sense of the lengths Sabzian will go to when he attempts to convince people of his status as a film director when everyone knows he is lying. Even more important, however, are those moments when Sabzian is merely waiting for the others to arrive to arrest him. People move just on the peripheral edge of our hearing (to mix yet another visual and aural concept), as they leave Sabzian to himself. The scene, emptied of all but the lowest perceivable sound, is the perfect complement to the aerosol can scene. There, we were asked to listen to one loud sound and consider its aural materiality. Here, however, we strain to hear conversations, which makes us all the more aware of the materiality of speech, particularly when given in hushed whispers between people in other rooms. Testimony is always out of reach, but as we reach for it (as we must), we learn what sound actually is when it is not merely the accompaniment for narrative.

In fact, we could contrast the relative silence of reenactments, which are almost always light on perceivable speech and heavy on nearly quiet, ambient noises (with the exception of a louder bus-ride scene), with the trial scenes. Here, the "grit" I mentioned earlier—what Chion calls the material sound index (*Audio-Vision* 114–17)—is much higher. We can hear the room tone, first of all, a common convention of documentary production, used to indicate an observational attitude from the filmmaker. In addition, the testimony itself registers the room's spatial dimensions through the echo that each voice produces. We are all too aware of the microphone's position; it often dips into the frame. None of this matters, of course, because we are meant to read this scene as documentary realism, and such slips are part of what we think of as cinéma vérité (or direct cinema, or observational cinema). With Kiarostami's film, however, the more we become convinced of the film's realism, the more we question what parts of the film are true and what parts are not. Although we feel the spatial dimensions of the room through its sounds, we are no more or less sure if the trial is actually occurring, a reenactment, or both. Kiarostami is constantly interrupting the judge, and there is a subtle joke here about who is actually in charge. Sabzian also often directs his testimony directly to the camera and microphone rather than the judge, again upsetting our sense of who is in charge. The interaction of sound we normally perceive to be "real" (that is, the noticeable room tone and voice echo) with these more complex interactions of characters makes us doubt any auditory perceptions as verifiable. What could be more appropriate for a film about someone whose motivations can never finally be verified?

In addition, the sounds of the trial not only function in a dialectic with the courtroom testimony, making it both more and less real, but sound more broadly functions as a larger structuring device. Remember that the reenactments (with the one exception of the bus ride) are typically more "quiet" than the courtroom scenes. Even the aerosol can is loud only in relation to the relative silence of the scene. The trial scenes, by contrast, are relatively noisy. Every scuff of the chair, cough, or other noise registers loudly on the microphone, and the room tone along with the vocal echo make for a great deal of ambient noise. Recall also that the film constantly switches back and forth between the trial scenes, which we read temporally as the relative present, with those events that the trial is about, which we read as the past. Since the film often cuts (rather than dissolving or fading) between these scenes, we often hear the jarring discontinuity between the noisy trial room and the quiet interior of the family home, for instance. This discontinuity alerts us to sound's structuring device, namely, the past is somehow quieter than the present. This simple structure can be read allegorically, as the past is necessarily more difficult to perceive (and, in the case of testimony, hear) than the noisy present. Of course, as if to caution us against reading this allegory too legibly, Kiarostami includes an exception: the initial meeting between the mother and Sabzian. Here, meeting on a bus for the first time, they must speak over the sound of the noisy engine. The past can be loud too, Kiarostami seems to say. We cannot rely too heavily on *any* trick of the filmmaker, even a larger structuring device such as this one.

Seeing the final scene in the context of so much self-awareness about sound should put any nagging doubts about Kiarostami's ethics out of my mind. I was never meant, in the first place, to interpret this scene overly legibly, even at the most basic formal level (that the sound had actually cut out). In fact, on reflection, I am surprised at how naive my initial response was. Indeed, why would the director and crew be miked as they sit parked across the street? By not paying attention to these basic factors of the filmmaking process, I let myself be taken in. But, of course, the irony is that what makes this scene most interesting is to imagine, if only briefly, that the microphone did fail at that moment. Kiarostami must know that the microphone accident functions just as much, if not more, as a crescendo to his film as the ethical treatment of someone who does not know he is being recorded. He even cues the orchestral score once we have gotten a long taste of this accidental lyricism. As a result, some part of me, that part that reads sound referentially and that same part that pays it less attention than the visuals, continues to believe that the microphone just happened to cut out the day they were filming, putting that final conversation just out of reach.

Should I therefore blame Kiarostami? I am not comfortable letting him off the hook completely, though that may have its roots in an adolescent shame of being duped. Still, one wonders, how many other viewers will see the film and assume the accident was real? Does such a situation present any ethical dilemmas? Perhaps it does. But we must also remember that Kiarostami's film is attempting to teach us about sound as much as use it for its own ends. Ironically, only by learning to hear the cinema more critically, a lesson his film seeks to teach, can we learn to critique the very film that seeks to give us this lesson. And here is the larger pedagogical lynchpin of so much intellectual development: in studying a subject, we learn how much more we have yet to learn. Given the still-early stages of film sound scholarship, this should be quite a lot indeed.

# 18

# Rethinking Point of Audition
# in *The Cell*

## ANAHID KASSABIAN

The opening sequence of *The Cell* (Tarsem Singh, USA, 2000) takes place in a desert, with sweeping vistas of dunes. The instrumentation of Howard Shore's musical cues includes *ghaita* (an Arab double-reed instrument), *lira* (a bamboo flute), and double-headed Moroccan drums. Parts of the music use microtonal modes typical of Arabic musics, suggesting in some of the basic musical materials that the setting is "Middle Eastern,"[1] even though we eventually learn that the landscape is in the mind of Edward, a psychotically disturbed young boy in southern California. But although the aural markers signifying the Middle East are clear, there are other musical sounds as well. In fact, the cue signals the unusual strategy of the entire score; it is a work that is as much sound design as composition.

*The Cell* is a clear example of the dissolution of boundaries among noise, sound, and music. Shore is well known for using this compositional strategy, but there are many other examples across styles and genres. Two brief ones: (1) at the onset of the terrible, incurable headaches that attack main character Max Cohen (Sean Gullette) in *Pi* (Darren Aronofsky, USA, 1998), what sounds like the beginning of a dance bass line becomes the throbbing of his head, mixed with sounds that may or may not be what Max hears and may or may not be music; (2) the opening sequence of *Hackers* (Iain Softley, USA, 1995) combines machine throbs, dog barks, a music box, and other sounds that both do and do not connect to visual materials, followed after the cut from slow-motion to real-time visuals by a thumping machine sound that could easily be a drum machine.

In this essay I am using *The Cell* as both a case study and a point of departure to consider this dissolution as one among several auditory markers of a shift in narrativity. And while there has been, at least since the work of Jean-François Lyotard, a discussion of the demise of linear narrativity, most models of power and culture still rely on notions of a fairly traditional form of narrativity. To my ears, then, *The Cell* and other films like it sound a call to rethink how to study power and culture, and this essay is an invitation to hear that call.

In the opening sequence of *The Cell,* sounds enter in the following order: wind, *lira,* jangling metal, whooshes, cello drone, horse galloping, cellos, Moroccan percussion, *ghaita.* And that's just the first minute or so of the cue. This layering of sound signals the strategy of *The Cell*'s score, exemplifying the evaporating segregations among aural materials. It suggests, among other things, a different register for the soundtrack, perhaps laying to rest once and for all the presumptions that sound and music are secondary to a film's visual materials.

The blurring of once apparently distinct sound registers is already drawing critical attention. For example, Daniel Falck has argued that speech in *The Thin Red Line* (Terrence Malick, USA, 1998) is often deprivileged, in some cases to the point of mumbling; he suggests that these moments can be thought of, after Michel Chion, as "emanation speech," speech that "becomes a kind of emanation of the characters, an aspect of themselves, like their silhouette is" (177). Falck makes the case that music, speech, and sound in *The Thin Red Line* are leveled out and used together to create a world not dependent a priori on the images. He highlights both the dissolving boundaries among kinds of sounds and the implications of their dissolution in which I am interested.

As I began to suggest in the opening paragraph of this essay, the soundtrack of *The Cell* is another example; it is not primarily, or even substantially, musical. Nor does it subordinate its aural world to its visual one. *The Cell* projects the interior world of its characters in hybrid sound forms as much as it envisions it in images. To return to the film's opening sequence, the overall effect is more layers of sounds than a through-realized cue. The invocation of West Asian or North African music (the musicians are Bachir Attar and the Master Musicians of Jajouka, from Morocco) serves an environmental rather than a narrative purpose; it seems mainly to signify dryness and desert (if anything at all). In fact, the same timbres and textures appear next alongside a sweep of dry, brown fields in central California, and follow a cut to our first view of Karl Rudolph Stargher (Vincent D'Onofrio), a serial killer, but the *lira* and *ghaita* do not accompany the subsequent cuts to the same terrain. They appear again not on the next cut to Stargher in the same setting, but when Catherine (Jennifer Lopez) dreams of Edward's internal desert scene, and then much later when she enters Karl's mind for the second time, a moment that shares nothing obvious with any of the prior Moroccan cues. In sum, the sounds of the Master Musicians of Jajouka do not appear regularly with either deserts or mindscapes, that is, not with any particular narrative or visual referents. The uses of the timbres and melodic materials of Jajouka follow no recognizable logic in the film; after the opening sequence, they are treated as another body of sounds available for inclusion in the score.

Many of the cues are built from layers of traditionally musical and tradition-ally sound-effect materials, utterly blurring the boundaries between the two. For example, Catherine's first entrance into Stargher's mind begins with a throbbing bass string sequence that clearly signifies "ominous." Layer upon layer is added to it: tympani, midrange strings, horns, a cash register, a baby crying, birds, the distorted sounds of a baptism, machine sounds. The sounds recede layer by layer and are replaced by extremely distorted sounds matched by surreal images: tight close-ups of a drop of blood falling in water and a grasshopper jumping, followed by a medium shot of a dog shaking off water and blood. The sound associated with each is extremely exaggerated and bears no resemblance to what one might hear from any of the visually represented spaces. Catherine finally meets up with the young boy Karl in a tiled room with a horse and a clock, with extreme echo in the ambient sound, pronounced ticking of several clocks, and increasing lay-ers of sounds and instrumental timbres.

This is neither music nor not music, but rather a textural use of sound that dis-regards most, if not all, of the "laws" of classic Hollywood film scoring technique. (On the basic rules of this practice, see, for example, Gorbman 73.) Instead, in this cue, the sound music is foregrounded for attention, it is not a signifier of emotion, nor does it provide continuity or unity. It is not subordinate to the narrative or the visuals; rather, it is on par with them in creating an affective world.

*The Cell* initiates a soundtrack of the unconscious, where the boundaries we are accustomed to recede in favor of a different logic. In an impressive early es-say on sound theory, Mary Ann Doane argues that synch sound in film is an important ideological force: "The drama played out on the Hollywood screen must be paralleled by the drama played out over the body of the spectator—a body positioned as unified and nonfragmented. The visual illusion of position is matched by an aural illusion of position. The ideology of matching is an obses-sion which pervades the practice of sound-track construction" ("Ideology and the Practice of Sound Editing and Mixing" 55). But *The Cell* actively strives to break that illusion, to mismatch visual and aural position, by using a range of techniques such as sound close-ups (for example, the horse's breathing in the opening sequence) and extreme echo (as with Catherine's and Edward's voices in the opening sequence) to signify perceived rather than objective sound. This is not a soundtrack of the conscious, material, rational world. It does not follow the logic outlined by Doane. It is driven instead by the logic of schizophrenia, of the loss of boundaries among unconscious, preconscious, and conscious. And not only are the boundaries of sound, noise, and music and those of unconscious, preconscious, and conscious erased, but the technology central to the film's plot also erases the boundary between subjects, so that we cannot distinguish between the unconscious sound worlds of the psychologist and the patient, or the cop and the patient.

Much of the film takes place inside the mind of Stargher, the serial killer, who has Waylon's Infraction[2] and whose mind Catherine has entered through an experimental technology in order to find his imprisoned, soon-to-be-next victim. As she meanders through the rooms and corridors of his strange mind, we are led through it by richly textured terrains of both aural and visual materials not organized by principles of narrative or of conscious, rational thought. It is reasonably clear that the topography we are exploring is continuous between Stargher's preconscious, that place where Freud says memories are stored, and his unconscious, the repository of that which is actively repressed, hidden behind the wall of censorship. In Stargher's mind, we cannot distinguish between what he is thinking—does he indeed think while in this coma?—and what Catherine sees that is stored away. Nor can we distinguish between repressed memories and available ones, though perhaps we could conjecture that the childhood scenes of abuse (which take place "outside," through a doorway, across a lawn, in what is presumably Stargher's childhood house) are stored elsewhere precisely because they are the repressed material of Stargher's unconscious.

Throughout these long sequences, it is impossible to determine the subject position being offered, because there is no physical boundary between Stargher and Catherine, nor is there a clear psychological one. One way that *The Cell* highlights the bizarre status of this domain is by problematizing point of audition. Point of audition is an idea that has been promulgated by Rick Altman and Michel Chion. Chion suggests that point of audition, like point of view, has two senses, not entirely overlapping: "1) spatial, the way that sound locates one in filmic space, and 2) subjective, the character through whom one hears" (*Audio-Vision* 89–92). In this way, Chion treats point of audition as a generalized phenomenon, like point of view or spectator position. All moments in film, then, have a point of audition; it is an auditory representational or narrative strategy.

For Rick Altman, however, point-of-audition sound is a tactic, one among many in the world of film sound. In his view, point-of-audition sound is sound that clearly articulates a particular spatial and subjective position ("Sound Space" 60–61). So, for a clear and lovely example from a contemporary film, there is a moment in *Garden State* (Zach Braff, USA, 2004) when the main character, Andrew Largeman (Zach Braff), meets his love interest, Sam (Natalie Portman), in a neurologist's office. She wants him to hear a track from the CD she is listening to, so she passes him her headphones. As he puts them on, we begin to hear the music, increasing in volume as the headphones get closer to his ears while room tone fades; when they are fitted on his head, there is no sound on the soundtrack except the music; as he removes the earphones, the process is reversed. This is an extreme point-of-audition sound experience of the sort Altman discusses; as is perhaps obvious, in these moments, Chion's spatial and subjective senses will coincide completely.

In narratology, Genette makes a distinction more similar to Chion's than to Altman's. He argues that the narrator and the character's point of view offered by the text are not necessarily the same. When they are the same, they offer an especially unified perspective and worldview.[3] Similarly, point of audition in both of Chion's senses can coincide—a moment Altman would also call point of audition—and sound then locates perceivers in a particular subject position. At least since the rise of the novel, such positioning has been one of the central requirements of narrative and narrativity. Shifts away from clear and consistent offers of such positions, then, indicate an erosion of narrative as we commonly experience it, and thus of many of the theoretical insights predicated on it.[4] Many feminist theories of film and of music, for example, begin with the basic presumptions of narrative as we generally think of it in modernity—a linear structure, a central individual character, and a beginning and an end. In this way, point of audition should be seen as a necessary guarantor of subject positionality and of narrativity, that is, of a film's offer of (the fantasy of) a unified, discrete subject. (For two defining moments in feminist theory in relation to narrative, see de Lauretis and McClary.)

Point of audition in *The Cell* is a strange phenomenon; it fails us in every sense of the term and therefore fails as a guarantor of subject positionality and narrativity. We are not placed in the film's space by sounds, nor do we have a clear subjective location. Sounds do not provide adequate information about spatial relations—there is no relationship between Catherine's location in the spatial world in Stargher's mind and the volume or placement of the sounds we hear. Moreover, the entire world is one point of audition, within which another character is "moving." Or put another way, the sound materials while Catherine and Peter (Vince Vaughn) are in Stargher's mind make it clear that we are hearing a very subjective point of audition; there is no relationship between echo or volume and space. But we can never be clear *whose* subjective point of audition we are occupying; is it that of the host (Stargher) or the guest (Catherine, Peter)? Thus, not only is sound space constructed according to unconscious rather than conscious principles, but it is also indistinguishable in terms of whose point of audition we are being asked to occupy, and in fact whether there is even more than one. Insofar as point of audition implies subject positioning, our placement in the sound world of the film is not of much help.

But if the scenes inside Stargher's mind destabilize point of audition, the two scenes of his schizophrenic episodes are even more challenging. When he is watching his next victim from his car and when he is in his bathtub, we hear a garble of sounds. In the first case, a passing motorcycle engine sounds as if it is in the car with him, then it is layered with high whines (possibly synthesized strings) and a distorted male voice that is extremely slowed down. It is clear that the voice (or voices) is speaking, but no words are comprehensible. In the second case, when Karl is in the bathtub—his last conscious moments—we hear a faint high drone, to which the same morphed voice is added, then factory sounds, and a sound like a car starting. On a cutaway from Karl's interior soundscape to the cops arriving at his house, repetitive string figures are added—the first overtly musical sounds in the cue. But whereas the audio in this sequence is a combination of noises and sounds, they are clearly treated compositionally—perhaps even instrumentally—as controlled and organized components of a score.

In some ways, this sequence is even more disorienting than the later one I already described. Whereas the earlier scene takes place in Stargher's mind, this one is filmed from outside his body, but the sounds we hear are clearly internal to Stargher's world. There is a clear mismatch between what we see and hear, between point of audition and point of view. There is, then, no point of audition that can be inhabited by the film's perceiver with any comfort or predictability. Here, as in almost every other important sequence, the soundtrack insists on being heard as not only subjective but also troubled, unreliable, and incoherent.

In such moments and throughout *The Cell*, point of audition loses its reliability as a marker of available or encouraged subject positionalities, suggesting shifts in narrative relations. Through a collapse of sound, noise, and music, point of audition is challenged, disrupted, problematized so that no comfortable subject positionality can be engaged. In this way, *The Cell* is a defining moment among a group of films, often science fiction films such as *Hackers* and *Lara Croft: Tomb Raider* (Simon West, UK/Germany/USA/Japan, 2001), and self-consciously postmodernist films such as *Lola rennt* (Run Lola Run) (Tom Tykwer, Germany, 1998), *Pi* and *Tank Girl* (Rachel Talalay, USA, 1995), in which the soundtrack, in all its complexities, demands enormous attention and priority, becoming a peer to the visual in every important sense. In these films, narrative structure and identification processes are de-emphasized in favor of a different organizing principle, one in which the moment supersedes the story. Such films demand new theories, based on the loss of boundaries, on the aural logics of dance music, iteration, and sound collisions and on the nonlinear narrative worlds of video games. As these new soundtracks test engagements between subjectivities and films, the many theoretical paradigms predicated on narrative and narrativity will be vexed, perhaps beyond repair.

## Notes

1. The music is actually a very old form of Moroccan Sufi music particular to the village of Jajouka that was introduced to the West by Brian Jones in his posthumous 1971 release *Brian Jones Presents the Pipes of Pan at Jajouka*. It has since drawn a wide range of European and North American artists, including the Rolling Stones, Ornette Coleman, DJ Talvin Singh, and Howard Shore.

2. According to the film, Waylon's Infraction is a form of severe schizophrenia that presents as catatonia. It does not appear in *DSM-IV*.

3. For a succinct discussion of this argument in Genette, see Webster 51–52.

4. This is one of the major points of departure between, for lack of better terminology, "entertainment" and "art" film critics (and our counterparts in literature and other arts). Whereas first-run theatrical releases are almost exclusively traditional linear narratives, many other kinds of films are not. This difference has had enormous influence on the development of theoretical work. As one small example, most theorizations of film music, my own included, have focused on narrative, "entertainment" films.

# Works Cited

Abel, Richard, and Rick Altman, eds. *The Sounds of Early Cinema*. Bloomington: Indiana University Press, 2001.

Acland, Charles R. *Screen Traffic: Movies, Multiplexes, and Global Culture*. Durham: Duke University Press, 2003.

Adorno, Theodor. *In Search of Wagner*. Trans. Rodney Livingstone. London: Verso, 1981.

Alexander, John. *The Films of David Lynch*. London: Letts, 1993.

Allan, Robin. *Walt Disney and Europe: European Influences on the Animated Feature Films of Walt Disney*. Bloomington: Indiana University Press, 1999.

Allen, Ioan. "The Dolby Sound System for Recording *Star Wars*." *American Cinematographer* 58.7 (July 1977): 709, 748, 761.

———. "Exhibs, Public Will Soon Eye Dolby Promos." *Variety* (12 January 1983): 19.

Altman, Rick. *The American Film Musical*. Bloomington: Indiana University Press, 1987.

———. "Cinema Sound at the Crossroads: A Century of Identity Crises." In *Le son en perspectives* [New Perspectives in Sound Studies]. Eds. Dominique Nasta and Didier Huvelle. Brussels: Peter Lang, 2004. 13–46.

———. "Deep-Focus Sound: *Citizen Kane* and the Radio Aesthetic." *Quarterly Review of Film and Video* 15.3 (1994): 1–33.

———. "De l'intermédialité au multimédia. Cinéma, médias, avènement du son." *Cinéma(s)* 10.1 (1999): 37–54.

———. "Film Sound—All of It." *iris* 27 (Spring 1999): 31–48.

———. "General Introduction: Cinema as Event." In *Sound Theory/Sound Practice*. 1–14.

———. "Introduction: Sound/History." In *Sound Theory/Sound Practice*. 113–25.

———. "Introduction: Sound's Dark Corners." In *Sound Theory/Sound Practice*. 171–77.

———. "The Material Heterogeneity of Recorded Sound." In *Sound Theory/Sound Practice*. 15–31.

------. "Moving Lips: Cinema as Ventriloquism." *Yale French Studies* 60 (1980): 67–79.

------. *Silent Film Sound*. New York: Columbia University Press, 2004.

------. "Sound Space." In *Sound Theory/Sound Practice*. 46–64.

------, ed. *Sound Theory/Sound Practice*. New York: Routledge, 1992.

------. "The State of Sound Studies." *iris* 27 (Spring 1999): 3–4.

------. "The Technology of the Voice." Pt. 1, *iris* 3.1 (1985): 3–20; Pt. 2, 4.1 (1985): 107–19.

------. "What It Means to Write the History of Cinema." In *Towards a Pragmatics of the Audiovisual*. Vol. 1. Ed. Jürgen E. Müller. Münster: Nodus, 1994/1995. 169–80.

Altman, Rick, with McGraw Jones and Sonia Tatroe. "Inventing the Cinema Soundtrack: Hollywood's Multiplane Sound System." In *Music and Cinema*. Eds. James Buhler, Caryl Flynn, and David Neumeyer. Hanover, N.H.: Wesleyan University Press, 2000. 339–59.

American Psychiatric Association. *Diagnostic and Statistical Manual of Mental Disorders DSM-IV*. Washington, D.C.: American Psychiatric Association, 2000.

Amundsen, Geir, et al. *Den Usynlige Stemmen: Om Bruk Av Voice-over I Reklamefilm*. Oslo: Department of Media and Communication, University of Oslo, 1999.

Andrew, Dudley. "Echoes of Art: The Distant Sounds of Orson Welles." In *Film in the Aura of Art*. Princeton, N.J.: Princeton University Press, 1984. 152–71.

Anzaldúa, Gloria. *Borderlands/La Frontera: The New Mestiza*. San Francisco: Aunt Lute Books, 1987.

Appadurai, Arjun. *Modernity at Large: Cultural Dimensions of Globalization*. Minneapolis: University of Minnesota Press, 1996.

Aristotle. "Sense and Sensibilia." Trans. J. I. Beare. *The Complete Works of Aristotle*, Vol. 1. Ed. Jonathan Barnes. Princeton, N.J.: Princeton University Press, 1984. 693–713.

Arnheim, Rudolf. "Perceptual Dynamics in Musical Expression." 1984. *New Essays on the Psychology of Art*. Berkeley: University of California Press, 1986. 214–27.

------. "The Rationalization of Color." 1974. *New Essays on the Psychology of Art*. Berkeley: University of California Press, 1986. 205–13.

Arthur, Paul. "Orson Welles, Beginning to End: Every Film an Epitaph." *Persistence of Vision* 7 (1989): 44–51.

Auslander, Philip. *Liveness: Performance in a Mediatized Culture*. New York: Routledge, 1999.

Auster, Albert. "*Saving Private Ryan* and American Triumphalism." *Journal of Popular Film and Television* 30.2 (June 2002): 98–104.

Ayfre, Amédée. "The Universe of Robert Bresson." Trans. Elizabeth Kingsley-Rowe. In *The Films of Robert Bresson*. Ed. Ian Cameron. New York: Praeger, 1969. 6–24.

Baldassare, Mark. "Human Spatial Behavior." *Annual Review of Sociology* 4 (1978): 29–56.

Bandy, Mary Lea, ed. *The Dawn of Sound*. New York: Museum of Modern Art, 1989.

Barnier, Martin. *En Route vers le parlant: Histoire d'un evolution technologique, économique et esthétique du cinéma (1926–1934)*. Liège: CÉFAL, 2002.

Bataille, Georges. *The College of Sociology (1937–39)*. Ed. Denis Hollier. Trans. Betsy Wing. Minneapolis: University of Minnesota Press, 1988.

Baxandall, Michael. *Painting and Experience in Fifteenth Century Italy.* Oxford: Oxford University Press, 1972.

Bazin, André. "An Aesthetic of Reality: Neorealism (Cinematic Realism and the Italian School of the Liberation)." 1948. In *What Is Cinema?* Vol. 2. Ed. and Trans. Hugh Gray. Berkeley: University of California Press, 1971. 16–40.

———. "The Ontology of the Photographic Image." 1945. In *What Is Cinema?* Vol. 1. Ed. and Trans. Hugh Gray. Berkeley: University of California Press, 1967. 9–16.

———. "Return to Hollywood: 'Using Up My Energy.'" In *Orson Welles: A Critical View.* 1972. Venice, Calif.: Acrobat Books, 1991. 122–35.

Bazin, André, Charles Bitsch, and Jean Domarchi. "Interview with Orson Welles." In *Touch of Evil: Orson Welles, Director.* Ed. Terry Comito. 199–212.

Beck, Jay. "A Quiet Revolution: Changes in American Film Sound Practices, 1967–1979." Ph.D. diss., University of Iowa, 2003.

Behlmer, Rudy. *Memo from Darryl F. Zanuck.* New York: Grove Press, 1993.

Belton, John. "1950s Magnetic Sound: The Frozen Revolution." In *Sound Theory/Sound Practice.* Ed. Rick Altman. 154–67.

———. *Widescreen Cinema.* Cambridge, Mass.: Harvard University Press, 1992.

Benjamin, Walter. *Charles Baudelaire: A Lyric Poet in the Era of High Capitalism.* Trans. Harry Zohn. London: Verso, 1973.

———. *Illuminations.* Ed. Hannah Arendt. Trans. Harry Zohn. New York: Schocken Books, 1969.

Berlant, Lauren. *The Queen of America Goes to Washington City: Essays on Sex and Citizenship.* Durham, N.C.: Duke University Press, 1997.

Bernstein, Leonard. "Why Don't You Run Upstairs and Write a Nice Gershwin Tune?" 1955. Republished in *The Joy of Music.* New York: Simon and Schuster, 1959. 52–64.

Besas, Peter. *Behind the Spanish Lens: Spanish Cinema under Franco and Democracy.* Denver: Arden Press, 1985.

Biesecker, Barbara. "Remembering World War II: The Rhetoric and Politics of National Commemoration at the Turn of the 21st Century." *Quarterly Journal of Speech* 88.4 (November 2002): 393–409.

"Bill's Tribute to 'The 5000 Fingers of Dr. T'" Web site. "Stills of Deleted Scenes." Last updated 11 January 2004. http://hometown.aol.com/seivadjl8/deleted.html.

Birren, Faber. *Color Psychology and Color Therapy: A Factual Study of the Influence of Color on Human Life.* New Hyde Park, N.Y.: University Books, 1961.

Bishop, Bainbridge. *A Souvenir of the Color Organ, with Some Suggestions in Regard to the Soul of the Rainbow and the Harmony of Light.* New Russia, N.Y.: De Vinne Press, 1893.

Blake, Larry. "Mixing Dolby Stereo Film Sound." *Recording Engineer/Producer* 12.1 (February 1981): 68, 70, 72, 74, 76–79. Reprinted in *Film Sound Today.* Hollywood, Calif.: Reveille Press, 1984. 1–10.

———. "A Sound Designer by Any Other Name . . ." *Mix* (October 2001): 226–27.

Bonitzer, Pascal. "Les silences de la voix." *Cahiers du Cinéma* no. 256 (February–March 1975): 22–33.

Bordwell, David. "Contemporary Film Studies and the Vicissitudes of Grand Theory." In *Post-Theory: Reconstructing Film Studies*. Eds. David Bordwell and Noël Carroll. Madison: University of Wisconsin Press, 1996. 3–36.

———. "The Introduction of Sound." In *The Classical Hollywood Cinema: Film Style and Mode of Production to 1960*. Eds. David Bordwell, Janet Staiger, and Kristin Thompson. New York: Columbia University Press, 1985. 298–308.

———. *On the History of Film Style*. Cambridge, Mass.: Harvard University Press, 1997.

Bordwell, David, and Noël Carroll, eds. *Post-Theory: Reconstructing Film Studies*. Madison: University of Wisconsin Press, 1996.

Bordwell, David, Janet Staiger, and Kristin Thompson, eds. *The Classical Hollywood Cinema: Film Style and Mode of Production to 1960*. New York: Columbia University Press, 1985.

Bordwell, David, and Kristin Thompson. *Film Art: An Introduction*. Seventh Edition. New York: McGraw-Hill, 2004.

Brady, Frank. *Citizen Welles: A Biography of Orson Welles*. New York: Charles Scribner and Sons, 1989.

Braziel, Jana Evans. "'In Dreams . . .': Gender, Sexuality and Violence in the Cinema of David Lynch." In *The Cinema of David Lynch*. Eds. Erica Sheen and Annette Davison. 107–18.

Bresson, Robert. *Notes on Cinematography*. Trans. Jonathan Griffin. New York: Urizen Books, 1977.

Brooks, Peter. *The Melodramatic Imagination: Balzac, Henry James, Melodrama, and the Mode of Excess*. New Haven, Conn.: Yale University Press, 1976.

Brown, Royal S. *Overtones and Undertones: Reading Film Music*. Berkeley: University of California Press, 1994.

Buckland, Warren. "A 'Sad, Bad Traffic Accident': The Televisual Prehistory of David Lynch's Film *Mulholland Drive*." *New Review of Film and Television Studies* 1.1 (November 2003): 131–47.

Buck-Morss, Susan. "Aesthetics and Anaesthetics: Walter Benjamin's Artwork Essay Reconsidered." *October* 62 (Fall 1992). 3–41.

Buhler, James. "Analytical and Interpretive Approaches to Film Music (II)." In *Film Music: Critical Approaches*. Ed. J. K. Donnelly. New York: Continuum, 2001. 39–61.

Bull, Michael, and Les Black, eds. *The Auditory Culture Reader*. Oxford: Berg, 2003.

Buñuel, Luis. *My Last Sigh*. Trans. Abigail Israel. New York: Knopf, 1983.

Čábelová, Lenka. *Radiojournal: Rozhlasové vysílání v Čechách a na Morav v letech 1923–1939*. Prague: Univerzita Karlova, 2003.

Calvo, Luz. "'Lemme Stay, I Want to Watch': Ambivalence in Borderlands Cinema." In *Latino/a Popular Culture*. Eds. Michelle Habell-Pallán and Mary Romero. New York: New York University Press, 2002. 73–81.

Carroll, Noël. "Prospects for Film Theory: A Personal Assessment." In *Post-Theory*. Eds. David Bordwell and Noël Carroll. 37–68.

Caruth, Cathy. *Unclaimed Experience: Trauma, Narrative, and History*. Baltimore: Johns Hopkins University Press, 1996.

Chatterji, Shoma. "Silence Juxtaposed against Sound in Contemporary Indian Cinema."

In *Soundscape: The School of Sound Lectures, 1998–2001.* Eds. Larry Sider, Diane Freeman, and Jerry Sider. 103–11.

Chion, Michel. "Audio-Vision and Sound." *Sound.* Ed. Henry Stobart and Patricia Kruth. Cambridge, U.K.: Cambridge University Press, 2000. 201–21.

———. *Audio-Vision: Sound on Screen.* Ed. and Trans. Claudia Gorbman. New York: Columbia University Press, 1994.

———. *David Lynch.* Trans. Robert Julian. London: British Film Institute, 1995.

———. "The Quiet Revolution and Rigid Stagnation." Trans. Ben Brewster. *October* 58 (Fall 1991): 69–80.

———. "The Silence of the Loudspeakers; or, Why with Dolby Sound It Is the Film That Listens to Us." In *Soundscape: The School of Sound Lectures, 1998–2001.* Eds. Larry Sider, Diane Freeman, and Jerry Sider. 150–54.

———. "Une esthétique dolby stéréo." *Cahiers du Cinéma* 329.18 (November 1981): xii–xiii.

———. *The Voice in Cinema.* Ed. and Trans. Claudia Gorbman. New York: Columbia University Press, 1999.

Chusid, Irwin. Phone interview with Tommy Rettig on WFMU Radio (East Orange, N.J.), c. 1987. Transcribed at http://hometown.aol.com/seivadj18/other2.html.

Cohen, Lisa. "The Horizontal Walk: Marilyn Monroe, CinemaScope, and Sexuality." *Yale Journal of Criticism* 11.1 (1988): 259–88.

Cohn, Carol, and Cynthia Weber. "Missions, Men and Masculinities: Carol Cohn Discusses *Saving Private Ryan* with Cynthia Weber." *International Feminist Journal of Politics* 1.3 (Autumn 1999): 460–75.

Collier, Gary. *Emotional Expression.* Hillsdale, N.J.: L. Erlbaum Associates, 1985.

Comito, Terry. "*Touch of Evil.*" In *Focus on Orson Welles.* Ed. Ronald Gottesman. Englewood Cliffs, N.J.: Prentice-Hall, 1976. 157–63.

———, ed. *Touch of Evil: Orson Welles, Director.* New Brunswick, N.J.: Rutgers University Press, 1985.

———. "Welles's Labyrinths: An Introduction to *Touch of Evil.*" In *Touch of Evil: Orson Welles, Director.* 3–33.

Conley, Tom. "Documentary Surrealism: On *Land without Bread.*" In *Dada and Surrealist Film.* Ed. Rudolf E. Kuenzli. New York: Willis Locker and Owens, 1987. 176–98.

Connor, Steven. *Postmodernist Culture: An Introduction to Theories of the Contemporary.* Second Edition. Oxford, U.K.: Blackwell, 1997.

Conrad, Randall. "'The Minister of the Interior Is on the Telephone': The Early Films of Luis Buñuel." *Cinéaste* 7.3 (1976): 2–11.

Conrad, Tony. *OxFF.* E-mail discussion list for topics relating to real-time video and digital image processing, sponsored by Columbia University School of Music, New York, 2003. http://music.columbia.edu/pipermail/oxff/.

Cowie, Peter. *The Cinema of Orson Welles.* New York: Da Capo Press, 1973.

Crafton, Donald. *The Talkies: American Cinema's Transition to Sound, 1926–1931.* History of the American Cinema, Vol. 4. Ed. Charles Harpole. New York: Charles Scribner and Sons, 1997.

Crary, Jonathan. *Techniques of the Observer: On Vision and Modernity in the Nineteenth Century.* Cambridge, Mass.: MIT Press, 1990.

Crowther, Bosley. "CinemaScope Seen at Roxy Preview." *New York Times* (25 April 1953): 10.

———. "Sound and (or) Fury: Stereophonic System Is Debated in Hollywood." *New York Times* (31 January 1954): 2:1.

Cubitt, Sean. "Pygmalion: Silence, Sound and Space." *Digital Aesthetics.* London: Sage, 1998. 92–121.

Culshaw, John. "The Record Producer Strikes Back." *High Fidelity/Musical America* 18.10 (October 1968): 68–71.

Dabashi, Hamid. *Close Up: Iranian Cinema, Past, Present and Future.* London: Verso, 2001.

Daston, Lorraine, and Peter Gallison. "The Image of Objectivity." *Representations* 40 (Fall 1992): 81–128.

Davison, Annette. "'Up in Flames': Love, Control and Collaboration in *Wild at Heart.*" In *The Cinema of David Lynch.* Eds. Erica Sheen and Annette Davison. 119–35.

Delamater, Jerome. *Dance in the Hollywood Musical.* Ann Arbor, Mich.: UMI Research Press, 1981.

de Lauretis, Teresa. *Alice Doesn't: Feminism, Semiotics, Cinema.* Bloomington: Indiana University Press, 1984.

De Pauw, Linda Grant. *Battle Cries and Lullabies: Women in War from Prehistory to the Present.* Norman: University of Oklahoma Press, 1998.

Denning, Michael. "The Politics of Magic: Orson Welles's Allegories of Anti-fascism." In *The Cultural Front: The Laboring of American Culture in the Twentieth Century.* London: Verso, 1996. 362–402.

Doane, Mary Ann. "Ideology and the Practice of Sound Editing and Mixing." In *The Cinematic Apparatus.* Eds. Teresa de Lauretis and Stephen Heath. London: Macmillan, 1980. 47–56. Reprinted in *Film Sound.* Eds. Elisabeth Weis and John Belton. 54–62.

———. "The Voice in the Cinema: The Articulation of Body and Space." *Yale French Studies* 60 (1980): 33–50. Reprinted in *Film Sound.* Eds. Elisabeth Weis and John Belton. 162–76.

"Dolby Encoded High-Fidelity Stereo Optical Sound Tracks." *American Cinematographer* 56.9 (September 1975): 1032–33, 1088–90.

Dolby Laboratories. *Bass Extension of Dolby Encoded 70mm Prints.* San Francisco: Dolby Laboratories, 1981.

———. *A Chronology of Dolby Laboratories: May 1965–May 1999.* San Francisco: Dolby Laboratories, 1999.

———. "Dolby Stereo: A Progress Report." *Variety* (16 August 1978): 7.

———. *Dolby Surround Mixing Manual.* San Francisco: Dolby Laboratories, 1998.

———. *Explore Our World* [DVD]. San Francisco: Dolby Laboratories, 2000.

———. "Licensing." http://dolbysearch.dolby.com/lic/. 2003.

———. "Trademark." http://dolbysearch.dolby.com/tm/. 2003.

Douglas, Ann. *Terrible Honesty: Mongrel Manhattan in the 1920s.* New York: Noonday, 1996.

Douglas, Susan J. *Inventing American Broadcasting, 1899–1922*. Baltimore, Md.: Johns Hopkins University Press, 1987.

———. *Listening In: Radio and the American Imagination*. Minneapolis: University of Minnesota Press, 2004.

Duguid, Brian. "Interview with Tony Conrad." June 1996. http://media.hyperreal.org/zines/est/intervs/conrad.html.

Eisenstein, Sergei M. "The Dramaturgy of Film Form [The Dialectical Approach to Film Form]." 1929. In *Film Form: Essays in Film Theory*. Ed. and Trans. Jay Leyda. New York: Harvest/Harcourt-Brace-Jovanovich, 1949. 45–63.

Eisler, Hanns, and Theodor Adorno. *Composing for the Films*. Freeport, N.Y.: Books for Libraries Press, 1947. Reprint 1971.

Elkus, Jonathan. Personal correspondence with Matthew Malsky. April 2004.

Elsaesser, Thomas. "Early Film History and Multi-media: An Archaeology of Possible Futures?" Paper presented at the Archaeology of Multi-media Conference. Brown University, Providence, R.I. 2–4 November 2000.

Elshtain, Jean Bethke. "Sovereignty, Identity and Sacrifice." In *Reimagining the Nation*. Ed. Marjorie Ringrose and Adam J. Lerner. Buckingham: Open University Press, 1993. 160–75.

———. *Women and War*. New York: Basic Books, 1987.

Erffmeyer, Thomas E. "20th Century-Fox Introduces CinemaScope: A Study of Technological and Organizational Innovation." *Film Reader* no. 6 (1985): 27–31.

Eysenck, M. W., and M. T. Keane. *Cognitive Psychology: A Student's Handbook*. London: Psychology Press, 1990.

Falck, Daniel. "Voyages on the Line." Unpublished manuscript, n.d.

Farahmand, Azadeh. "Perspectives on Recent (International Acclaim for) Iranian Cinema." In *The New Iranian Cinema: Politics, Representation and Identity*. Ed. Richard Tapper. New York: I. B. Tauris, 2002. 86–108.

Felman, Shoshana, and Dori Laub, M.D. *Testimony: Crises of Witnessing in Literature, Psychoanalysis, and History*. New York: Routledge, 1992.

Feuer, Jane. *The Hollywood Musical*. Second Edition. London: Macmillan, 1993.

———. "The Self-Reflexive Musical and the Myth of Entertainment." In *Film Genre Reader II*. Ed. Barry Keith Grant. Austin: University of Texas Press, 1995. 441–55.

Figgis, Mike. "Silence: The Absence of Sound." In *Soundscape: The School of Sound Lectures, 1998–2001*. Ed. Larry Sider, Diane Freeman, and Jerry Sider. 1–14.

Finlay, Victoria. *Color: A Natural History of the Palette*. New York: Random House Trade Paperbacks, 2004.

Flaherty, Robert J. "How I Filmed *Nanook of the North*." *World's Work* (October 1922): 632–40. Reprinted in *Film Makers on Film Making*. Ed. Harry M. Geduld. Bloomington: Indiana University Press, 1967. 56–64.

Flinn, Caryl. *Strains of Utopia: Gender, Nostalgia, and Hollywood Film Music*. Princeton: Princeton University Press, 1992.

Foucault, Michel, and Jay Miskowiec. "Of Other Spaces." *Diacritics* 16.1 (Spring 1986): 22–27.

Freedman, Estelle B. "Uncontrolled Desires: The Response to the Sexual Psychopath, 1920–1960. *Journal of American History* 74.1 (June 1987): 83–106.

Frejka, Jiří. "Divadlo a film." *Národní Osvobození* 6.218 (10 August 1929): 1–2.

———. "O zvukovém filmu a trochu o filmu na dálku." *Národní Osvobození* 6.242 (2 September 1929): 1–2.

———. "První zvukový film v Praze." *Národní Osvobození* 6.223 (15 August 1929): 2.

———. "Rozhlasový žurnalismus." *Radiojournal* 8.1 (1930): 6.

Frith, Simon. "Art versus Technology: The Strange Case of Popular Music." *Media, Culture and Society* 8 (1986): 263–79.

———. *Sound Effects: Youth, Leisure, and the Politics of Rock 'n Roll*. New York: Pantheon Books, 1981.

Frith, Simon, Andrew Goodwin, and Lawrence Grossberg, eds. *Sound and Vision: The Music Video Reader*. New York: Routledge, 1993.

Frow, George L. *The Edison Disc Phonographs and the Diamond Discs: A History with Illustrations*. Sevenoaks, U.K.: George L. Frow, 1982.

Fussell, Paul. *Wartime: Understanding and Behavior in the Second World War*. New York: Oxford, 1989.

Gage, John. *Color and Culture: Practice and Meaning from Antiquity to Abstraction*. Berkeley: University of California Press, 1993.

———. *Color and Meaning: Art, Science, and Symbolism*. Berkeley: University of California Press, 1999.

Gamwell, Lynn. *Exploring the Invisible: Art, Science, and the Spiritual*. Princeton, N.J.: Princeton University Press, 2002.

Garity, William E., and J. N. A. Hawkins. "Fantasound." *Journal of the Society of Motion Picture Engineers* 37 (August 1941): 127–46.

Gelatt, Roland. *The Fabulous Phonograph, 1877–1977*. New York: Macmillan, 1977.

Genette, Gerard. *Narrative Discourse: An Essay in Method*. 1972. Ithaca, N.Y.: Cornell University Press, 1983.

Goldfarb, Phyllis. "Orson Welles's Use of Sound." *Take One* 3.6 (July–August 1971): 10–14. Reprinted in *Focus on Orson Welles*. Ed. Ronald Gottesman. Englewood Cliffs, N.J.: Prentice Hall, 1976. 85–95.

Gomery, Douglas. *Shared Pleasures: A History of Movie Presentation in the United States*. Madison: University of Wisconsin Press, 1992.

Goodwin, Andrew. *Dancing in the Distraction Factory: Music Television and Popular Culture*. Minneapolis: University of Minnesota Press, 1992.

———. "Fatal Distractions: MTV Meets Postmodern Theory." In *Sound and Vision*. Eds. Simon Frith, Andrew Goodwin, and Lawrence Grossberg. 45–66.

Gorbman, Claudia. *Unheard Melodies: Narrative Film Music*. Bloomington: Indiana University Press, 1987.

Gössel, Gabriel. *Fonogram: Prakticky průvodce historií záznamu zvuku*. Prague: Radioservis, 2001.

Grajeda, Tony. "Machines of the Audible: A Cultural History of Sound, Technology, and a Listening Subject." Ph.D. diss., University of Wisconsin–Milwaukee, 2001.

Greenewalt, Mary Hallock. *Nourathar: The Fine Art of Light Color Playing*. Philadelphia: Westbrook, 1946.

Grossberg, Lawrence. "The Media Economy of Rock Culture: Cinema, Postmodernity and Authenticity." In *Sound and Vision*. Eds. Simon Frith, Andrew Goodwin, and Lawrence Grossberg. 185–209.

Gunning, Tom. "Doing for the Eye What the Phonograph Does for the Ear." In *The Sounds of Early Cinema*. Eds. Richard Abel and Rick Altman. 13–31.

Habermas, Jürgen. *The Structural Transformation of the Public Sphere*. Trans. Thomas Burger. Cambridge, Mass.: MIT Press, 1993 [1962].

Hall, Edward T. "Proxemics." *Current Anthropology* 9.2–3 (1968): 83–108.

———. *The Hidden Dimension: Man's Use of Space in Public and Private*. London: Bodley Head, 1969.

Hammond, Paul. *The Shadow and Its Shadow: Surrealist Writings on Cinema*. New York: Columbia University Press, 1981.

Handel, Stephen. *Listening: An Introduction to the Perception of Auditory Events*. Cambridge, Mass.: MIT Press, 1989.

Hansen, Miriam Bratu. "Benjamin and Cinema: Not a One-Way Street." *Critical Inquiry* 25.2 (Winter 1999): 306–43.

———. "The Mass Production of the Senses: Classical Cinema as Vernacular Modernism." *Modernism/Modernity* 6.2 (April 1999): 55–77. Reprinted in *Reinventing Film Studies*. Eds. Christine Gledhill and Linda Williams. London: Arnold, 2000. 332–50.

Harvith, John, and Susan Edwards Harvith, eds. *Edison, Musicians, and the Phonograph: A Century in Retrospect*. New York: Greenwood Press, 1987.

Hay, James. "Piecing Together What Remains of the Cinematic City." *The Cinematic City*. Ed. David B. Clarke. New York: Routledge, 1997. 209–29.

Heath, Stephen. "Film and System: Terms of Analysis, Part I." *Screen* 16:1 (Spring 1975): 7–77.

———. "Film and System: Terms of Analysis, Part II." *Screen* 16:2 (Summer 1975): 91–113.

Higginbotham, Virginia. *Luis Buñuel*. Boston: Twayne Publishers, 1979.

———. *Spanish Films under Franco*. Austin: University of Texas Press, 1988.

Hilmes, Michele. *Hollywood and Broadcasting: From Radio to Cable*. Urbana: University of Illinois Press, 1990.

———. "Is There a Field Called Sound Culture Studies? And Does It Matter?" *American Quarterly* 57.1 (March 2005): 249–59.

Hincha, Richard. "Selling CinemaScope, 1953–1956." *Velvet Light Trap* 21 (Summer 1985): 44–53.

Hindes, Andrew. "Dolby Is Digital Leader." *Variety.Com* (database online). Los Angeles: Reed Business Information, 1998.

———. "Theaters Surround Dolby Ex." *Variety.Com* (database online). Los Angeles: Reed Business Information, 1998.

———. "Universal Finally Says Yes to Dolby Format." *Variety.Com* (database online). Los Angeles: Reed Business Information, 1998.

Hirdman, Anja. *Tv-Reklam I Sverige 1990 Och 1995: Maskulinitet, Femininitet Och Etnicitet*. Stockholm: Konsumentverket, 1995.

Hodge, Bob, and Gunther Kress. *Social Semiotics*. Cambridge, U.K.: Polity Press, 1988.

Hogeland, Lisa Maria. Afterword to *In a Lonely Place*. Dorothy B. Hughes. New York: Feminist Press, 2003. 225–50.

Hollaender, Friedrich. *Von Kopf bis Fuss: Mein Leben Mit Text und Musik*. Munich: Kindler, 1965.

"The Hollywood Revue." *Harrison's Reports* (24 August 1929): 135.

Holman, Tomlinson. *Sound for Film and Television*. Boston: Focal Press, 1997.

Honzl, Jindřich. "K diskusi o řeči ve filmu." *Slovo a slovesnost* no. 1 (1935): 38–40.

Hopewell, John. *Out of the Past: Spanish Cinema after Franco*. London: British Film Institute, 1986.

Hošek, Arne. "Komposice, umění, film a divadlo." *Magazin DP* no. 9 (1933–1934): 277–79.

———. "Poznámky k estetice filmu." *Program D 40* (24 October 1939): 15–18.

Hubert, Henri, and Marcell Mauss. *Sacrifice: Its Nature and Function*. Trans. W. D. Halls. Chicago: University of Chicago Press, 1964.

Hughes, David. *The Complete David Lynch*. London: Virgin, 2001.

Huhtamo, Erkki. "From Kaleidoscomaniac to Cybernerd: Notes toward an Archeology of Media." In *Electronic Culture: Technology and Visual Representation*. Ed. Timothy Druckrey. New York: Aperture, 1996. 296–303.

Husserl, Edmund. *The Phenomenology of Internal Time-Consciousness*. Ed. Martin Heidegger. Trans. James S. Churchill. Bloomington: Indiana University Press, 1964.

Hutcheon, Linda. *A Poetics of Postmodernism: History, Theory, Fiction*. New York: Routledge, 1988.

Huysmans, Joris-Karl. *Against Nature*. Trans. Robert Baldick. Baltimore: Penguin Books, 1966.

Inglis, Ian, ed. *Popular Music and Film*. London: Wallflower Press, 2003.

"In 70mm and 6–Track Dolby Stereo." *International Newsletter about 70mm Film* 13.62 (September 2000): 8–13.

Isaacs, Hermine Rich. "New Horizons: *Fantasia* and Fantasound." *Theatre Arts* 25.1 (January 1941): 55–61.

Jackson, Kenneth T. *Crabgrass Frontier: The Suburbanization of the United States*. New York: Oxford University Press, 1985.

James, David E. *Allegories of Cinema: American Film in the Sixties*. Princeton: Princeton University Press, 1989.

———. *Power Misses: Essays across (Un)Popular Culture*. London: Verso, 1996.

Jameson, Fredric. "Postmodernism and Consumer Society." In *The Anti-aesthetic: Essays on Postmodern Culture*. Ed. Hal Foster. Port Townsend, Wash.: Bay Press, 1983. 111–25.

———. *Postmodernism; or, The Cultural Logic of Late Capitalism*. Durham, N.C.: Duke University Press, 1991.

Jasper, Bill. "A Message from Dolby President." *Dolby News* (Fall 1998): 1.

Jay, Martin. *Downcast Eyes: The Denigration of Vision in Twentieth-Century French Thought*. Berkeley: University of California Press, 1993.

Jenkins, Henry. "'No Matter How Small': The Democratic Imagination of Doctor Seuss." *Hop on Pop: The Politics and Pleasure of Popular Culture*. Eds. Henry Jenkins, Tara McPherson, and Jane Shattuc. Durham, N.C.: Duke University Press, 2002. 187–208.

———. "Tales of Manhattan: Mapping the Urban Imagination through Hollywood." In *Imaging the City: Continuing Struggles and New Directions*. Eds. Lawrence Vale and Sam Bass Warner. New Brunswick, N.J.: Center for Urban Policy Research, 2001. 179–212.

Johnson, Heather. *Roy de Maistre: The English Years, 1930–1968*. Roseville East, New South Wales, U.K.: Craftsman House, 1995.

Johnston, William A. "The Public and Sound Pictures." *Journal of the Society of Motion Picture Engineers* 12.35 (1928): 614–19.

Jones, James. *The Thin Red Line*. New York: Charles Scribner's Sons, 1962.

Julesz, Bela, and Ira J. Hirsh. "Visual and Auditory Perception: An Essay of Comparison." In *Human Communication: A Unified View*. Eds. E. E. David Jr. and P. B. Denes. New York: McGraw-Hill, 1972. 283–340.

Kahn, Douglas. *Noise, Water, Meat: A History of Sound in the Arts*. Cambridge, Mass.: MIT Press, 1999.

Kaleta, Kenneth C. *David Lynch*. New York: Twayne, 1993.

Kalinak, Kathryn. "Max Steiner and the Classical Hollywood Film Score." In *Film Music 1*. Ed. Clifford McCarty. New York: Garland Publishing, 1989.

Kandinsky, Wassily. *Kandinsky: Complete Writings on Art*. Eds. Kenneth C. Lindsay and Peter Vergo. New York: Da Capo Press, 1994.

Karten, Harvey S. Review of *Touch of Evil* (1998). http://reviews.imdb.com/Reviews/141/14199.

Kassabian, Anahid. "At the Twilight's Last Scoring." In *Keeping Score: Music, Disciplinarity, Culture*. Charlottesville: University Press of Virginia, 1997. 258–74.

———. *Hearing Film: Tracking Identifications in Contemporary Hollywood Film Music*. New York: Routledge, 2001.

Katz, Mark. *Capturing Sound: How Technology Has Changed Music*. Berkeley: University of California Press, 2004.

Kauffmann, Stanley. "Welles in the Underworld." *New Republic* (28 September 1998): 30–31.

Keightley, Keir. "'Turn It Down!' She Shrieked: Gender, Domestic Space and High Fidelity, 1948–1959." *Popular Music* 15.2 (May 1996): 149–77.

Kelman, Ken. "The Other Side of Realism." In *The Essential Cinema: Essays on the Films in the Collection of Anthology Film Archives, Volume One, Anthology Film Archives, Series 2*. Ed. P. Adams Sitney. New York: Anthology Film Archives and New York University Press, 1975.

Kenny, Tom. "Walter Murch: The Search for Order in Sound and Picture." *Mix* (April 1998): Sound for Picture Supplement, 12–24.

———. "Cruising with David Lynch Down the 'Lost Highway.'" *Sound for Picture: Film Sound through the 1990s*. Vallejo, Calif.: MixBooks, 2000. 128–38.

Kittler, Friedrich A. *Discourse Networks, 1800/1900*. 1985. Trans. Michael Metteer with Chris Cullen. Foreword by David Wellbery. Stanford, Calif.: Stanford University Press, 1990.

———. *Gramophone, Film, Typewriter*. Trans. Geoffrey Winthrop-Young and Michael Wutz. Stanford, Calif.: Stanford University Press, 1999.

Klimek, Mary Pat. "Imagining the Sound(s) of Shakespeare: Film Sound and Adaptation." In *Sound Theory/Sound Practice*. Ed. Rick Altman. 204–16.

Klimeš, Ivan. "Zvuk na procenta." *Kino-Ikon* 7.2 (2002): 119–30.

Klinger, Barbara. "The New Media Aristocrats: Home Theater and the Domestic Film Experience." *Velvet Light Trap* 42 (1998): 4–19.

Kocourek, Franta. "Žalosti a radosti zvukového filmu." *Studio* 3.4 (1931): 97–98.

Koszarski, Richard. "On the Record: Seeing and Hearing the Vitaphone." In *The Dawn of Sound.* Ed. Mary Lea Bandy. 15–21.

Kozloff, Sarah. *Overhearing Film Dialogue.* Berkeley: University of California Press, 2000.

Kracauer, Siegfried. *From Caligari to Hitler: A Psychological History of the German Film.* Princeton, N.J.: Princeton University Press, 1947.

———. *Theory of Film: The Redemption of Physical Reality.* New York: Oxford University Press, 1960.

Kramer, Stanley, with Thomas Coffey. *A Mad, Mad, Mad, Mad World: A Life in Hollywood.* New York: Harcourt Brace, 1997.

Krauss, Rosalind. "Paul Sharits." *Film Culture* nos. 65–66 (1978): 89–102.

Kreilkamp, Ivan. "A Voice without a Body: The Phonographic Logic of *Heart of Darkness.*" *Victorian Studies* 40 (1997): 211–43.

Krueger, Eric M. "*Touch of Evil*: Style Expressing Content." *Cinema Journal* 12:1 (Fall 1972): 57–63.

Kubelka, Peter. "The Theory of Metrical Film." In *The Avant-Garde Gilm: A Reader of Theory and Criticism.* Ed. P. Adams Sitney. New York: Anthology Film Archives, 1987.

Kučera, Jan. "Forma zvukového filmu." *Studio* 1.8 (1929–1930): 242–45.

Kushner, Marilyn S. *Morgan Russell.* New York: Hudson Hills Press, 1990.

Lacan, Jacques. "The Agency of the Letter in the Unconcious or Reason since Freud." In *Écrits: A Selection.* Trans. Alan Sheridan. New York: Norton, 1977. 146–75.

———. "On Jouissance." In *The Seminar of Jacques Lacan: Book XX.* Ed. Jacques-Alain Miller. Trans. Bruce Fink. New York: Norton, 1998. 1–13.

Lacasse, Serge. "'Listen to My Voice': The Evocative Power of Vocal Staging in Recorded Rock Music and Other Forms of Vocal Expression." Ph.D. thesis, University of Liverpool, 2000.

Landy, Marcia. "'American under Attack': Pearl Harbor, 9/11, and History in the Media." In *Film and Television after 9/11.* Ed. Wheeler Winston Dixon. Carbondale: Southern Illinois University Press, 2004. 79–100.

Lastra, James. "Reading, Writing, and Representing Sound." In *Sound Theory/Sound Practice.* Ed. Rick Altman. 65–86.

———. *Sound Technology and the American Cinema: Perception, Representation, Modernity.* New York: Columbia University Press, 2000.

Lawson, Ted W. *Thirty Seconds over Tokyo.* Washington, D.C.: Brassey's, 2002.

Leeper, Jill. "Crossing Musical Borders: The Soundtrack for *Touch of Evil.*" In *Soundtrack Available: Essays on Film and Popular Music.* Eds. Pamela Robertson Wojcik and Arthur Knight. Durham, N.C.: Duke University Press, 2001. 226–43.

Legendre, Maurice. *Las jurdes: Études de geographie humaine.* Paris: Feret and Fils, 1927.

Le Grice, Malcolm. "Colour Abstraction—Painting—Film—Video—Digital Media," 1995. *Experimental Cinema in the Digital Age.* London: British Film Institute, 2001.

Le Guin, Ursula K. "Tangents: Film—*Close Encounters of the Third Kind* and *Star Wars.*" *Parabola* 3.1 (Winter 1978): 92–94.

Leslie, Esther. *Hollywood Flatlands: Animation, Critical Theory and the Avant-Garde.* London: Verso, 2002.

Levy, Shawn. "Touch of the Master." *Portland Oregonian* (16 October 1998). http://www.oregonlive.com/ent/movies/9810/mv981016_touch.html.

Lindsay, Vachel. *The Art of the Moving Picture.* New York: Liveright, 1970.

Linhart, Lubomír. "L." *Malá abeceda filmu.* Prague: Pokrok, 1930. 44–59.

LoBrutto, Vincent. *Sound-on-Film: Interviews with Creators of Film Sound.* Westport, Conn.: Praeger, 1994.

Lynch, David. "Action and Reaction." In *Soundscape: The School of Sound Lectures, 1998–2001.* Eds. Larry Sider, Diane Freeman, and Jerry Sider. 49–53.

Lyotard, Jean-François. *The Postmodern Condition: A Report on Knowledge.* Minneapolis: University of Minnesota Press, 1984.

Maasø, Arnt. *"Se-Hva-Som-Skjer!": En Studie Av Lyd Som Kommunikativt Virkemiddel I Tv.* Acta Humaniora Nr. 132. Oslo: Universitetet i Oslo, 2002.

Maeterlinck, Maurice. "Silence." In *The Treasure of the Humble.* Trans. Alfred Sutro. New York: Dodd, Mead, 1914. 1–21.

Malcolm, Janet. *Diana and Nikon: Essays on the Aesthetic of Photography.* Boston: Godine, 1980.

Mancini, Marc. "Sound Thinking." *Film Comment* 19.6 (November–December 1983): 40–43, 45–47. Reprinted as "The Sound Designer" in *Film Sound: Theory and Practice.* Eds. Elisabeth Weis and John Belton. 361–68.

Marchetti, Gina. "Documenting Punk: A Subcultural Investigation." *Film Reader* no. 5 (1982): 269–84.

Margulies, Ivone. "Exemplary Bodies: Reenactment in *Love in the City, Sons,* and *Close Up.*" *Rites of Realism: Essays on Corporeal Cinema.* Ed. Ivone Margulies. Durham, N.C.: Duke University Press, 2003. 217–44.

Marmorstein, Gary. *Hollywood Rhapsody: Movie Music and Its Makers 1900 to 1975.* New York: Schirmer Books, 1997.

Marvin, Carolyn. *When Old Technologies Were New: Thinking about Electric Communication in the Late Nineteenth Century.* New York: Oxford University Press, 1988.

Matossian, Nouritza. *Xenakis.* New York: Taplinger, 1986.

Maur, Karin von. *The Sound of Painting: Music in Modern Art.* Munich and London: Prestel, 1999.

Maynard, Patrick. "Talbot's Technologies: Photographic Depiction, Detection, and Reproduction." *Journal of Aesthetics and Art Criticism* 47.3 (Summer 1989): 263–76.

McBride, Joseph. *Orson Welles.* London: Secker and Warburg, in association with the British Film Institute, 1972.

McClary, Susan. *Feminine Endings: Music, Gender and Sexuality.* Minneapolis: University of Minnesota Press, 1991.

McCracken, Allison. *Real Men Don't Sing: Crooning and American Culture, 1928–1933.* Durham, N.C.: Duke University Press, 2008.

McGowan, Todd. "Lost on *Mulholland Drive*: Negotiating David Lynch's Panegyric to Hollywood." *Cinema Journal* 43.2 (Winter 2004): 67–89.

McGuinness, Patrick. *Maurice Maeterlinck and the Making of Modern Theatre*. Oxford: Oxford University Press, 2000.

Merleau-Ponty, Maurice. "The Film and the New Psychology." In *Sense and Non-sense*. Evanston, Ill.: Northwestern University Press, 1964. 48–59.

———. *Phenomenology of Perception*. Trans. Colin Smith. London: Routledge and Kegan Paul, 1962.

Metz, Christian. "Aural Objects." In *Film Sound: Theory and Practice*. Eds. Elisabeth Weis and John Belton. 154–61.

Miklitsch, Robert. "Cinephilia/Scopophilia/Audiophilia: Audiovisual Pleasure and Narrative Cinema in *Jackie Brown*." *Screen* 45.4 (Winter 2004): 287–304.

Míšková, Alena. "Národní filmová galerie." *Iluminace* 2.2 (1990): 71–95.

Moor, Elisabeth. "Branded Spaces: The Scope of 'New Marketing.'" *Journal of Consumer Culture* 3.1 (2003): 39–60.

Morgan, Judith, and Neil Morgan. *Dr. Seuss and Mr. Geisel: A Biography*. New York: Random House, 1995.

Moritz, William. *Optical Poetry: The Life and Work of Oskar Fischinger*. Bloomington: Indiana University Press, 2004.

Morris, Paul J. "Making Music More Musical." *Musician* 21.5 (May 1916): 263, 265.

Morton, David. *Off the Record: The Technology and Culture of Sound Recording in America*. New Brunswick, N.J.: Rutgers University Press, 2000.

Müller, Corinna. *Vom Stummfilm zum Tonfilm*. Munich: Wilhelm Fink, 2003.

Mulvey, Laura. "Afterword." In *The New Iranian Cinema: Politics, Representation and Identity*. Ed. Richard Tapper. New York: I. B. Tauris, 2002. 254–61.

———. "Visual Pleasure and Narrative Cinema." *Screen* 16.3 (Autumn 1975): 6–18.

Mumford, Lewis. *The Culture of Cities*. New York: Harcourt, Brace, 1938.

———. *Technics and Civilization*. New York: Harcourt, Brace, 1934.

Mundy, John. *Popular Music on Screen: From Hollywood Musical to Music Video*. Manchester, U.K.: Manchester University Press, 1999.

Murch, Walter. Foreword to *Audio-Vision: Sound on Screen*. Michel Chion. vii–xxiv.

———. "Restoring the Touch of Genius to a Classic." *New York Times* (6 September 1998): Arts and Leisure Section, 1, 16–17.

———. "Sound Design: The Dancing Shadow." In *Projections* 4. Eds. John Boorman, Tom Luddy, David Thompson, and Walter Donohue. London: Faber and Faber, 1995. 237–51.

———. Telephone interview with Jay Beck. 23 January 2001.

———. "10: Walter Murch—Subject: Designing Sounds for *Apocalypse Now*." In *Projections* 6. Ed. John Boorman and Walter Donohue. New York: Faber and Faber, 1996.

———. "Touch of Silence." In *Soundscape: The School of Sound Lectures, 1998–2001*. Eds. Larry Sider, Diane Freeman, and Jerry Sider. 83–102.

Murphy, John. "What Is Branding?" In *Brands: The New Wealth Creators*. Eds. Susannah Hart and John Murphy. Basingstoke, U.K.: Palgrave, 1998. 1–12.

Murray, Leo. "*Un condamne a mort s'est echappe* (A Man Escaped)." In *The Films of Robert Bresson*. Ed. Ian Cameron. New York: Praeger, 1969. 68–81.

Naremore, James. *The Magic World of Orson Welles*. Dallas: Southern Methodist University Press, 1989.

Natale, Richard. "Three Companies Wage Battle for the Hearts—and Ears—of U.S. Moviegoers." *Los Angeles Times* (5 September 1995): D1.

Neale, Steve. *Genre and Hollywood*. London: Routledge, 2000.

Nelson, Chris. "Lip-Synching Gets Real: The New Technology and Etiquette of Faking It." *New York Times* (1 February 2004): Arts and Leisure Section, 30.

Nericcio, William Anthony. "Of Mestizos and Half-Breeds: Orson Welles's *Touch of Evil*." In *Chicanos and Film: Representation and Resistance*. Ed. Chon A. Noriega. Minneapolis: University of Minnesota Press, 1992. 47–58.

Newton, Sir Isaac. *Opticks; or, A Treatise on the Reflections, Refractions, Inflections & Colours of Light*. Based on the fourth edition, London, 1730. New York: Dover Publications, 1952.

Nochimson, Martha. "'All I Need Is the Girl': The Life and Death of Creativity in *Mulholland Drive*." In *The Cinema of David Lynch*. Eds. Erica Sheen and Annette Davison. 165–81.

———. "Mulholland Drive." *Film Quarterly* 56.1 (2002): 37–45.

Nornes, Abé Mark. "Cherry Trees and Corpses: Representations of Violence from WWII." In *The Japan/America Film Wars: World War II Propaganda and Its Cultural Contexts*. Eds. Abé Mark Nornes and Fukushima Yukio. Chur, Switzerland: Harwood Academic Publishers, 1994. 147–61.

O'Brien, Charles. *Cinema's Conversion to Sound: Technology and Film Style in France and the U.S.* Bloomington: Indiana University Press, 2005.

Oliver, Kelly, and Benigno Trigo. "The Borderlands of *Touch of Evil*." In *Noir Anxiety*. Minneapolis: University of Minnesota Press, 2003. 115–36.

Olsen, Eric J. "Fresh Star Sounds." *Variety.Com* (database online). Los Angeles: Reed Business Information, 1998.

Ondaatje, Michael. *The Conversations: Walter Murch and the Art of Editing Film*. New York: Alfred A. Knopf, 2002.

Ong, Walter J. *Orality and Literacy: The Technologizing of the Word*. London: Routledge, 1982.

O'Pray, Michael. *Derek Jarman: Dreams of England*. London: British Film Institute, 1996.

Patzaková, A. J. "Vývoj Československého rozhlasu." In *Prvních deset let eskoslovenského rozhlasu*. Ed. A. J. Patzaková. Prague: Radiojournal, 1935. 13–655.

Peacock, Kenneth. "Instruments to Perform Color Music: Two Centuries of Technological Experimentation." *Leonardo* 21.4 (1988): 397–406.

Peck, A. P. "What Makes 'Fantasia' Click." *Scientific American* 164.1 (January 1941): 28–30.

Phillip, Robert. *Performing Music in the Age of Recording*. New Haven: Yale University Press, 2004.

Picker, John M. *Victorian Soundscapes*. Oxford: Oxford University Press, 2003.

Plumb, E. H. "The Future of Fantasound." *Journal of the Society of Motion Picture Engineers* 39 (July 1942): 16–21.

Rádl, Otto. "Pro mluvený film." *Studio* 2 (1930–1931): 65–68.

Ragona, Melissa. "Hidden Noise: Strategies of Sound Montage in the Films of Hollis Frampton." *October* 109 (Summer 2004): 96–118.

Read, Oliver, and Walter L. Welch. *From Tin Foil to Stereo: Evolution of the Phonograph.* Indianapolis: Bobbs-Merrill, 1976.

Rice, Elmer. *Minority Report: An Autobiography.* New York: Simon and Schuster, 1963.

———. *Street Scene: A Play in Three Acts.* New York: Samuel French, 1929.

Richter, Hans. "Easel—Scroll—Film." *Magazine of Art* 45.2 (February 1952): 78–86.

Riviere, Francis. "Rebirth Next: An Exclusive Interview with Francis Ford Coppola." *L.A. Weekly* (28 November 1979): 40.

Rodley, Chris. *Lynch on Lynch.* London: Faber and Faber, 1997.

Rogin, Michael. *Ronald Reagan, the Movie.* Berkeley: University of California Press, 1987.

Rubinstein, E. "Visit to a Familiar Planet: Buñuel among the Hurdanos." *Cinema Journal* 22.4 (Summer 1983): 3–17.

Ruby, Jay. "The Image Mirrored: Reflexivity and the Documentary Film." In *New Challenges for Documentary.* Ed. Alan Rosenthal. Berkeley: University of California Press, 1988. 64–77

Ryder, Loren L. "Magnetic Sound Recording in Motion Picture and Television Industries." *JSMPTE* 85.7 (July 1976): 528–30.

Saeed-Vafa, Mehrnaz, and Jonathan Rosenbaum. *Abbas Kiarostami.* Urbana: University of Illinois Press, 2003.

Sánchez Vidal, Agustín. *Luis Buñuel: Obra cinematografica.* Madrid: Ediciones J. C., 1984.

———. Unpublished personal correspondence. 1990–1991.

Sanjek, Russell. *American Popular Music and Its Business: The First Four Hundred Years, Volume III: From 1900 to 1984.* New York: Oxford University Press, 1988.

Scarry, Elaine. *The Body in Pain: The Making and Unmaking of the World.* New York: Oxford University Press, 1985.

Schafer, R. Murray. *The Tuning of the World.* New York: Alfred A. Knopf, 1977. Reprinted as *The Soundscape: Our Sonic Environment and the Tuning of the World.* Rochester, Vt.: Destiny Books, 1994.

Schatz, Thomas. "World War II and the 'War Film.'" In *Refiguring American Film Genres: History and Theory.* Ed. Nick Browne. Berkeley: University of California Press, 1998. 89–128.

Schreger, Charles. "Altman, Dolby, and the Second Sound Revolution." In *Film Sound: Theory and Practice.* Eds. Elisabeth Weis and John Belton. 348–55.

Schulte-Sasse, Jochen. "Imagination and Modernity; or, The Taming of the Human Mind." *Cultural Critique* 5 (1986–87): 23–48.

Sennett, Richard. *Classic Essays on the Culture of Cities.* Englewood Cliffs, N.J.: Prentice-Hall, 1969.

———. *The Fall of Public Man.* New York: Alfred A. Knopf, 1977.

Sergi, Gianluca. "A Cry in the Dark: The Role of Post-classical Film Sound." In *Contem-*

*porary Hollywood Cinema*. Eds. Steve Neale and Murray Smith. London: Routledge, 1998. 156–65.

———. *The Dolby Era: Film Sound in Contemporary Hollywood*. Manchester: Manchester University Press, 2004.

———. "The Sonic Playground: Hollywood Cinema and its Listeners." In *Hollywood Spectatorship*. Eds. Melvyn Stokes and Richard Maltby. London: British Film Institute, 2001. 121–31.

Sharits, Paul. "Cinema as Cognition: Introductory Remarks." *Film Culture* nos. 65–66 (1978): 76–78.

———. "A Cinematics Model for Film Studies in Higher Education." *Film Culture* nos. 65–66 (1978): 43–68.

———. "Hearing: Seeing." *Film Culture* nos. 65–66 (1978): 69–75.

———. "Postscript as Preface." *Film Culture* nos. 65–66 (1978): 1–6.

———. "Statement Regarding Multiple Screen/Sound 'Locational' Film Environments—Installations." *Film Culture* nos. 65–66 (1978): 79–80.

———. "-UR(i)N(ul)LS:TREAM:S:S:ECTION:S:S:ECTIONED: (A) (lysis), JO: '1968–79'." *Film Culture* nos. 65–66 (1978): 7–28.

———. "Words Per Page." *Afterimage* no. 4 (Autumn 1972): 27–43. Reprinted in *Film Culture* nos. 65–66 (1978): 29–42.

Sheen, Erica, and Annette Davison, eds. *The Cinema of David Lynch*. London: Wallflower Press, 2004.

Sider, Larry, Diane Freeman, and Jerry Sider, eds. *Soundscape: The School of Sound Lectures, 1998–2001*. London: Wallflower Press, 2003.

Silverman, Kaja. *The Acoustic Mirror: The Female Voice in Psychoanalysis and Cinema*. Bloomington: Indiana University Press, 1988.

———. *The Subject of Semiotics*. New York: Oxford University Press, 1983.

Singer, Ben. "Modernity, Hyperstimulus, and the Rise of Popular Sensationalism." In *Cinema and the Invention of Modern Life*. Eds. Leo Charney and Vanessa Schwartz. Berkeley: University of California Press, 1995. 72–99.

Sitney, P. Adams. *Visionary Film: The American Avant-Garde, 1943–2000*. New York: Oxford University Press, 2002.

Small, Christopher. *Music of the Common Tongue: Survival and Celebration in Afro-American Music*. New York: Riverrun Press, 1987.

Smith, Jeff. *The Sounds of Commerce: Marketing Popular Film Music*. New York: Columbia University Press, 1998.

Smrž, Karel. "Stín, který promluvil." *Kinorevue* 1.22 (1934): 421–26.

Sobchack, Vivian. "Synthetic Vision: The Dialectical Imperative of Luis Buñuel's *Las Hurdes*." In *Documenting the Documentary: Close Readings of Documentary Film and Video*. Eds. Barry Keith Grant and Jeannette Sloniowski. Detroit: Wayne State University Press, 1998. 70–82.

———. "When the Ear Dreams: Dolby Digital and the Imagination of Sound." *Film Quarterly* 58.4 (2005): 2–15.

Spelke, Elisabeth S., and Alexandra Courtelyou. "Perceptual Aspects of Social Knowing:

Looking and Listening in Infancy." In *Infant Social Cognition*. Eds. Michael E. Lamb and Lonnie R. Sherrod. Hillsdale, N.J.: Lawrence Erlbaum Associates, 1981. 61–84.

[Sponable, Earl]. Research and Development Division of the Twentieth Century-Fox Film Corporation. *CinemaScope Information for the Theatre: Equipment, Installation Procedures, Maintenance Practices, Operating Considerations, Demagnetization.* Third Revision. Los Angeles: Twentieth Century-Fox, 1954.

Spoto, Donald. *Stanley Kramer: Film Maker.* 1978. Hollywood: Samuel French, 1990.

Staiger, Janet. "Sophistophobia: *Mulholland Drive* as a Remake of *Meshes of the Afternoon.*" Paper presented at the Society for Cinema and Media Studies Conference, Denver, May 2002. 1–12.

Steiner, Frederick. *The Making of an American Film Composer: A Study of Alfred Newman's Music in the First Decade of the Sound Era.* Ph.D. diss., University of Southern California, 1981.

Steiner, Max. "Scoring the Film." In *We Make the Movies.* Ed. Nancy Naumberg. New York: W. W. Norton, 1937.

Sterne, Jonathan. *The Audible Past: Cultural Origins of Sound Reproduction.* Durham, N.C.: Duke University Press, 2003.

Stoklas, Eugen. "Úvaha o hře rozhlasové." *Radiojournal* 8.14 (1930): 1–2.

*Světlo a výtvarné umění v díle Zdenka a Jöny Pešánkových na transformační stanici Edisonově v Praze.* Prague: Elektrické podniky Hlavního města Prahy, 1930.

Symes, Colin. *Setting the Record Straight: A Material History of Classical Recording.* Middletown, Conn.: Wesleyan University Press, 2004.

Szczepanik, Petr. "Speech and Noise as the Elements of Intermedia History of the Early Czech Sound Cinema." In *MLVs, Cinema and Other Media.* Ed. Veronica Innocenti. Pasian di Prato: Campanotto, 2006. 175–90.

Taylor, Timothy D. *Strange Sounds: Music, Technology and Culture.* New York: Routledge, 2001.

Teige, Karel. "Divadlo a zvukový film." *Nová Scéna* 1.2 (1930): 36–41.

———. "Film II: Optofonetika." *Pásmo* 1.7–8 (1924–1925): 11.

Théberge, Paul. *Any Sound You Can Imagine: Making Music/Consuming Technology.* Middletown, Conn.: Wesleyan University Press, 1997.

Theweleit, Klaus. *Male Fantasies.* Vol. 1, *Women, Floods, Bodies, History.* Vol. 2, *Male Bodies: Psychoanalyzing the White Terror.* Trans. Stephen Conway in collaboration with Erica Carter and Chris Turner. Minneapolis: University of Minnesota Press, 1987.

Thom, Randy. "Designing a Movie for Sound." *iris* 27 (1999): 9–20.

Thomas, Bob. "Purchase of Columbia Gives Sony Some [of] America's Most Treasured Movies." *Associated Press.* 28 September 1989.

Thompson, Emily. "Machines, Music, and the Quest for Fidelity: Marketing the Edison Phonograph in America, 1877–1925." *Musical Quarterly* 79 (Spring 1995): 131–71.

———. *The Soundscape of Modernity: Architectural Acoustics and the Culture of Listening in America, 1900–1933.* Cambridge, Mass.: MIT Press, 2002.

Thompson, Stacy. "Punk Cinema." *Cinema Journal* 34.2 (Winter 2004): 47–66.

THX. "Press release." http://www.thx.com/news. 2003.

Tille, Václav. "Film spasí divadlo." *Rozpravy Aventina* 7.10 (1931): 73–74.

Tomblin, Barbara Brooks. *G.I. Nightingales: The Army Nurse Corps in World War II.* Lexington: University Press of Kentucky, 1996.

Truppin, Andrea. "And Then There Was Sound: The Films of Andrei Tarkovsky." In *Sound Theory/Sound Practice.* Ed. Rick Altman. 235–48.

Tsivian, Yuri. *Early Cinema in Russia and Its Cultural Reception.* Trans. Alan Bodger. London: Routledge, 1994.

Uhlig, Ronald E. "Stereophonic Photographic Soundtracks." *Journal of the SMPTE* 82.4 (April 1973): 292–95.

Urban, Jaroslav A. "V továrně na slova v Joinvillu." *Studio* 2 (1930–1931): 204–6.

van Leeuwen, Theo. *Speech, Music, Sound.* London: Macmillan, 1999.

Vančura, Vladislav. "K diskusi o řeči ve filmu." *Slovo a slovesnost* no. 1 (1935): 40–42.

Vaughan, Dai. *For Documentary.* Berkeley: University of California Press, 1999.

Walló, Kimi. "Sny, které z dálky šeptají." *Magazin DP* (1935–1936): 78–80.

Webster, Roger. *Studying Literary Theory: An Introduction.* London and New York: Edward Arnold, 1990.

Weingart, Miloš. "Zvukový film a řeč. Čtyři zásadní kapitoly." In *Abeceda filmového scenaristy a herce. Soubor přednášek scenaristického kursu Filmového studia.* Ed. Karel Smrž. Prague: Filmová knihovna, 1935. 39–72.

———. "Zvuková kultura českého jazyka." In *Spisovná čeština a jazyková kultura.* Eds. B. Havránek and M. Weingart. Prague: Melantrich, 1932.

Weis, Elizabeth. *The Silent Scream: Alfred Hitchcock's Sound Track.* Rutherford, N.J.: Fairleigh Dickinson University Press, 1982.

Weis, Elizabeth, and John Belton, eds. *Film Sound: Theory and Practice.* New York: Columbia University Press, 1985.

Weiss, Allen S., ed. *Experimental Sound and Radio.* New York and Cambridge, Mass.: New York University and MIT Press, 2001.

Welles, Orson, and Peter Bogdanovich. *This Is Orson Welles.* Ed. Jonathan Rosenbaum. New York: De Capo Press, 1998.

White, Susan. "Male Bonding, Hollywood Orientalism, and the Repression of the Feminine in Kubrick's *Full Metal Jacket.*" In *Inventing Vietnam: The War in Film and Television.* Ed. Michael Anderegg. Philadelphia: Temple University Press, 1991. 204–30.

Whitney, John. "Audio-Visual Music: Color Music—Abstract Film." 1944. In *Digital Harmony: On the Complementarity of Music and Visual Art.* Peterborough, N.H.: Byte Books, 1980. 138–43.

Wilde, Oscar. *The Picture of Dorian Gray.* Ed. Isobel Murray. London: Oxford University Press, 1974.

Wilfred, Thomas. "Light and the Artist." *Journal of Aesthetics and Art Criticism* 5.4 (June 1947): 247–55.

Williams, Mark. "Figuring the Representations: Inter-medial Borders and L.A.-*Frontera* Subjectivity." Paper presented at the American Studies Conference, Houston, November 2002. 1–9.

Wittgenstein, Ludwig. *Remarks on the Foundations of Mathematics.* 1956. Trans. G. E.

M. Anscombe. Eds. G. H. von Wright, R. Rhees, and G. E. M. Anscombe. Cambridge, Mass.: MIT Press, 2001.

Wolf, Michael J. *The Entertainment Economy.* New York: Penguin Books, 1999.

Wolfe, Charles. "On the Track of the Vitaphone Short." In *The Dawn of Sound.* Ed. Mary Lea Bandy. 35–41.

Wollen, Peter. "Foreign Relations: Welles and *Touch of Evil.*" *Sight and Sound* 6.10 (October 1996). 20–23.

———. *Singin' in the Rain.* BFI Film Classics. London: British Film Institute, 1992.

Woram, John M. *Sound Recording Handbook.* Oslo: Vett and Viten AS, 1995.

Wurtzler, Steve J. *Electric Sounds: Technological Change and the Rise of Corporate Mass Media.* New York: Columbia University Press, 2006.

———. *"She Sang Live, but the Microphone Was Turned Off: The Live, the Recorded and the Subject of Representation."* In *Sound Theory/Sound Practice.* Ed. Rick Altman. 87–103.

Zeman, Pavel. "Idea filmového archivu v echách." *Iluminace* 7.1 (1995): 43–58.

Zizek, Slavoj. *Looking Awry: An Introduction to Jacques Lacan through Popular Culture.* Cambridge, Mass.: October–MIT Press, 1991.

———. *Organs without Bodies: On Deleuze and Consequences.* New York: Routledge, 2004.

# Contributors

**JAY BECK** is an assistant professor of media and cinema studies in the College of Communication at DePaul University. He has published widely on film sound topics in *Southern Review, iris, Journal of Popular Film & Television, Moving Image, Kino-Ikon, Scope,* and *Iluminace.* His dissertation, "A Quiet Revolution: Changes in American Film Sound Practices, 1967–1979," received the 2004 Society for Cinema and Media Studies Dissertation Award.

**JOHN BELTON** is a professor of English and film at Rutgers University. In 2005–6, he received a Guggenheim Fellowship to research a book on digital cinema. He is the author of five books, including *Widescreen Cinema,* winner of the 1993 Kraszna Krausz prize for books on the moving image, and *American Cinema/American Culture,* a textbook written to accompany the PBS series *American Cinema.* He has edited three books, most recently *Alfred Hitchcock's "Rear Window,"* and he edits a series of books on film and culture for Columbia University Press.

**CLARK FARMER** is an assistant professor of film studies at the University of Colorado–Boulder.

**PAUL GRAINGE** is an associate professor of film studies at the University of Nottingham. He is the author of *Brand Hollywood: Selling Entertainment in a Global Media Age* and *Monochrome Memories: Nostalgia and Style in Retro America.* He is also coauthor of *Film Histories: An Introduction and Reader* and editor of *Memory and Popular Film.* His work on contemporary film and media culture has been published in journals including *Screen, Cultural Studies, American Studies,* and the *International Journal of Cultural Studies.*

**TONY GRAJEDA** is an associate professor of cultural studies in the Department of English at the University of Central Florida. He is coeditor (with Timothy Taylor and Mark Katz) of *The Social Life of Early Sound Technologies: A History in Documents, 1878–1945.* His work has appeared in such journals as *Jump Cut, Film Quarterly, Social Epistemology, Chain, Journal of Popular Music Studies,* and *disClosure,* as well as several anthologies, most recently *Rethinking Global Security: Media, Popular Culture, and the "War on Terror."*

**DAVID T. JOHNSON** is an assistant professor of English at Salisbury University. He is also coeditor of *Literature/Film Quarterly.*

**ANAHID KASSABIAN** is the James and Constance Alsop Chair of Music at the University of Liverpool. Her research and teaching focus on ubiquitous musics and music, sound, and moving images. She is the author of *Hearing Film,* coeditor of *Keeping Score,* and a past editor of the *Journal of Popular Music Studies.* With Ian Gardiner, she recently cofounded the journal *Music, Sound and the Moving Image.* She has also written, with David Kazanjian, on Armenian diasporic film. She has curated Armenian film festivals in San Francisco and New York, and she has taught in media studies, women's studies, literary studies, and Middle East studies.

**DAVID LADERMAN** is a professor of film at the College of San Mateo and is the author of *Driving Visions: Exploring the Road Movie.* He has published in the *Journal of Film and Video, Film Quarterly* and *Cinema Journal* and has taught at San Francisco State University and at the University of California–Davis.

**JAMES LASTRA** is an associate professor of English and cinema and media studies at the University of Chicago where he teaches courses devoted to film sound and music on a regular basis. He is the author of *Sound Technology and the American Cinema: Perception, Representation, Modernity.*

**ARNT MAASØ** is an associate professor in the Department of Media and Communication at the University of Oslo. He has published Norwegian-language books and articles on sound in film and television since the mid-1990s. A few publications are also available in German, French, and English. He currently serves on the editorial board of *Music, Sound and the Moving Image,* and since 2005 he has been director of studies and vice head at the Department of Media and Communication, where he also teaches a course in music and media.

**MATTHEW MALSKY** is a composer on the music faculty at Clark University where he oversees the computer music and recording studios and serves as the director of the Communication and Culture Program. His current creative project is an evening-length electroacoustic soundtrack for the classic silent film *Berlin, Symphony of a Great City,* which will use his recent sound recordings of Berlin street life. He has published widely on issues related to music, technology, and culture, including articles published by Wesleyan University Press, *Leonardo Music Journal,* and the *Kurt Weill Gesellschaft.*

**BARRY MAUER** earned his doctorate in cultural studies from the University of Florida's English Department. He is an associate professor at the University of Central Florida in Orlando. His primary responsibility at UCF is to the Texts and Technology Ph.D. Program, which is housed in the English Department. In addition to his work as a teacher and researcher, Mauer also writes and records music. His CDs are available through CDBaby and Amazon.

**ROBERT MIKLITSCH** is a professor in the English Department at Ohio University. He is the author of *From Hegel to Madonna: Towards a General Economy of "Commodity Fetishism"* and *Roll Over Adorno: Critical Theory, Popular Culture, Audiovisual Media.* He is also the editor of *Psycho-Marxism.* His work has appeared in *Screen, Camera Obscura, Film Quarterly,* and the *New Review of Film and Television Studies,* and is forthcoming in the Wallflower collection *Neo-Noir.* He is presently writing a book on sound in film noir, *Audio Noir: Genre, Performance, Audiovisuality.*

**NANCY NEWMAN** is an assistant professor of music at the University at Albany, State University of New York. After earning her Ph.D. at Brown University, she taught at Clark, Wesleyan, and Tufts. She is currently completing a book about the nineteenth-century Germania Musical Society titled *Good Music for a Free People.* Articles on the orchestra have appeared in the *Yearbook of German-American Studies* and the *Institute for Studies in American Music Newsletter.* Other projects include articles on Clara Wieck Schumann, Björk's music for *Dancer in the Dark,* and an essay in the book *Musical Savants.* A former piano teacher, she most recently performed in the Extensible Toy Piano Festivals in Albany and Worcester.

**MELISSA RAGONA** is currently an assistant professor of art at Carnegie Mellon University. Her critical and creative work focuses on sound design, film theory, and new media practice and reception. By forging approaches from the disciplines of film studies, art history, and new media technologies, her work has sought to present a more complex aesthetic, theoretical, and historical foundation for the analysis of contemporary time-based arts. Her current book project, *Readymade Sound: Andy Warhol's Recording Aesthetics,* examines Warhol's tape recording projects from the mid-1960s until the late 1970s in light of audio experiments in modern art as well as contemporary practices of pattern matching and information visualization. She has published essays that explore the nexus between sound and image in the films of Hollis Frampton and Marie Menken. She has also published in monographs on the work of artists Heike Mutter and Ulrich Genth as well as Christian Jankowski.

**PETR SZCZEPANIK** is an assistant professor in the Department of Film and Audiovisual Culture at Masaryk University, Brno, and researcher in the National Film Archive, Prague. He is an editor in chief of the Czech film studies journal *Iluminace.* He has published essays in Czech, English, and German on the coming of sound in the context of broader media culture, on the history of industrial film, and on film's interrelations with other technical media, and he edited or coedited three books on the history of film thought and on film historiography. In 2005–6 he was a postdoc member of Graduiertenkolleg Mediale Historiographien in Weimar.

**PAUL THÉBERGE** is a professor and a Canada Research Chair at the Institute for Comparative Studies in Literature, Art and Culture at Carleton University. He has published widely on music, technology, and culture, and is the author of *Any Sound You Can Imagine: Making Music/Consuming Technology.* His current research is concerned with the history of sound technology and recording practices across a variety of media during the 1960s and 1970s. He has published articles on the uses of sound in film, television, and multimedia, including sound and music in the films of David Cronenberg.

**DEBRA WHITE-STANLEY** is an assistant professor of film studies at Indiana University/Purdue University, Indianapolis. Her research interests include the war film and memoir, film sound, and the representation of gender and race. Her publications include "'God Give Me Strength': The Melodramatic Soundtracks of Allison Anders" and forthcoming articles on *The Manchurian Candidate* and the F/X television series *Over There.* She is currently working on a book focusing on military nurses in war films and literature.

# Index

The University of Illinois Press
is a founding member of the
Association of American University Presses.

---

Composed in 10.5/13 Adobe Minion Pro
with Frutiger display
by Jim Proefrock
at the University of Illinois Press
Manufactured by Cushing-Malloy, Inc.

University of Illinois Press
1325 South Oak Street
Champaign, IL 61820-6903
www.press.uillinois.edu